Eco-Resilience Biol

The book *Eco-Resilience Biology: An Exploration of Clima* offers comprehensive information on the dynamic interaction systems and changing climate regimes. As climate change ac ing the resilience mechanisms within forest biomes becomes biodiversity, ecosystem services, and sustainable human livelih to bridge the knowledge gap by exploring eco-resilience from b and climate perspectives, providing an integrated framework f ers, and policymakers.

With an emphasis on interdisciplinary insights, this volume est species adapt and respond to climatic stressors through ph functional trait shifts, and ecological feedback loops. It investi silience and the potential for enhancing adaptive capacity thro tions like forest management, conservation planning, and restor existing works, this book uniquely blends empirical case studies and cutting-edge scientific findings to present a holistic view of

Key Features:

- Explores plant functional traits and species adaptation m climate stress.
- Addresses forest-climate feedback, carbon sequestration stability.
- Integrates traditional ecological knowledge with m approaches.
- Discusses resilience assessment tools and strategies for f planning.
- Case studies from diverse forest ecosystems, including tro and boreal zones.

Eco-Resilience Biology is ideal for a wide range of readers, incl tal scientists, forest ecologists, conservation biologists, climate re dents of ecological and environmental sciences. It is also suitabl graduate, postgraduate, and Ph.D. students in fields such as agricu physiology, forestry, biotechnology, environmental science, and a

Eco-Resilience Biology

An Exploration of Climate-Forest Connection

Edited by
Amit Kumar and Rajesh Kaushal

CRC Press
Taylor & Francis Group
Boca Raton London New York

CRC Press is an imprint of the
Taylor & Francis Group, an **informa** business

Designed cover image: Shutterstock

First edition published 2026
by CRC Press
2385 NW Executive Center Drive, Suite 320, Boca Raton FL 33431

and by CRC Press
4 Park Square, Milton Park, Abingdon, Oxon, OX14 4RN

CRC Press is an imprint of Taylor & Francis Group, LLC

ISBN: 978-1-032-96212-2 (hbk)
ISBN: 978-1-032-96214-6 (pbk)
ISBN: 978-1-003-58858-0 (ebk)

DOI: 10.1201/9781003588580

Typeset in Times
by KnowledgeWorks Global Ltd.

Contents

Preface

In response to the rapid pace of climate change, forest ecosystems are now at a critical juncture when their resilience is not only an ecological issue but also a global concern. Often referred to as the "lungs of our planet," forests are essential for preventing climate change because they store carbon, control the water cycle, and protect biodiversity. But the effects of changing precipitation patterns, intense weather events, and warming temperatures are making it harder for them to survive and remain stable. The idea of this book has been emerged in response to the pressing need to comprehend, document, and share information about how forest ecosystems respond to and adapt to these unprecedented climate changes.

The book is a collective effort of researchers, ecologists, climate scientists, and environmental thinkers who have contributed insights grounded in empirical data, field-based evidence, and theoretical models. It seeks to build a bridge between ecological theory and practical strategies for enhancing forest resilience, drawing from diverse biogeographic regions and ecological contexts. From the microscopic dynamics of soil microbial interactions to the large-scale analysis of species distribution and functional traits, the content encompasses a multi-layered view of eco-resilience. This work also aims to inspire a broader appreciation for the role of forests in sustaining life and fostering climate stability. We believe that a clear understanding of forest resilience mechanisms can guide more informed decision-making, foster adaptive management strategies, and support the development of future-proof conservation policies.

We sincerely hope this book serves as a valuable resource for students, researchers, practitioners, and policymakers working toward a sustainable and climate-resilient future.

Editors
Amit Kumar
Rajesh Kaushal

About the Editors

Amit Kumar is a distinguished researcher, renowned author, and an expert in the fields of plant physiology, agroforestry, and climate change. With an insatiable curiosity about the natural world and a deep-rooted passion for environmental conservation, he has dedicated his life to understanding the intricate relationships between plants, ecosystems, and the ever-changing climate. He earned his Ph.D. in Forestry from Forest Research Institute, Dehradun, India. His research is primarily focused on plant functional traits of forestry ecosystem. Earlier, he worked as Assistant Professor in many organizations, i.e., Dolphin P.G. Institute of Biomedical and Natural Sciences, Dehradun; Alpine Group of Institution, Dehradun, Uttarakhand. He has published several research articles, books, and book chapters in peer-reviewed journals with high impact factors. He has edited several books with international publishers. He has received several awards from prestigious organization, i.e., Young Scientist Award from UCOST Dehradun, Uttarakhand, Emerging Scientist Award from VDGOOD Foundation, Kolkata, and Best Research Paper Award from Indian Association of Soil and Water Conservationists. He has participated in more than 30 national and international conferences, symposia, and seminars. He has published more than 30 research papers in well-reputed national and international journals. Currently, he is working as a scientist at CSIR-Institute of Himalayan Bioresource Technology, Palampur – HP.

Rajesh Kaushal has more than 20 years of experience in the field of Conservation Forestry and NRM. He obtained his B.Sc. (Forestry), M.Sc., and Ph.D. (Agroforestry) from Dr. Y.S. Parmar University of Horticulture and Forestry, Nauni, Solan, HP. He began his career as an assistant scientist (Agroforestry) at G.B. Pant University of Agriculture & Technology, Pantnagar, Uttarakhand, in 2003. In 2011, he joined the ICAR-Indian Institute of Soil and Water Conservation (IISWC), Dehradun, as a senior scientist (Forestry) and has been serving as a principal scientist (Forestry) since 2017. He has made significant contributions to the field of natural resource management, particularly in the Himalayan region. He has developed and demonstrated several sustainable land management technologies for the rehabilitation of degraded lands. His work encompasses the design and implementation of agroforestry systems that promote resource conservation, enhance ecological stability, and support livelihood security for local communities. He has also established standardized propagation protocols for commercially important bamboo and agroforestry tree species, facilitating large-scale cultivation and adoption. Furthermore, he has developed predictive models to assess the carbon sequestration potential of various bamboo and agroforestry species, contributing to climate change mitigation strategies. He has been recognized by International Bamboo and Rattan Organization (INBAR) as resource person for capacity-building activities of African and South American countries as part of the South-South knowledge transfer program. He visited Kenya, Ethiopia, Tanzania, Madagascar, and Ecuador for providing training to scientists and Field Officers on resource conservation aspect of bamboo. He was also Visiting Scholar at The Ohio State University, USA in 2014 where he underwent specialized training on carbon sequestration. He also handled additional responsibilities of Assistant Director, Agroforestry Research Centre and Assistant Coordinator, Bamboo Coordinating Centre at Pantnagar. He is a member of the editorial board of the *Indian Journal of Soil Conservation*. He is a recipient of the University Gold Medal, Gold Medal, and Fellow of the Indian Society of Agroforestry, as well as a recipient of the Gold Medal and D.N. Puri Award from the Indian Association of Soil and Water Conservationists. He has handled ten externally funded projects and has published more than 75 research articles and 14 books to his credit.

Contributors

Ajay
Bioscience Division, Institute of Pesticide Formulation Technology (IPFT)
Gurugram, India

Darshita
Division of Genetics and Tree Improvement, Forest Research Institute (Deemed to be University)
Dehradun, India

Garima
Dr Y S Parmar University of Horticulture and Forestry, College of Horticulture and Forestry
Mandi, India

Vatsala
Forest Research Institute (Deemed to be University)
Dehradun, India

Avantika Adhruj
Indian Institute of Forest Management
Bhopal, India

Varun Attri
Regional Research Station (Punjab Agricultural University), Ballowal Saunkhri
Balachaur, India

Sumantra Basu
ICFRE-Forest Research Institute
Dehradun, India

D. R. Bhardwaj
College of Forestry, Dr. Y. S. Parmar University of Horticulture and Forestry
Solan, India

Jigyasa Bisaria
Centre for Climate Change Studies, Indian Institute of Forest Management
Bhopal, India

Kunal Biswas
Division of Genetics and Tree Improvement, Forest Research Institute (Deemed to be University)
Dehradun, India

A. K. Chanchal
Nalanda College of Horticulture
Nalanda, India

Rajdeep Chanda
Department of Forestry, School of Natural Resources and Management, Mizoram University
Aizawl, India

Lhingneivah Chongloi
Bioscience Division, Institute of Pesticide Formulation Technology (IPFT)
Gurugram, India

Suresh Choudhary
Bioscience Division, Institute of Pesticide Formulation Technology (IPFT)
Gurugram, India

Khoisnam Vramari Devi
Department of Forestry, School of Natural Resources and Management, Mizoram University
Aizawl, India

Meera Devi
Krishi Vigyan Kendra Kandaghat
Solan, India

Ningthoujam Ranjana Devi
Department of Forestry, School of Natural Resources and Management, Mizoram University
Aizawl, India

Suruchi Devi
Centre for Climate Change Studies, Indian Institute of Forest Management
Bhopal, India

Ram Gopal
North Eastern Hill University
Tura Campus, Meghalaya, India

Sandhya Goswami
Department of Forestry, Dolphin (PG) Institute of Biomedical & Natural Sciences
Dehradun, India

Suraj Goswami
Indian Institute of Forest Management
Bhopal, India

Akrita Gupta
Centre for Climate Change Studies, Indian Institute of Forest Management
Bhopal, India

Deepak Kumar Jaiswal
Bioscience Division, Institute of Pesticide Formulation Technology (IPFT)
Gurugram, India

Sagrika Kalta
Forest Research Institute (Deemed to be) University
Dehradun, India

Alisha Kaprate
College of Forestry, Dr. Y. S. Parmar University of Horticulture and Forestry
Solan, India

Shubhi Kulshrestha
ICFRE-Institute of Forest Biodiversity
Hyderabad, India

Aakash Kumar
School of Advanced Engineering, University of Petroleum and Energy Studies
Dehradun, India

Abhay Kumar
KVK, Jehanabad, Bihar Agricultural University
Bhagalpur, India

Abhishek Kumar
Graphic Era Deemed to be University
Dehradun, India

Dhirender Kumar
College of Forestry, Dr. Y. S. Parmar University of Horticulture and Forestry
Solan, India

Manish Kumar
Department of Forestry, Dolphin (PG) Institute of Biomedical & Natural Sciences
Dehradun, India

Mukesh Kumar
Botany Department, Government College Bahadurgarh
Jhajjar, India

Jaya Kumari
Nalanda College of Horticulture
Nalanda, India

Shilpa Kumari
Division of Genetics, Indian Council of Agricultural Research, Indian Agriculture Research Institute
New Delhi, India

Sunitā Kumari
Department of FPU
Faculty of Forestry, Birsa Agricultural University
Ranchi, Jharkhand, India

Yourmilla Kumari
Dr Y S Parmar University of Horticulture and Forestry, College of Horticulture and Forestry
Mandi, India

Dinesh Mahto
KVK, Jehanabad, Bihar Agricultural University
Bhagalpur, India

M. S. Malik
Department of Silviculture & Agroforestry
Birsa Agricultural University
Ranchi, Jharkhand, India

Yalal Mallesh
Rani Lakshmi Bai Central Agricultural University
Jhansi, India

Aruna Mehta
Dr Y S Parmar University of Horticulture and Forestry, College of Horticulture and Forestry
Mandi, India

Gargi V. Navpute
Forest Research Institute
Dehradun, India

Shivansh Negi
Division of Genetics and Tree Improvement, Tropical Forest Research Institute
Jabalpur, India

P. R. Oraon
Department of Silviculture & Agroforestry
Birsa Agricultural University
Ranchi, Jharkhand, India

Muneshwar Prasad
KVK, Jehanabad, Bihar Agricultural University
Bhagalpur, India

Komal Rani
Division of Genetics and Tree Improvement, ICFRE-Tropical Forest Research Institute
Jabalpur, India

Vikaspal Singh Rawat
Department of Forestry, Dolphin (PG) Institute of Biomedical & Natural Sciences
Dehradun, India

Amitava Roy
Division of Genetics and Tree Improvement, ICFRE Rain Forest Research Institute
Jorhat, India

Vipul Sharma
Forest Research Institute
Dehradun, India

Swati Shabnam
School of Agriculture, Rural and Tribal Development, Ramakrishna Mission Ashrama
Morabadi, Ranchi, Jharkhand, India

Gurwinder Singh
College of Forestry, Dr. Y. S. Parmar University of Horticulture and Forestry
Solan, India

Hariom Singh
Bioscience Division, Institute of Pesticide Formulation Technology (IPFT)
Gurugram, India

Manisha Singh
Centre for Climate Change Studies, Indian Institute of Forest Management
Bhopal, India

Ngangbam Somen Singh
Department of Forestry, School of Natural Resources and Management, Mizoram University
Aizawl, India

Rohit Kumar Singh
Bioscience Division, Institute of Pesticide Formulation Technology (IPFT)
Gurugram, India

Salam Suresh Singh
Department of Forestry, School of Natural Resources and Management, Mizoram University
Aizawl, India

Sanjay Singh
Centre of Excellence on Sustainable Land Management, Indian Council of Forestry Research and Education
Dehradun, India

Swati Singh
Department of Environmental Science, Graphic Era (Deemed to be University)
Dehradun, India

Yashaswi Singh
Graphic Era Deemed to be University
Dehradun, India

Bhaskar Sinha
Centre for Climate Change Studies, Indian Institute of Forest Management
Bhopal, India

Rishabh Srikar
Centre for Climate Change Studies, Indian Institute of Forest Management
Bhopal, India

Abhishek Srivastav
Bioscience Division, Institute of Pesticide Formulation Technology (IPFT)
Gurugram, India

Pavin Praize Sunny
Department of Forest Products and Utilization, College of Forestry, Kerala Agricultural University
Thrissur, India

Ajay Thakur
Forest Research Institute
Dehradun, India

C. L. Thakur
College of Forestry, Dr. Y. S. Parmar University of Horticulture and Forestry
Solan, India

Jagriti Thakur
Chaudhary Sarwan Kumar Himachal Pradesh Krishi Vishvavidyalaya
Palampur, India

Ruchi Thakur
College of Forestry, Dr. Y. S. Parmar University of Horticulture and Forestry
Solan, India

Shri Kant Tripathi
Department of Forestry, School of Natural Resources and Management, Mizoram University
Aizawl, India

Keshav Kumar Upadhyay
Department of Forestry, School of Natural Resources and Management, Mizoram University
Aizawl, India

David C. Vanlalfakawma
Department of Forestry, School of Natural Resources and Management, Mizoram University
Aizawl, India

Ritika Verma
Forest Research Institute
Dehradun, India

Kashish Walia
Dr Y S Parmar University of Horticulture and Forestry, College of Horticulture and Forestry
Thunag, Mandi, India

Shambhavi Yadav
Forest Research Institute
Dehradun, India

1 Introduction to Resilience of Forest Ecosystems and Its Role in Climate Change

Pavin Praize Sunny

1.1 INTRODUCTION

1.1.1 Ecological Resilience

Ecological resilience was first put up by C.S. Holling, who understood that systems that were disrupted beyond their ability to recover may change into a different state, also known as a different regime. The degree of disruption or disturbance that a system can tolerate before reaching a critical threshold is known as its resilience (Holling, 1973). Ecoresilience is the growing ability of ecosystems to provide desired services in the face of environmental changes and human exploitation (Folke et al., 2005). The primary objective is to enhance ecosystems' capacity to withstand disturbances and maintain the necessary natural structures, feedback, and processes. According to Marchese et al. (2018), resilience can also be defined as an ecosystem's capacity to withstand threats, absorb effects, and even recover and adapt after ongoing stressors or disruptive events. Resilience is defined as "jumping or restoring to the primary state" in one of the earliest definitions in this context (Holling, 1973). Notably, resilience does not necessarily return to the primary balance; instead, it can determine survival rates, future alterations, and the likelihood of adaptation and transformation in the current state (Folke et al., 2010). Because of their disciplinary background, these descriptions demonstrate resilience in several ways (Bone et al., 2016).

1.1.2 Key Components of Ecological Resilience

Furthermore, two essential components of resilience are thresholds and adaptability (Gunderson, 2000). Three fundamentals are revealed by this ability: adaptive capacity, recuperation or rearrangement, and resistance (specifically, thresholds) (Boyd & Folke, 2011; Elmqvist et al., 2003; Folke, 2006; Folke et al., 2010). Each principle also refers to thoughtful realities: adaptive capacity indicates the ability of ecosystems to withstand, depending on their diversity and complexity as well as a learning mechanism (Angeler & Allen, 2016; Angeler et al., 2019; Ferro-Azcona et al., 2019; Gunderson & Allen, 2010; Gunderson et al., 2010), recovery or reorganization introduces alternative regimes following disturbances and resistance embraces thresholds that ecosystems can withstand (Baho et al., 2017; Carpenter et al., 2001;

DOI: 10.1201/9781003588580-1

Ferro-Azcona et al., 2019; Folke et al., 2004; Gunderson & Allen, 2010; Gunderson et al., 2010).

1.1.3 Forest Resilience Studies

The idea of forest resilience is still evolving and encompasses many different topics. In order to control tree resilience in forest parks, Hale et al. (2015) identified a method based on a system with many dependencies, co-increments, and tensions. They also highlighted the specific shortcomings and noted the functional capacities and resilience of urban forests. Furthermore, Siedl et al. (2016) investigated the connection between resilience and ecosystem-associated services and came to the conclusion that ecological disturbance may have an effect on these services and that resilience may be a potent strategy to mitigate the effects of disturbance regimes and current changes. Bowditch et al. (2019) employed a combination of techniques to map the spatial resilience of forests and further examined forest resilience through public perceptions. In this regard, certain elements, including forest development and carbon marketing, may have a big impact on public opinion. Public partnership investments may also promote the resilience of forests. Similarly, Albrich et al. (2020) examined simulation models for forest resilience and discovered that little research had taken resilience processes into account, necessitating additional development of the used models. Tree growth after recovery may be impacted by the non-linear trajectories of the link between growth resilience and intense drought, as shown by Ovenden et al. (2021). The resilience of forest trees to water stress gradients was investigated by Pérez-Luque et al. (2021) in a related study. It was discovered that the resilience did not follow a uniform pattern because of differences in ecological conditions. Furthermore, Mina et al. (2021) used landscapes as functional networks to evaluate resilience and came to the conclusion that climate change had a negative impact on forest resilience indicators and decreased the ability of forests to adapt to global changes. Ecoresilience has also been the subject of numerous studies. Baho et al. (2017) noted that the lack of operational frameworks constituted a gap between science and practice and that the application of useful quantitative resilience metrics was difficult. Additionally, the majority of recent studies have only examined a few resilience qualities. Consequently, these evaluations fall short in capturing the system's overall resilience. However, there is an alternate method that measures ecological outcomes rather than assessing the factors influencing resilience. This method is similarly based on the idea that if an ecosystem is resilient, its results would reflect resilience. Consequently, rather than emphasizing partial parameters, which raise uncertainty, this method uses a holistic framework. The alternative approach's five primary parameters—eco-volume, eco-height, eco-climax, basal area, and bio-volume—are also used to calculate resilience (Torrico & Janssens, 2010). Since resilience is an emergent phenomenon of complex systems, it cannot be inferred from the system's constituent elements. The system may break down and restructure if a critical threshold is crossed. The original system may simply renew as a result of reorganization, or it may reorganize around new drivers, in which case a new system that is possibly very different may arise. In terms of the provision of the ecological goods and services generated, which are essential to human existence, the new system (an alternative system state) may be less appealing to humans when reorganization takes place around new drivers. Thus, preserving systems in favorable states, avoiding critical thresholds, and boosting resilience are frequently in the best interests of humanity. On the other hand, when a system is in an undesired condition, it might be required to intentionally change it to a more desirable one by eroding resilience. In either scenario, it is critical to be able to evaluate a system's resilience so that, based on the value of the current system state and the possible exposure and severity of present and future drivers of alternative state change, humanity can either promote or undermine resilience (Chaffin et al., 2016).

1.1.4 Complex Systems and Non-linear Dynamics

The ability of a social-ecological system to absorb and react to a disturbance while retaining its fundamental structure and functions is known as the system property of resilience, and resource managers frequently rely on it. Unlike engineering resilience, which emphasizes bounce-back and presumes stationarity, equilibrium, and frequently the occurrence of a single, idealized state, ecological resilience is a more comprehensive term. These presumptions work best for completely designed systems operating on modest scales for brief periods of time. According to ecological resilience, a focus on bounce-back ignores the emergence of destabilizing feedbacks that keep pushing the system in the direction of another state. A key emphasis of social-ecological systems study today is the capacity of a state to either absorb disturbance and persist or to self-organize and re-emerge after collapse caused by disturbance. This capacity is also essential for managers seeking to preserve or improve resilience as a management goal or approach. Resource managers may be able to prevent or facilitate regime shifts in order to preserve or restore desired ecosystem services if they have a better understanding of the boundaries of resilience, the thresholds that divide one state of a system from alternative states, and the ways in which social and ecological system components interact to reinforce or degrade resilience. Assessing the vulnerability of systems to perturbation requires combining this information with knowledge of the likelihood of a regime change and the possibility that other drivers will emerge and cause alternative state changes after catastrophic disruption (Holling, 1973). It is commonly accepted that complex systems are open, self-organizing, and characterized by emergent phenomena, non-linear dynamics, and intrinsic unpredictability and uncertainty. Because there is no central authority controlling system components, complex systems self-organize; complexity develops gradually from numerous simple interactions. Because complex systems exhibit non-linear dynamics, their behavior cannot be deduced or anticipated by adding up its constituent parts. Small changes that have a disproportionately significant effect are known as non-linear dynamics. For instance, a lake's phosphorus levels may gradually increase over time without seeming to have any negative effects on the lake until they surpass a certain threshold and the lake suddenly shifts into a new regime, becoming eutrophic, vulnerable to algal blooms, and less appealing to people. Changing conditions cause ecosystems and other complex systems to alter, which affects our understanding of and approach to managing ecosystems. Because humans depend on the steady supply of ecological goods and services—such as drinking water, crop pollination, soil restoration and regeneration, a wealth of marine life for food, carbon dioxide storage, and more—it is critical to comprehend the resiliency of ecosystems. When an ecosystem's resilience is surpassed and the system transitions into a different state, the system may become less desirable from a human perspective and produce fewer goods and services, which can be dangerous for those impacted. This is because risk is the uncertain negative (or undesirable) consequence of an event or activity with regard to something that humans value. Therefore, risk approaches should concentrate on identifying critical thresholds.

The non-linear dynamics of complex systems make it hard to forecast when an ecosystem's resistance will be surpassed, although it is frequently evident when the system observably changes, such as when a lake turns from clear to turbid. Small systemic changes could have disproportionately big effects, and vice versa. It is important to distinguish ecological resilience from engineering resilience, which focuses on a system's capacity to carry out a given duty reliably and consistently as well as to promptly resume operation in the event of a disruption. Because complex adaptive systems cannot be viewed as an engineering system with predictable and consistent outputs, the implications of applying this way of thinking to the management of complex systems—particularly social-ecological systems—are troublesome (Liu et al., 2023). When competing forces are in balance, ecosystems do not have equilibrium. Instead, an ecosystem occurs inside a regime, and within a given regime, the composition and quantity of the constituent elements may vary over time in a highly dynamic manner. This implies that in a risk scenario, combined disruptions brought on by catastrophes or global change may have unanticipated and extremely unpredictable consequences for people, societies, and natural resources. Economic crises, conflicts, human health issues, and systemic hysteresis can all have these effects, as can regime changes, which may cause environmental degradation and lock a social-ecological system into an undesirable system state that may be hard to escape. The idea that surprise and uncertainty are part of any complex social-ecological system is a fundamental one. It is challenging to achieve and manage trade-offs between lowering risks and maximizing benefits for mankind, and this approach falls short in addressing surprise and uncertainty in system behavior. As a result, social-ecological system governance and management have largely struggled to maintain human-desired ecological regimes in terms of reliable supply of ecological goods and services during system change. The long-term misuse of natural resources has frequently resulted in regime shifts, such as the demise of commercial fisheries or agricultural systems (Rahimi et al., 2023; Seekell et al., 2022).

1.1.5 Risk, Thresholds, and Undesirable States

Resilience-based management reduces risk in two ways because risk and undesirability are identical ideas (risk is the unwanted result of uncertainty). First, management can concentrate on developing and strengthening the resilience required to sustain this regime when systems are in a state that people find desirable. This can be achieved by making sure that functional qualities pertinent to processes that provide ecosystem services are varied and imbricated. This lowers the possibility that the system will change into an undesired condition if it experiences unmanageable stress, which could result in the system's inability to continue providing the services for which it was designed. Prioritizing resilience in planning and management, however, has frequently proven challenging (Chambers et al., 2019). Management has long placed a high priority on maximizing resource use and reducing variability or uncertainty in harvest use or efficiency. While these goals are crucial, they frequently weaken resilience by restricting the system's use of state space, which may narrow the basin of attraction for a desired state and increase the likelihood of collapse. Second, even if systems in undesirable states can be extremely resilient, they nonetheless pose hazards to the resources or individuals that depend on them (such as those trapped in poverty). Systems in undesirable states that are resistant to change—that is, when an undesirable state is resilient—may need to have their resilience decreased in order to induce a shift to a more desirable regime. The system may then need to be managed in order to support the regime of this desirable state.

1.1.6 Transformability and Governance

Walker et al. (2004) proposed the concept of "transformability" as a characteristic of complex systems that characterized the possible dynamics and future course of a system. The ability to "create a fundamentally new system when ecological, economic, or social conditions make the existing system untenable" is how the authors defined transformability in this context. According to Folke et al. (2010), a transformation at the Social-Ecological System (SES) level is a purposeful, socially driven process that involves "a phased introduction of one or more new state variables" in order to push a system over a threshold. Other definitions of transformation assert that there are two kinds of transformative change: unintended or unforeseen due to a process or event, and intentional or planned. A social-ecological system experiences a regime transition and a threshold is crossed in each of these situations, whether deliberate or not. However, we define a transition as occurring when the regime shift is a direct result of human planning and activity, human agency. Transformation is a multilayer, scale-dependent process that can occur at the system level or at the organizational, personal, or sub systemic levels. As a result, a transformation necessitates intentionally surpassing a system's resistance in an unwanted state and then encouraging a reorganization into a more desired state by human intervention.

The key to comprehending transformation is the idea of panarchy (Allen et al., 2014). It is very "dependent on the nature and extent of adaptive actions being taken at other interacting scales" and is probably the result of a layered system that is impacted by cross-scale interactions from above and below. Larger-scale change, for instance, might only happen when individual or personal modifications are intensified to create the group's ability to effect change. In this sense, directive or purpose transformation has also been used to describe intentional transformation. In any case, transformation refers to alterations in the prevailing mechanisms and systems that sustain a complex social-ecological system within a specific state. These variables can include processes and related feedbacks like biophysical cycles, ecological hierarchies, human activity, or social institutions; however, they differ for every SES at a given scale. Therefore, transformative governance increases society's ability to change these processes to support a new regime in social-ecological systems that fulfills societal goals. An emerging method for managing complex systems that has the ability to both lead and catalyze regime transformations in systems at various scales is transformative governance. To displace the established forms of environmental governance and create room for innovation, the transformation process necessitates upending internal or external drivers or introducing new ones. This will promote a fundamentally positive change in the nature of the impacted social-ecological system.

1.2 CHALLENGES, FUTURE DIRECTIONS, AND CONCLUSION

The fact that it is fairly easy to identify a system that has experienced a regime change yet very challenging to identify when a system's resilience has been undermined presents a challenge to resilience theory. Although there have been recent developments in identifying early warning of impending regime shifts, the quantification of system resilience is still in its infancy and is still underdeveloped. These developments typically center on increasing variance in important ecological system parameters (Angeler & Allen, 2016). The creation of leading indicators is essential to resilience management and, consequently, to good environmental management. In understanding the interconnectedness of ecosystems and human societies, this highlights the urgent need to embrace ecoresilience as a guiding framework, paving the way for sustainable solutions that harmonize environmental, social, and economic systems in the face of ongoing challenges.

REFERENCES

Albrich, K., Ramme, W., Turner, M. G., Ratajczak, Z., Braziunas, K. H., Hansen, W. D., & Seidl, R. (2020). Simulating forest resilience: A review. *Global Ecology and Conservation*, *29*(12), 2082–2096.

Allen, C. R., Angeler, D. G., Garmestani, A. S., Gunderson, L. H., & Holling, C. S. (2014). Panarchy: Theory and application. *Ecosystems*, *17*(3), 578–589. Reviews the development, use and evidence for panarchical structure in social, ecological and social-ecological systems.

Angeler, D. G., & Allen, C. R. (2016). Quantifying resilience. *Journal of Applied Ecology*, *53*(3), 617–624.

Angeler, D. G., Fried-Petersen, H. B., Allen, C. R., Garmestani, A., Twidwell, D., Chuang, W. C., & Wonkka, C. L. (2019). Adaptive capacity in ecosystems. *Advances in Ecological Research*, *60*, 1–24.

Baho, D. L., Allen, C. R., Garmestani, A. S., Fried-Petersen, H. B., Renes, S. E., Gunderson, L. H., & Angeler, D. G. (2017). A quantitative framework for assessing ecological resilience. *The Journal of Integrative Environmental Sciences*, *22*(3), 1–26.

Bone, C., Moseley, C., Vinyeta, K., & Bixler, R. P. (2016). Employing resilience in the United States forest service. *Land Use Policy*, *52*, 430–438.

Bowditch, E. A., McMorran, R., Bryce, R., & Smith, M. (2019). Perception and partnership: Developing forest resilience on private estates. *Economic Policy Journal*, *99*, 110–122.

Boyd, E., & Folke, C. (2011). *Adapting Institutions: Governance, complexity and social-ecological resilience*. Cambridge University Press.

Carpenter, B., Walker, J. M., Anderies, N., & Abel, 2001. From metaphor to measurement: Resilience of what to what? *Ecosystems*, *4*(8): 765–781.

Chaffin, B. C., Garmestani, A. S., Gunderson, L. H., Benson, M. H., Angeler, D. A., Arnold, C., & Allen, C. R. (2016). Transformative environmental governance. *Annual Review of Environment and Resources*, *41*, 399–423.

Chambers, J. C., Allen, C. R., & Cushman, S. A. (2019). Operationalizing ecological resilience concepts for managing species and ecosystems at risk. *Frontiers in Ecology and Evolution*, *7*, 241.

Elmqvist, T., Folke, C., Nyström, M., Peterson, G., Bengtsson, J., Walker, B., & Norberg, J. (2003). Response diversity, ecosystem change, and resilience. *Frontiers in Ecology and the Environment, 1*(9), 488–494.

Ferro-Azcona, H., Espinoza-Tenorio, A., Calderón-Contreras, R., Ramenzoni, V. C., País, M. D. L. M. G., & Mesa-Jurado, M. A. (2019). Adaptive capacity and social-ecological resilience of coastal areas: A systematic review. *Ocean & Coastal Management, 173*, 36–51.

Folke, C. (2006). Resilience: The emergence of a perspective for social–ecological systems analyses. *Global Environmental Change, 16*(3), 253–267.

Folke, C., Carpenter, S., Walker, B., Scheffer, M., Elmqvist, T., Gunderson, L., & Holling, C. (2004). Regime shifts, resilience, and biodiversity in ecosystem management. *Annual Review of Ecology and Systematics, 35*(1), 557–581.

Folke, C., Carpenter, S. R., Walker, B., Scheffer, M., Chapin, T., & Rockstrom, J. (2010). Resilience thinking: Integrating resilience, adaptability and transformability. *Ecology and Society, 15*(4).

Folke, C., Hahn, T., Olsson, P., & Norberg, J. (2005). Adaptive governance of social-ecological systems. *Annual Review of Environment and Resources, 30*, 441–473.

Gunderson, L., & Allen, C. (2010). Why resilience? Why now. In L. Gunderson, C. Allen, & C. Holling (Eds.), *Foundations of ecological resilience* (pp. xiii–xxv). Island Press.

Gunderson, L. H. (2000). Ecological resilience in theory and application. *Annual Review of Ecology, Evolution, and Systematics, 31*(1), 425–439.

Gunderson, L. H., Allen, C. R., & Holling, C. S. (2010). *Foundations of ecological resilience*. Island Press.

Hale, J. D., Pugh, T. A., Sadler, J. P., Boyko, C. T., Brown, J., Caputo, S., & Hales, C. (2015). Delivering a multi-functional and resilient urban forest. *Sustainability, 7*(4), 4600–4624.

Holling, C. S. (1973). Resilience and stability of ecological systems. *Annual Review of Ecology, Evolution, and Systematics*, 1–23.

Liu, F., Dai, E., & Yin, J. (2023). A review of social–ecological system research and geographical applications. *Sustainability, 15*(8), 6930.

Marchese, D., Reynolds, E., Bates, M. E., Morgan, H., Clark, S., & Linkov, I. (2018). Resilience and sustainability: Similarities and differences in environmental management applications. *Science of the Total Environment, 613*, 1275–1283.

Mina, M., Messier, C., Duveneck, M., Fortin, M. J., & Aquilué, N. (2021). Network analysis can guide resilience-based management in forest landscapes under global change. *Ecological Applications, 31*(1), e2221.

Ovenden, T. S., Perks, M. P., Clarke, T. K., Mencuccini, M., & Jump, A. S. (2021). Life after recovery: Increased resolution of forest resilience assessment sheds new light on post-drought compensatory growth and recovery dynamics. *Journal of Ecology, 9*, 3157–3170.

Pérez-Luque, A. J., Gea-Izquierdo, G., & Zamora, R. (2021). Land-use legacies and climate change as a double challenge to oak forest resilience: Mismatches of geographical and ecological rear edges. *Ecosystems, 24*(4), 755–773.

Rahimi, M., Ahmadaali, K., & Goli, A. M. (2023). Revealing regime shifts and their impact on social-ecological systems: A case study on Iran. *Water Conservation Science and Engineering, 8*(1), 56.

Seekell, D. A., Pace, M. L., Heffernan, J. B., & Holbrook, S. J., 2022. Nonlinear dynamics, resilience, and regime shifts in aquatic communities and ecosystems: An overview. *Limnology and Oceanography, 67*(S1).

Seidl, R., Spies, T. A., Peterson, D. L., Stephens, S. L., & Hicke, J. A., 2016. Searching for resilience: addressing the impacts of changing disturbance regimes on forest ecosystem services. *Journal of Applied Ecology, 53*(1), 120–129.

Torrico, J. C., & Janssens, M. J. (2010). Rapid assessment methods of resilience for natural and agricultural systems. *Anais da Academia Brasileira de Ciencias, 82*(4), 1095–1105.

Walker, B., Holling, C. S., Carpenter, S. R., & Kinzig, A. (2004). Resilience, adaptability and transformability in social–ecological systems. *Ecology and Society, 9*(2), 5.

2 Exploring the Dynamics of Forests in Climate Change

Insights and Impacts

Varun Attri

2.1 INTRODUCTION

Forest dynamics involve changes in stand structure, species composition, and interactions, either caused or affected by the type and severity of disturbances that occur. These dynamics are then modified by interactions between disturbances and landforms across various spatial and temporal scales. Climate change is a complex phenomenon influenced by both natural and anthropogenic factors. While natural processes like solar variations, volcanic activity, and tectonic shifts have historically driven changes over long timescales, human activities now dominate, accounting for over 90% of recent changes (IPCC, 2007). The combustion of fossil fuels, deforestation, and changes in landuse significantly increased greenhouse gas (GHG) concentrations, intensifying global warming (NAS, 2001). Global climate changes due to rising temperature and GHG concentrations are expected to profoundly impact forests dynamics by altering forest ecosystem processes like establishment, tree growth, survival, and reproductive patterns. This influence extends to the changes in community structure, composition, species distribution, and overall function of the forest's ecosystem (IPCC, 2007). As climate conditions shift beyond historical baselines, plant species may shift their ranges either latitudinally or altitudinally to find more favorable environments. These climate-driven changes reshape forest characteristics, affecting forest structure, composition, and ecological functions. For example, boreal forests might be expanding into tundra regions whereas temperate forests and grasslands are encroaching upon them (Neilson et al., 1998). Such changes in terrestrial vegetation have profound implications, such as altering water quality and supply, disturbance regimes, and biodiversity (Gitay et al., 2001). This redistribution disrupts long-established ecological interactions, such as mutualistic and competitive relationships, leading to the reorganization of forest communities. This may lead to higher probabilities of extinction for some species that cannot adapt or move away, thus threatening the functional resilience and stability of ecosystems (Thomas et al., 2004). Climate-induced shifts in phenology can alter competitive interactions, resource supply and availability, and ecosystem dynamics, often favoring earlier flowering and advancing other related processes (Bajpai et al., 2017). Many invasive species thrive in disturbed areas such as post-fire landscapes, which are anticipated to become

DOI: 10.1201/9781003588580-2

more prevalent with changing climate (Yadav et al., 2016). Natural disturbances and human activities significantly impact forest dynamics, introducing changes in ecological, social, and economic functions (Wang et al., 2023). Hence, it is imperative to examine the forests under such factors of climate change because projections indicate increased counts of record warm nights and record cold days, increased frequencies of drought, longer periods of highly intense as well as more frequent hot spells (He et al., 2022). Other climatic parameters, such as temperature, CO_2, ozone, nitrogen deposition, and extreme events, as well as land-use changes, influence forest dynamics via species composition, growth, and mortality rate changes which in turn define the forest's ability to withstand and recover from future episodic disturbances (Mina et al., 2022).

Forest dynamics in response to climate change are closely linked to changes in composition and structure through its influence on forest growth such as dominant plant species, size and height of the trees, crown density and stem density (Reyes-Palomeque et al., 2021). As climate variables viz., temperature, precipitation, and drought frequency shift, these changes directly affect the growth rates, survival, and reproduction of tree species undergo changes due to site potential, climatic regimes, and even competition for light, nutrients, and water where the interrelations among these are affected by seasonal, annual, and long-term changes in the environmental conditions (Hao et al., 2020). Under favorable conditions, such as sufficient water availability, a short and moderate dry period, and elevated temperatures, forests can experience enhanced growth rates. However, such conditions may not be uniformly beneficial, as the vulnerability of tree species to extreme climatic variations plays a critical role in their ability to thrive and regenerate. Forests play a dual role in the global carbon cycle, acting as both significant sources and sinks of GHGs due to natural processes like fires, insect infestations, and tree growth, as well as anthropogenic activities such as harvesting, afforestation, and deforestation. To mitigate climate change, managing forests effectively is key to reducing their potential as GHG sources and enhancing their capacity as carbon sinks. This involves strategies like limiting deforestation, minimizing emissions from forestry operations, reducing disturbances during harvesting, and facilitating faster tree re-establishment. Additionally, afforestation on abandoned agricultural land, wastelands, or non-forested areas and promoting growth in existing forests further enhance their role as carbon sinks (Watson et al., 2000). Agroforestry is one of the most promising agroecological approaches to climate change adaptation because many agroforestry systems provide multiple cobenefits in addition to climate change adaptation (Tschora & Cherubini, 2020). Adaption practices such as resistance to change, evolutionary adaptation, phenotypic plasticity, genetic diversity facilitated through regeneration and assisted migration, silvicultural management through selective cutting and thinning i.e. optimizing tree density and managing growing stock, these practices reduce vulnerability to biotic stresses, including pests and diseases, as well as abiotic pressures like harsh weather events. It also promotes diverse growth forms and increases ecological resilience which is important in addressing the impacts of climate change, focusing on forest resilience. Adapting forest management practices to changing climate requires an integrated understanding of how climate influences the dynamics of forests, industries, and communities, as well as the ability to predict how these impacts may change through time. Incorporating such knowledge into management decisions will involve using diverse forms of expertise and innovative approaches. As these adaptive strategies are implemented, they are expected to gradually change the structure and composition of forest communities. Change in species distributions, growth patterns, and disrupted ecological processes will likely lead to changes in the dynamics of forests. This chapter, therefore, explores the dynamics of forests in climate change: insights and impacts in the context of a changing climate. By examining how climate change is influencing distributions, forest growth pattern, ecosystem processes and functions, this chapter aims to provide a comprehensive understanding of the impacts faced by forest ecosystems and the potential pathways for mitigation and adaptation in changing climatic scenario.

2.2 CAUSES OF CLIMATE CHANGE

Climate change anchored various forcing mechanisms, or "climate forcings" (IPCC, 2007), including natural processes like variability in solar radiation, changes in Earth's orbit, and geological shifts such as mountain and continental drift, along with changing concentrations of GHGs. These forcings can be intensified or weakened by climate feedbacks, which differ in response speed; for example, oceans and ice caps respond slowly, while other systems are faster to react. Climate forcings are classified as either internal or external. Internal forcings include natural climate variability, such as oceanic changes driven by phenomena like El Niño, the Pacific decadal oscillation, and thermohaline circulation, atmospheric circulation such as monsoon, cyclone etc., sea ice and snow cover changes, carbon cycle fluctuation which redistributes heat across the ocean and land but only affects climate over the short term. External forcings, however, encompass both natural and human-induced factors. Solar variation, including the 11-year solar cycle, influences climate through changes in solar intensity, while volcanic eruptions can temporarily cool the climate by blocking solar radiation; for instance, the 1991 eruption of Mount Pinatubo lowered global temperatures by about 0.5°C (IPCC, 2007; Oppenheimer, 2003). Tectonic shifts, which influence climate over millions of years, impact ocean and atmospheric circulation patterns (Figure 2.1); for example, the formation of the Isthmus of Panama altered ocean currents and likely contributed to Northern Hemisphere glaciation (Forest, 1999). Today, human activity is the major driver of climate change, responsible for over 90% of recent changes. The fossil fuels burning, pollution, population, industries deforestation, and agricultural practices have led to increased emissions of GHGs like CO_2, CH_4, and N_2O, which trap heat in the atmosphere (IPCC, 2007). Land-use changes further exacerbate CO_2 levels by converting forests to agricultural land. These GHGs vary in their warming potential and atmospheric lifetimes, amplifying their contribution to global warming (NAS, 2001).

2.3 IMPACTS OF CLIMATECHANGE ON FOREST DYNAMICS

Forest dynamics processes like establishment, growth, and reproduction are meticulously linked to environmental conditions, making them sensitive to climate change. Changes in temperature and precipitation pattern altered species dynamics and competition (IPCC, 2007). Species dynamics typically respond to climate change through increased growth, local extinctions, or dispersal to favorable sites, often observed as range shifts, phenological changes (Figure 2.2) and invasive species spread (Sykes, 2009).

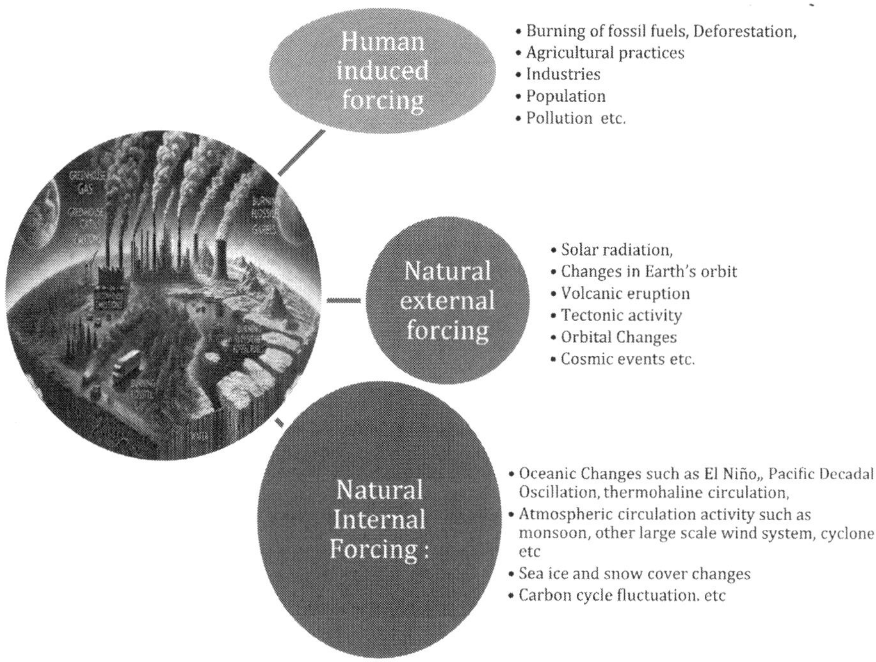

FIGURE 2.1 Drivers of climate change.

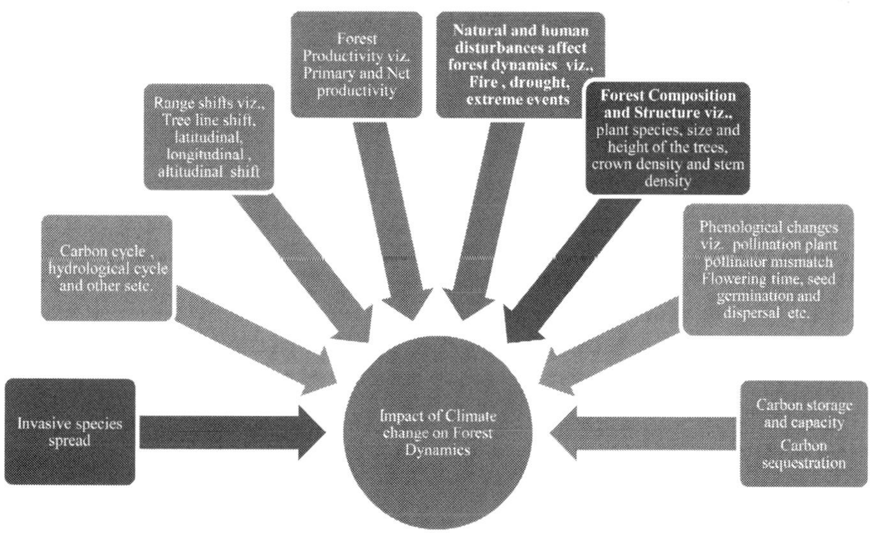

FIGURE 2.2 Influence of climate change on forest growth and productivity.

2.3.1 Shifts in Range

Biogeographic regions, molded by unique species compositions and evolved traits, are increasingly impacted by climate change. The rapid environmental shifts outpace species' adaptive capacities, leading to significant biome alterations such as forest expansion into tundra, montane forest invasions into warmer areas, and forest loss. Factors like temperature, rainfall, CO_2 levels, niche filling, and environmental filters play an important role in reshaping ecosystems (Elsen et al., 2022). In India future scenarios predict a 70 percent shift of Indian forests toward moist types under increased CO_2 levels and changing climate regimes. Himalayan forests, undergoing rapid warming, face threats to species like Himalayan birch due to dependency on snowmelt (Chaturvedi et al., 2011). Invasive species like *Prosopis juliflora* in arid grasslands and *Lantana camara* in the Western Ghats further stress ecosystems (Hiremath & Sundaram, 2005). High latitudes in boreal forests indicate rapid biomass increases, enhancing carbon sequestration and productivity under warming trends (Pillay et al., 2022). Tree-line shifts are driven by warming temperatures and prolonged growing seasons, causing species to move to higher altitudes or latitudes. Examples includelatitudinal shifts of *Ilex aquifolium* expanded northward in response to the 0°C isotherm shift, while thermophilic species have proliferated in Europe, replacing cold-tolerant species (Colwell et al., 2008). Altitudinal shifts of treelines which have moved upslope in Vermont (91–115 m) and southern California (65 m) due to warming (Kelly & Goulden, 2008). In Costa Rica, upslope movements are partially offset by lowland species' arrival. In Europe's mountain regions, endemic species are being replaced by invasive species highlighting complex interspecies dynamics (Dainese et al., 2014). Soil conditions, water availability, and topography interact with inter- and intra-species competition to determine plant distribution. Species ranges shift as plants respond to climate changes through competition, resource availability, and tolerance limits (Sykes, 2009). This dynamic interplay underscores the urgency of monitoring and modeling biogeographic shifts under climate change to mitigate ecological and biodiversity impacts.

2.3.2 Phenology of Plants

Phenology, the timing of events such as flowering, fruiting, and leaf emergence, serves as a sensitive biological indicator of climate change (Calinger et al., 2013). Changes in flowering phenophases significantly impact plant health, biodiversity, and ecosystem functions by altering processes like pollination, germination, and seed dispersal, and increasing risks of plant-pollinator mismatches (Franco-Cisterna et al., 2020). Flowering time is influenced by factors such as plant growth, flower color, pollination types (Wang et al., 2020), and environmental changes, including warming temperature and precipitation (Rafferty et al., 2020). Climate-induced phonological shift can alter competitive interactions, resource supply and availability, and ecosystem dynamics, often favoring earlier flowering and advancing other related processes (Bajpai et al., 2017). Global trends in flowering and leafing episodes have advanced of 2.5 days per decade, while leaf fall by 0.2 days per decade (Menzel et al., 2006). In temperate regions of the UK, 385 species showed a 4.5-day earlier flowering per decade in the 1990s (Fitter & Fitter, 2002). In North

America, spring phenology advanced by 5–6 days from 1959 to 1993 due to rising temperatures (Schwartz & Reiter, 2000). Studies on alpine and temperate mountains, including the Alps and Nanling Mountains, show consistent advancements in spring-flowering, often linked to temperature increases (Song et al., 2021). Species-specific responses *viz.*, Rhododendron flowering advanced by 2.27 days/°C rises in annual temperature (Hart et al., 2014). Eucalyptus species in Australia exhibited mixed flowering responses to temperature and rainfall changes (Keatley et al., 2002). In the Gongga Mountains, flowering times were held up over the last 70 years but varied by elevation and season, with spring-flowering plants blooming earlier while summer/autumn plants showed delays (Zu et al., 2023). In the north-western Indian Himalayas, *Rhododendron arboreum* showed a shift in blooming onset compared to historical herbarium records (Rana et al., 2024). These studies accent that many plant species exhibit shifts in phenological responses to climate change; the extent and nature of these changes vary across species, regions, and climatic conditions.

2.3.3 Expansion of Invasive Plant Species

Invasive alien species characterized by rapid vegetative and reproductive expansion, often gain a competitive edge in elevated atmospheric CO_2 (Bradley et al., 2010). This citability allows them to increase their range beyond that of native species and migrate to higher elevations, particularly under conditions of increased human activity and global climate change. Many invasive species thrive in disturbed areas such as post-fire landscapes, which are more common with climate change (Yadav et al., 2016). Expanding risk areas for invasives by shifts in precipitation and temperature alters land areas vulnerable to invasion. Higher precipitation levels can encourage the proliferation of non-native grasses in the western United States (Martin-R et al., 1995). Rising temperatures enable the northward spread of invasive species in the southeastern United States (Rogers & McCarty, 2000). Species like yellow starthistle (*Centaurea solstitialis*) and tamarisk (*Tamarix* spp.) are showing expansion trends, while others, including cheatgrass (*Bromus tectorum*) and spotted knapweed (*Centaurea biebersteinii*), may be undergoing range shifts that could involve either expansion or contraction (Bradley et al., 2010). Weedy plants cause threats to man-managed systems such as agriculture, rangelands, and forests which lead to great financial costs (Pimentel et al., 2000). In the Indian Himalayan Region, 297 naturalized alien plant species from 65 families have been identified, with Himachal Pradesh leading at 78.1%, followed by Jammu and Kashmir at 64.6%, and Uttarakhand at 60.9% (Pathak et al., 2019). Species such as *L. camara, Ageratina adenophora*, *Parthenium hysterophorus*, and *Ageratum conyzoides* have also invaded other ecosystems. *A. adenophora's* expansion to a limit of 2900 m elevation has surpassed its previous pattern, demonstrating how climate change can exacerbate the encroachment pattern over alien taxa over the higher Himalayas.

2.3.4 Forest Composition and Structure

Forests have distinct composition and structure formed by dominant plant species, size and height of the trees, crown density and stem density (Reyes-Palomeque et al., 2021).

The height and diameter of trees undergo changes due to site potential, climatic regimes, and even competition for light, nutrients, and water where the interrelations of these are affected by seasonal, annual, and long-term changes in the environmental conditions (Hao et al., 2020). Other climatic parameters, such as temperature, CO_2, ozone, nitrogen deposition, and extreme events, as well as land-use changes, influence forest dynamics via species composition, growth, and mortality rate changes which in turn define the forest's ability to withstand and recover from future episodic disturbances (Mina et al., 2022). Moreover, activities of logging coupled with reforestation influence the aging of the forests and their density and carbon sequestration in a manner that end up affecting the global carbon cycle (Astigarraga et al., 2020). The thermal niches that define the optimal condition of a particular species are not static, but rather shift as temperature remains constantly changing, moving tree species in latitudinal or altitudinal gradients. However, the migration rates are limited by seed dispersal and other environmental factors that restrict them (Coelho et al., 2023). In tropical forests, rising temperatures and water stress reduce species richness. Prolonged dry seasons and intensified competition for soil moisture favor early successional species with deeper roots capable of accessing seasonal water stores (Chaudhary et al., 2021). These findings point out the complex interplay of climate change and ecological factors on forest dynamics, influencing their composition, structure, and long-term resilience.

2.3.5 Natural and Human Disturbances Affect Forest Dynamics

Natural disturbances and human activities significantly impact forest dynamics, introducing changes in ecological, social, and economic functions (Wang et al., 2023). Climate change necessitates examining forests, as projections indicate an increase in climatic anomalies such as record warm nights, unusually cold days, heightened drought frequency, prolonged periods of extreme intensity, and recurring heatwaves. (He et al., 2022). Climate change's effects on wildfires are expected to be severe due to extreme weather events (Keith et al., 2009). Such changes in climate may ultimately cause a radical shift in the size, intensity, and frequency of fires, leading to increased risks of fire occurrence, extended fire seasons, and more calamitous effects. The associated risks and frailties caused by the fire are microclimate change, floods, infrastructure damage, economic loss, and human loss (WRI, 2024). Wildfire may activate other environmental perturbations accelerating transformed vegetation regimes, degradation and desertification of land, and perturbations in hydrological cycles. Somewhat modest climate changes have been known to spark catastrophic regime shifts in ecosystems where human exploitation erodes their resilience (SCBD [Secretariat of the Convention on Biological Diversity], 2001). Wildfires contribute to a feedback loop in the climate system, wherein the emission of carbon and other GHGs exacerbates climate warming. This warming creates more favorable conditions for wildfires, which in turn lead to further emissions, perpetuating the cycle (Abatzoglou & Kolden, 2013). Fire is becoming an increasingly significant driver of global tree cover loss. While historically responsible for less than 20% of tree cover loss, fires now account for approximately 33%. A significant proportion, nearly 70%, of fire-related tree cover loss between 2001 and 2023 occurred in boreal regions. In these regions, fire-induced tree cover loss has risen by approximately 138,000 ha annually (around 3.6%) over the past 23 years, representing about half of the total global increase in fire-related tree cover loss during this period (WRI, 2024). Ahmad and Goparaju (2018) analyzed forest fire patterns in Himachal Pradesh and Uttarakhand, India, and concluded that Pauri Garhwal had the highest incidences of forest fires, 22.4%, followed by Nainital at 16.4% and Tehri Garhwal at 8.5%. Deciduous broadleaf forests, covering 17.2% of the forest area, constituted 44.7% of the total fire events. Most forest fires in both regions occurred in April and May. April recorded the most frequent forest fires of about 79%. Climate analysis revealed that rising temperatures and reduced precipitation, particularly in April as projected for 2050, are likely to exacerbate forest fire risks in high-incidence areas. Sagar et al. (2024) identified significant forest fire hotspots in India, with incidents rising sharply in Mizoram by 2022 (averaging 2614.78 ± 1519.36). The authors emphasize the need for proactive, interdisciplinary strategies to address ecological and public health impacts determined in the findings to be strongly correlated with temperature and forest fires. Mina et al. (2023) documented a sharp increase in forest fires in Uttarakhand between 2002 and 2019, attributing the trend to rising temperatures associated with climate change. These changes not only heighten fire risks but also disrupt carbon sequestration processes. Similarly, in the western United States, earlier spring warming, increased vapor pressure deficit, and reduced summer precipitation have extended the fire season and expanded the burned area (Dennison et al., 2014).

Insects and diseases, due to their shorter generational cycles, can adapt more rapidly to changing climatic conditions compared to trees. However, if insect populations migrate to new environments faster than tree species can, certain segments of the tree population may experience less impact in the future (Regniere, 2009). The interaction between pests, diseases, and fire is another critical factor, particularly for ecosystems such as western white pine (*Pinus monticola*) in Montana, United States. Loehman et al. (2011) suggested that rising temperatures could benefit western white pine over climax and shade-tolerant species, primarily due to the increased frequency and extent of wildfires under warmer conditions, which support the regeneration of these species. Climate change has increased length, severity, and extreme events frequencies such as droughts, wildfires, and heatwaves that affect the integrity of ecosystems and make them less resilient (Jay et al., 2018). Increased droughts and shifted precipitation have disturbed forest ecosystems mainly across the United States. West, Southeast, and Lake States, where warmer and variable precipitation conditions have amplified drought severity, lowered tree growth, and increased tree mortality over the past two decades (Clark et al., 2016). Such droughts weaken tree defenses, making them more susceptible to disturbances, including pests, pathogens, invasive species, and fires (Trottier et al., 2017). Drought directly impacts long-term ecosystem structure; it also facilitates immediate changes through these disturbances (Loehman et al., 2017). In the North Atlantic, rising ocean temperatures and sea levels have intensified storms, leading to stronger cyclones and increased rainfall (Patricola & Wehner, 2018). This heightened storm intensity poses risks through extreme flooding, erosive waves, and higher storm surges, complicating recovery for affected ecosystems and communities. Ocean warming has forced marine heatwaves, such as those in the Northwest Atlantic in 2012 and 2016, and the northeast

Pacific in 2014–2015, which brought about coral bleaching, loss of coral cover, modified fish communities, and enhanced coastal exposure to waves (Hobday et al., 2018). Climatic changes exert both positive and negative effects on forests. Positive impacts include enhanced vigor and growth due to CO_2 fertilization, improved water use efficiency, and longer growing seasons. Conversely, negative effects involve reduced growth, increased stress, and heightened mortality driven by climatic changes and their influence on forest insect and pathogen dynamics (Lloyd & Bunn, 2007). Forests also endure significant human-induced stresses, such as elevated ground-level ozone and atmospheric deposition (Ollinger et al., 2008). Modeling the effects of these processes and other stressors on large-scale tree mortality, referred to as "forest mortality," remains highly uncertain under changing climatic conditions (Bugmann et al., 2001). Despite this uncertainty, recent increases in tree mortality linked to drought and elevated temperatures suggest that climate change may already be exacerbating forest mortality in some regions. Tree mortality has been observed across diverse forest types and regions, driven primarily by drought, heat, and associated stressors. In southern Europe, mortality rates have risen significantly (Bigler et al., 2006), while temperate and boreal forests in western North America have experienced considerable increases in background mortality over recent decades. Drought-induced mortality has been documented in tropical forests, such as in Uganda (Lwanga, 2003), as well as in mountain acacia in Zimbabwe (Tafangenyasha, 2001) and mesic savanna trees in South Africa's Kruger National Park (Viljoen, 1995). Atlas cedar (*Cedrus atlantica*) mortality has occurred during severe droughts in Morocco and Algeria (Bentouati, 2008). In Asia, droughts have caused significant tree die-offs, such as in *Pinus tabuliformis* across 0.5 million ha and *Pinus yunnanensis*in China, *Abies koreana*in South Korea (Lim et al., 2008), and *Juniperus procera* in Saudi Arabia. Similar impacts have been observed in pine and fir species in Turkey. Tropical regions have also experienced mortality events linked to severe El Niño-induced droughts, affecting the forests of Borneo (van Nieuwstadt & Sheil, 2005) and dry forests in India (Khan et al., 1994). In northeastern Australia, multi-year droughts have led to die-offs of Eucalyptus and Corymbia species (Fensham & Fairfax, 2007), along with impacts on acacia woodlands (Fensham & Fairfax, 2005). Mediterranean forests have not been spared; *Abies cephalonica* in Greece suffered from severe droughts (Tsopelas et al., 2004), while drought-resistant Mediterranean pines faced mortality in eastern Greece (Korner et al., 2005). In eastern Europe and northern regions, combined drought and other stressors have driven mortality in *Quercus robur* in Poland (Siwecki & Ufnalksi, 1998), *Picea abies* in Norway (Solberg, 2004), and *Picea obovata* in Russia (Kauhanen et al., 2008). Large-scale mortality has also occurred in North America, including over 20 million hectares of spruce in Alaska since 1997 (Berg et al., 2006), *Pinus contorta* in British Columbia (Kurz et al., 2008), and *Populus tremuloides* in Saskatchewan and Alberta (Hogg et al., 2008). Seasonal droughts associated with ENSO events have further intensified tree mortality in tropical forests globally (Chazdon et al., 2005).

2.3.6 Forest Productivity

Primary producers, especially photosynthetic organisms, are the basis for nearly all life on Earth, constituting the root of most food webs and food chain, providing oxygen, and playing a regulatory role in carbon cycling and sequestration. Climate change has altered primary production differently over various regions and timescales, resulting in a tendency for effects often to cascade to higher trophic levels and interactively affect ecosystem functioning (Lipton et al., 2018). Globally, terrestrial primary production increased during the late 20th and early 21st century due to aspects such as enhanced atmospheric CO_2, nutrient inputs from human activities, extended growing seasons, and forest regrowth (Domke et al., 2018). But regional trends vary widely because climate change can have direct opposing effects on productivity: while increased CO_2 concentration may enhance vegetation growth, productivity could be limited by factors such as nutrient imbalances, water scarcity, and pollution (Norby et al., 2010). Alterations of CO_2 and temperature also influence belowground processes, carbon and nitrogen cycling, and overall productivity (Melillo et al., 2017). Climate-driven changes in forest productivity also vary with forest type and elevation. Primary production could decline in water-limited forests during growing seasons but could rise in energy-limited forests as snow and cold constrain growth periods. However, even in these energy-limited regions, droughts and extreme temperatures could mitigate any productivity increases (Hember et al., 2017). The sustainability of fertilization effects over time also remains uncertain, especially as forests age (Norby et al., 2010). In this respect, climate change influences the net primary productivity (NPP) of forests, which are estimated to sequester an average of 835 g C m^{-2} $year^{-1}$ under the current scenario. According to the A2 GHG scenario, this value is expected to more than double and, in B2, 73% higher due to CO_2 fertilization. Tropical forests have NPP increases of 1.35–1.57 times, whereas temperate forests observe smaller gains. A Hybrid-V4.1 model simulation of carbon stored in vegetation is estimated to rise by 290 GtC from 1860 to 2100, while global NPP rises from 45 to 50 GtC/year in the 1990s to 65 GtC/year by the 2080s (White et al., 1999).

2.3.7 Carbon Sequestration/Carbon Storage Capacity

The impact of the increase in CO_2 and climatic change on the carbon balance in forests has major implications for the global carbon cycle (Morichetti et al., 2024). A near-doubled concentration of atmospheric CO_2 has a direct stimulating effect on carbon assimilation in trees; this may lead to potential gains in growth, productivity, and carbon storage. This effect of elevated carbon dioxide has been well-illustrated in different types of forests; each specific ecosystem responds in a different way to the optimality of other resources such as light, water, and nutrients. For instance, long-term observations of trees have indicated that the biomass of the majority of forest has enhanced during the last few decades, simultaneously with a general rise in temperature and the CO_2 concentration (Schepaschenko et al., 2021). Other compounding stressors, however, include climate change. These might be positive and create a warming effect, but as shown by recent massive droughts and forest dieback, the frequency and intensity of these climatic factors may result in higher compositional shifts, productivity declines, and also longer-lasting carbon losses in forests (Vacek et al., 2023).

Forests play a crucial role in carbon storage, sequestering carbon for decades to millennia. However, human-driven deforestation and land-use change, especially for agriculture, have released significant GHGs from these natural carbon reservoirs, a process that began early in the Holocene epoch. Efforts to conserve forests have increased recently, and scientific knowledge regarding carbon storage, along with the physiology, genetics, and evolutionary patterns of thousands of tropical tree species, continues to grow. Notably, tropical forests contribute significantly to global carbon storage: equatorial and tropical forests hold 60% of the world's terrestrial carbon, with the Amazon rainforest alone containing 43%, and monsoonal forests in Southeast Asia adding 22% (Walker et al., 2022).The importance of carbon storage as a climate mitigation strategy is expected to rise with accelerating climate change. While national forest areas have generally increased since 2000, it remains uncertain if carbon storage from afforestation can continuously offset emissions from ongoing deforestation. Additionally, carbon storage capacity in many forests may decline due to higher temperatures, increased water stress, disturbances, and the lower CO_2 uptake in aging forests compared to younger, regrowth forests (Pugh et al., 2019). Coastal wetlands also play a vital role in carbon storage and act as natural defenses against erosion, flooding, and storm surges (Howard et al., 2017). However, the degradation of these ecosystems, whether due to human development or rising sea levels, reduces their ability to provide these essential services.

2.4 MITIGATION AND ADAPTATION

Forests are dynamic ecosystems and active components in the global carbon cycle and changing climate. They are sensitive to, and collectively have strong feedbacks on, climate-induced changes as they naturally sequester carbon, but may at the same time, due to enhanced climate stresses, shift from serving as a significant carbon sink to becoming a substantial carbon source. Forest adaptation to climate change involves managing forests to enhance their resilience, ensuring they continue to provide ecosystem services despite shifting climatic conditions. Climate change brings a mix of threats and challenges to forest ecosystems, such as altered temperature and precipitation patterns, increased pest and disease pressures, extreme weather events, and a higher frequency of wildfires. Effective mitigation and adaptation strategies help forests respond to these changes in ways that maintain or even enhance biodiversity, productivity, and ecosystem functions. Here are some key approaches:

2.4.1 Mitigation

Forests are key to carbon sequestration through afforestation, reforestation, and sustainable forest management. These actions not only store carbon but also reduce the risks of significant carbon release through fires, pests, or diseases (Krankina & Harmon, 2006).

2.4.1.1 Expanding Forest Areas: Afforestation and Reforestation

Expanding forest areas through reducing deforestation and enhancing afforestation has importance, as forests act as a long-term sink capturing carbon dioxide from the atmosphere. Restoration and expansion of forest ecosystems could be completely counterbalanced by emissions caused by human and natural disturbances, settling within the targets imposed by the Paris Agreement on climate change. There has been much interest in the environmental and socio-economic benefits of sustainable forest management, especially in afforestation and reforestation efforts. Global efforts have resulted in 294 million hectares of plantation forests, representing 7% of the total forest area of the world, as of 2020, with an addition of 123 million hectares since 1990; however, growth rates have been tapering off. Global trends in planted forest distribution are notable, with Asia leading, where 22% of forests are planted through human effort. Only 2% of the total forests are reported as planted in Africa and South America, while in North and Central America, it is 6%. In Europe, excluding Russia, it stands at 30% but goes down to 7% when Russian forests are included. These variations symbolize various emphases and strategies for forest management (Food and Agriculture Organization of the United Nations [FAO], 2020). Urban tree-planting could substantially reduce emissions. For example, global urban tree planting alone could sequester 82.4 ± 25.7 million metric tons of CO_2 each year, offsetting about 25% of urban emissions (Teo etal., 2021). According to a study by Bastin et al. (2019), global forest expansion could potentially cover 0.9 billion hectares, sequestering up to 205 gigatons of carbon, though later critiques suggest this estimate might be optimistic due to factors such as pre-existing vegetation and soil carbon levels. However, large-scale afforestation initiatives may present challenges to food security by competing with agricultural land. A substantial increase in forest cover could sequester significant amounts of CO_2 but might also increases food prices, with potential impacts on global food security, especially in regions like Africa and South Asia. Projections suggest that expanding forests by 1.1 billion hectares to achieve the 2°C warming target could negatively impact the food security of up to 441 million people by the end of the century. As a result, any forest-based climate strategy must carefully balance these food security risks (Doelman et al., 2019).

2.4.1.2 Agroforestry

Agroforestry is one of the climatic adaptation strategies in increasing the resilience of farmers and agricultural systems against climate risk, providing multiple biophysical and socio-economic benefits. According to IPCC (2022), agroforestry refers to integration of trees or shrubs with crops and/or livestock on a same unit of land, plot, farm, and/or landscape scale. As recent Intergovernmental Panel on Climate Change (IPCC) reports have stressed, agroforestry is one of the most promising agroecological approaches to climate change adaptation because many agroforestry systems provide multiple cobenefits in addition to climate change adaptation, for example, synergies with climate change mitigation through carbon sequestration, improved food security and income opportunities, provisioning of ecosystem services, and biodiversity conservation (Tschora & Cherubini, 2020). Agroforestry thus offers a climate mitigation and adaptation solution. Further strategies can be applied at the stand level and landscape level to enhance carbon storage. Partial harvesting, reduced residue burning, and regeneration management will increase carbon density within individual stands. At the landscape level, extension of rotation periods, the conservation zone, and the management of disturbance ensure sustained carbon storage across broader forested areas.

2.4.1.3 Adopting Wood-Based Biofuels and Bioenergy

Wood-based bioenergy has some advantages over fossil fuels and certain renewable energy sources because it has a relatively low fossil fuel input and a smaller carbon footprint (Malmsheimer et al., 2008). It is carbon-neutral because the carbon released during wood combustion is offset by forest regrowth and carbon sequestration (Johnson, 2009). Evaluating the carbon impacts of bioenergy requires assessing the full lifecycle of energy generation, including the long-term land-use changes associated with energy production compared to conventional fossil fuel use (Millar et al., 2012). Studies indicate that sustainably produced forest bioenergy would provide long-term carbon advantages, but these are extremely susceptible to the factors like the initial conditions of forests, productivity, energy inputs arising from harvesting and transport activities, and the kind of fossil fuel substituted (Eriksson et al., 2007). A comprehensive analysis should compare forest carbon stocks under bioenergy scenarios versus no-bioenergy alternatives and include a lifecycle assessment of emissions from bioenergy production and displaced fossil fuels (Ter-Mikaelian et al., 2015). Long-term use of wood products, recycling, biofuels, and bioenergy can enhance carbon storage and reduce dependence on fossil fuels, supporting a sustainable, low-carbon future. Integrating these strategies in forest conservation and expansion efforts could address multiple objectives, including mitigating climate change, enhancing food security, and preserving biodiversity.

2.4.1.4 Reduce Emissions

Reducing carbon emissions from wildfires and pest outbreaks involves practices like thinning to improve forest health and resilience, prescribed fires, or managed wildfire suppression. Using harvested biomass for energy generation or long-term storage further reduces emissions (Stephens & Moghaddas, 2005).

2.4.1.5 Prioritizing Management Practices

Species' responses to climate change are individualistic, and forest management must prioritize interventions based on sensitivity, urgency, and resource availability. Using triage approaches can help managers allocate resources effectively, especially in high-demand situations (Fitzgerald, 2000). This prioritization process must be dynamic, adapting as environmental conditions and management needs change.

2.4.2 Adaptations

2.4.2.1 Resistance to Change

One approach is to manage forest ecosystems to resist climate change impacts, forestalling unwanted changes (Parker et al., 2000). This is often applied where maintaining the status quo is necessary, such as in high-value plantations or endangered species' habitats. Practices to enhance forest defenses include reducing fire, insect, and disease risks, such as establishing fuel breaks around high-risk areas, intensive removal of invasive species, and deploying targeted interventions like resistance breeding, pheromone treatments, and herbicides (Agee & Skinner, 2005). These strategies may ultimately fail if environmental changes become too drastic, but they are useful short-term options for forests of high value or low sensitivity to climate (Harris et al., 2006).

2.4.2.2 Promote Resilience to Change

Resilient forests are better able to recover from disturbances with minimal intervention, making resilience-building a popular adaptive strategy. Techniques to enhance resilience include seed-banking and intensive management during the establishment phase, particularly where young plants are sensitive to climate shifts (Betancourt et al., 2004). These methods are generally applied to ecosystems with high value or low sensitivity to climate change.

2.4.2.3 Enable Forests to Respond to Change

This strategy promotes gradual adaptation by facilitating species' natural processes, such as migration and community shifts. Practices include assisting species migrations, modifying silvicultural treatments to anticipate new climates, and increasing genetic diversity through relaxed guidelines (Spittlehouse & Stewart, 2003). Assisted migration of species to suitable new habitats is an emerging concept, although it must be carefully considered given the unpredictability of future climates (McLachlan et al., 2007).

2.4.2.4 Shifting Geographical Distribution, Timing of Growth, and Reproduction

Natural populations have responded to climate change by shifting their geographic distribution and changing the timing of growth and reproduction. But these shifts, in turn, reshape community composition and change species interactions. For many populations, though, these responses have been inadequate to the speed and magnitude of climate change, leaving a few groups exposed to further decline and extinction.

2.4.2.5 Evolutionary Adaptation

Extinction can be avoided by population shifts to favorable habitats, plastic changes in species traits, or evolutionary adaptation to overcome stress (Williams et al., 2008). Rapid evolutionary changes have been recorded in species invading new areas and native species responding to biotic invasions, emphasizing the role of evolution in countering climate change. Threatened species may survive through evolutionary adaptation or human-mediated translocation to suitable habitats (Hoffmann & Sgro, 2011). Adaptive changes can help species take advantage of such favorable conditions as CO_2 enrichment and longer favorable seasons. Climate change, however, is happening at the same time as habitat fragmentation, species invasions, and other pressures that can change gene flow and introduce new genotypes through hybridization (Bajpai et al., 2015, 2018; Chaturvedi et al., 2017). Favorable conditions may create new contact zones between lineages that increase interspecific competition and hybridization, which can sometimes enhance adaptation to new environments (Donovan et al., 2010).

2.4.2.6 Phenotypic Plasticity

Plant species adapt to new conditions through phenotypic plasticity, the ability of a single genotype to express different based on environmental factors. Adaptive plasticity enhances fitness, while non-adaptive plasticity arises from physical processes or resource limitations. Both types influence plant responses to climate change. Genetic variation plays a crucial role, enabling plants to sense environmental changes and produce plastic responses. For instance, variation in genes encoding temperature sensors and transcription factors aids adaptation to temperature shifts. Plasticity thus buffers against rapid climate changes and facilitates adaptation (Lande, 2009).

2.4.2.7 Increase Redundancy and Buffers

Introducing species across varied environments rather than limiting them to historical ranges spreads ecological risks and increases resilience. Monitoring replicated forest plantations and other plantings helps track species' responses to climate shifts, providing insight for future actions (Betancourt et al., 2004). Past climate analogues offer guidance for planting species outside their current ranges, where future conditions may favor them. For example, Monterey pine has adapted to non-native sites in California, showing the feasibility of establishing such "neo-native" stands in suitable areas (Millar, 1998).

2.4.2.8 Expand Genetic Diversity

Traditional genetic management aimed at preserving local gene pools might shift toward incorporating genes from broader environmental ranges. Adjusting seed sources to account for warming trends, while carefully documenting these changes, could yield better-adapted forests in the long term (Millar & Brubaker, 2006). Seed transfer zones assume optimal climatic limits for species and provenances, but the climate change will alter its geographic extent, generally shifting the ranges upward in elevation and northward in latitude within the Northern Hemisphere, leading to new species assemblages (Hansen et al., 2001). These changes require determining species and genotype responses to climate, establishing dynamic climate-based seed zones (Parker et al., 2000), and breeding for pest resistance and broader tolerance to climate stresses (Namkoong, 1984). Existing seed orchards may face unsuitable climates, increasing reliance on wild-stand seeds. Rare and endangered species, with low genetic diversity and specific requirements, may require preservation in artificial reserves or arboreta (Hansen et al., 2001; Parker et al., 2000).

2.4.2.9 Managing Disturbances in Changing Forest Landscapes

Maintaining forest landscape connectivity supports species migration and reduces fragmentation, aiding natural adaptation to climate change (Noss, 2001). Rapid shifts in forest age and landscape patterns may result from increased disturbances, including fire, wind, ice storms, and pests, with interactions between insect outbreaks, warmer climates, and fire (Fleming et al., 2002). Adaptive actions for forest fire regimes include protecting high-value areas, allowing fire to run its course elsewhere (Parker et al., 2000), altering forest structure to reduce disturbance risk (Dale et al., 2001), using prescribed burns to manage fuel (Wheaton, 2001), and creating fire-smart landscapes through strategic harvesting and regeneration. For instance, planting fire-resistant species like aspen among flammable conifers in boreal forests could mitigate fire risks in longterm.

Climate change may increase thermal and moisture stress on forests, making aging trees more vulnerable to pests and diseases. Adaptive measures include partial cutting to boost stand vigor, sanitation cuts to remove infected trees, shorter rotation lengths to reduce vulnerability, transitioning to suitable species (Lindner et al., 2000), and using pest-resistant genotypes (Namkoong, 1984). Insecticides and fungicides may be necessary where silvicultural methods are ineffective (Parker et al., 2000).

2.4.2.10 Anticipate Surprises and Threshold Effects

Managers must be prepared for unexpected ecological shifts or extreme events, such as year-round fires or unprecedented drought-induced diebacks. Proactive management can help forestall or mitigate such threshold effects (Breshears et al., 2005). In heavily altered environments, restoration efforts should align with present and anticipated climates, rather than aiming to restore historical conditions. This approach helps maintain ecological function in changing conditions (Harris et al., 2006).

2.4.2.11 Silvicultural Management

Silviculture has transformed from viewing forests as timber sources mainly to more comprehensive ecological functions (Gustafson et al., 2020). Adaptive management must be maintained on forests to ensure their resilience. It will focus on risk identification, vulnerability assessment, and appropriate, responsive interventions subject to ongoing monitoring (Szmyt & Dering, 2024). Their strategies for forest sustainability and continuity in the face of global environmental changes can be expressed best by active, planned adaptation. Diversity in species composition enhances resilience since different species vary in sensitivities to climatic and site conditions (Cosofret & Bouriaud, 2019). Mixed forests have ecological insurance and are superior to monocultures in term of resilience against species-specific disturbances, such as insects. In general, mixed forests, especially conifer-broadleaved mixtures, display significantly higher resilience and productivity, compensating for economic losses (Haberstroh & Werner, 2022). Species selection should focus on functional traits, such as drought tolerance and regenerative ability (Forrester & Bauhus, 2016). Silvicultural best practices should include diverse systems, for instance selection and shelterwood, artificial planting with resilient species, enhancement of natural regeneration and assisted migration. Intensive thinning in fertile sites enhances growth of shade-tolerant understory, whereas an effective population control of herbivores supports regeneration. Canopy gaps are provided by selection cutting, uneven-aged shelterwood cutting, and thinning, leading to structural complexity as well as resistance to stresses such as pests, disease, drought, and wind (Dusenge et al., 2019). Underplanting of monocultures improves structural diversity, though management in fire-prone areas has to be precautionary. Thinning treatments result in structural diversity, enhance wind resistance, and reduce the vulnerability of forests to pests and pathogens (Jactel et al., 2017). Other common practices, such as lower stocking and intense thinning, result in vigorous, hardy trees. Low-growing stock helps avoid financial losses due to hurricanes and wildfires, where tree competition and fuel loads are reduced significantly, especially in regions with droughts (Puettmann, 2011). Adaptive practices like short rotation cycles, high-density thinning, and diverse species composition reduce the impact of climate change. Climate change may temporarily increase northern forest productivity but is constrained by nutrient availability and acclimation (Cohen &Miller, 2001). Warmer sites may be exposed to greater respiration, and drier zones may experience grassland encroachment. Adaptive management measures for declining stands encompass selective thinning, underplanting, reduced rotation age, and planting better-adapted species (Dale et al., 2001; Lindner et al., 2000; Papadopol, 2000).

2.5 CLIMATE-SMART FOREST MANAGEMENT

Climate-smart forestry expands sustainable forest management to provide more benefits to more people in a changing climate. Since forests can offset a significant portion of the nation's GHG emissions, there has been investment in forest research and conservation for more than a century, paving the way for new climate applications that make landscapes more resilient to the impacts of climate and weather extremes. Particular emphasis is placed on the pivotal roles of forest trees and soils in the carbon and water cycles. By maintaining forest product markets, diverting more small-diameter trees to bioenergy, and reusing slash, we constrain fossil fuel emissions while completing the fuel treatment work needed on narrow treatment windows (Wade et al., 2022). All of our natural and working landscapes, from agricultural soils to urban forests, interact with our planet's climate system. Each landscape has the potential to support climate adaptation. There is a clear understanding of what actions can offset climate change, recover extinction-prone species, and build resilience to severe weather. Our goal is to better coordinate management actions to reach multiple goals in a rapidly changing climate. We have proposed a climate-smart forestry platform that can provide new incentives to improve the permanence of conservation, restoration, and sustainable forest management efforts (Outhwaite et al., 2022)

2.6 COMMUNITY AND STAKEHOLDER ENGAGEMENT

The role of local communities, stakeholders, and policymakers in sustainable forest management and conservation activities is emphasized to improve understanding and raise awareness of the value of forests in mitigating and adapting to climate change. Stakeholder organizations, researchers, and practitioners need to engage with local communities and policymakers, especially in developing countries, to incorporate their concerns, catalyze joint benefits, and validate community-based monitoring of the impacts of climate change. Such collaborative research can bridge gaps in the translation of scientific knowledge to policy and practices for sustainable, climate-resilient landscape management, goal achievement, and land-use planning. Discovering conservation, land restoration, and the reduction of deforestation and degradation as multiple benefits for mitigating and adapting to climate change, national and international governments, NGO signatories, and the private sector have evolved several initiatives (Bosela et al., 2024). However, ex situ conservation efforts, such as seed micropropagation and cryopreservation, gene banks, zoos, and botanic gardens, have attracted less attention and funding from these initiatives as a climate change adaptation strategy.

2.7 FUTURE RESEARCH DIRECTIONS AND EMERGING TECHNOLOGIES

The increasing pace of information generation and synthesis of research questions pertinent to forests in the current Anthropocene era of rapid global change actually demanded. The state of current technological approaches to monitor forest dynamics changes due to climate using emerging technologies such as remote sensing, artificial intelligence advances in allometric relationships, and modeling. The synthesis of advanced remote sensing tools with artificial intelligence, a discovery in the field of machine learning, together with new insights about the structure and functioning of forest ecosystems alter due to climate can provide cost-effective advances in monitoring and modeling forests.

2.8 CONCLUSION

Forests and vegetation dynamics play a critical role in mitigating the impacts of global climate change by capturing atmospheric carbon and providing vast numbers of bio-products. Forests are seen to be playing an increasingly vital role in climate change adaptation, but there are large uncertainties associated with predicting exactly how these essential ecosystem resources will change in the future. Forest dynamics in response to shifting climatic conditions highlights the intricate and complex interactions between species and their environments under climate change. Rising temperatures, altered precipitation patterns, and increasing climate-driven disturbances are causing significant range shifts in forest, phenophases, invasive species expansion, tree mortality due to extreme events, structure and composition and species distribution. These changes are affecting biodiversity, disrupting ecosystem distribution and function, tree diversity interactions, and impacting essential forest services, including carbon storage, water cycle, and forest soil stability. Understanding these processes is essential for designing mitigation and adaptation strategies that enhance species resilience and support overall forest health. The variability in future forest response is closely related to the timing and magnitude of changes in climate, in addition to how different forests are likely to respond to the same changes, and whether a threshold climate tolerance has been exceeded.

REFERENCES

Abatzoglou, J. T., & Kolden, C. A. (2013). Relationships between climate and macroscale area burned in the western United States, *International Journal of Wildland Fire*, *22*, 1003–1020. https://doi.org/10.1071/WF13019

Agee, J. K., & Skinner, C. N. (2005). Basic principles of fuel reduction treatments. *Forest Ecology and Management*, *211*, 83–96.

Ahmad, F., & Goparaju, L. (2018). Climate change and its impact on forest fire in the state of Himachal Pradesh and Uttarakhand states of India: Remote sensing and GIS analysis. *Contemporary Trends in Geosciences*, *7*(2), 229–246. https://doi.org/10.2478/ctg-2018-0016

Astigarraga, J., Andivia, E., Zavala, M. A., Gazol, A., Cruz-Alonso, V., Vicente-Serrano, S. M., & Ruiz-Benito, P. (2020). Evidence of non-stationary relationships between climate and forest responses: Increased sensitivity to climate change in Iberian forests. *Global Change Biology*, *26*(9), 5063–5076.

Bajpai, O., Dutta, V., Chaudhary, L. B., & Pandey, J. (2018). Key issues and management strategies for the conservation of the Himalayan Terai forests of India. *International Journal of Conservation Science*, *9*(4), 749–760.

Bajpai, O., Kushwaha, A. K., Srivastava, A. K., Pandey, J., & Chaudhary, L. B. (2015). Phytosociological status of a monotypic genus *Indopiptadenia*: A near threatened tree from the Terai-Bhabar region of Central Himalaya. *Research Journal of Forestry*, *9*(2), 35–47.

Bajpai, O., Pandey, J., & Chaudhary, L. B. (2017). Periodicity of different phenophases in selected trees from Himalayan Terai of India. *Agroforestry Systems, 91*, 363–374.

Bastin, J.-F., Finegold, Y., Garcia, C., Mollicone, D., Rezende, M., Routh, D., Zohner, C. M., & Crowther, T. W. (2019). The global tree restoration potential. *Science, 365*, 76–79.

Bentouati, A. (2008). La situation du cèdre de l'Atlas dans les Aurès (Algérie). *Foret Méditerranéenne, 29*, 203–208.

Berg, E. E., Henry, J. D., Fastie, C. L., De Volder, A. D., & Matsuoka, S. M. (2006). Spruce beetle outbreaks on the Kenai Peninsula, Alaska, and Kluane National Park and Reserve, Yukon Territory: Relationship to summer temperatures and regional differences in disturbance regimes. *Forest Ecology and Management, 227*, 219–232.

Betancourt, J., Breshears, D., & Mulholland, P.(2004). Ecological impacts of climate change. Report from a NEON Science Workshop. August 24–25, 2004, Tucson, AZ. American Institute of Biological Sciences, Washington, D.C., USA.

Bigler, C., Braker, O. U., Bugmann, H., Dobbertin, M., & Rigling, A. (2006). Drought as an inciting mortality factor in Scots pine stands of the Valais, Switzerland. *Ecosystems, 9*, 330–343.

Bosela, M., Larocque, G. R., Baycheva, T., Valbuena, R., & Lier, M. 2024. Criteria and indicators of sustainable forest management. In *Ecological forest management handbook* (pp. 356–385). CRC Press.

Bradley, B. A., Blumenthal, D. M., Wilcove, D. M., & Ziska, L. H. (2010). Predicting plant invasions in an era of global change. *Trends in Ecology & Evolution, 25*(5), 310–318.

Breshears, D., et al. (2005). Regional vegetation die-off in response to global-change-type drought. *Proceedings of the National Academy of Sciences (USA), 102*, 15144–15148.

Bugmann, H. K. M., Wullschleger, S. D., Price, D. T., Ogle, K., Clark, D. F., & Solomon, A. M. (2001). Comparing the performance of forest gap models in North America. *Climatic Change, 51*, 349–388.

Calinger, K. M., Queenborough, S., & Curtis, P. S. (2013). Herbarium specimens reveal the footprint of climate change on flowering trends across north-central North America. *Ecology Letters, 16*, 1037–104

Chaturvedi, R. K., Gopalakrishnan, R., Jayaraman, M., Bala, G., Joshi, N. V., Sukumar, R., & Ravindranath, N. H. (2011). Impact of climate change on Indian forests: A dynamic vegetation modeling approach. *Mitigation and Adaptation Strategies for Global Change, 16*(2), 119–142.

Chaturvedi, R. K., Raghubanshia, A. S., Tomlinson, K. W., & Singh, J. S. (2017). Impacts of human disturbance in tropical dry forests increase with soil moisture stress. *Journal of Vegetation Science, 28*(5), 997–1007.

Chaudhary, C., Richardson, A. J., Schoeman, D. S., & Costello, M. J. (2021). Global warming is causing a more pronounced dip in marine species richness around the equator. *Proceedings of the National Academy of Sciences, 118*(15), e2015094118.

Chazdon, R. L., Brenes, A. R., & Alvarado, B. V. (2005). Effects of climate and stand age on annual tree dynamics in tropical second-growth rain forests. *Ecology, 86*, 1808–1815.

Clark, J. S., Iverson, L., Woodall, C. W., Allen, C. D., Bell, D. M., Bragg, D. C., D'Amato, A. W., Davis, F. W., Hersh, M. H., Ibanez, I., Jackson, S. T., Matthews, S., Pederson, N., Peters, M., Schwartz, M. W., Waring, K. M., & Zimmermann, N. E. (2016). The impacts of increasing drought on forest dynamics, structure, and biodiversity in the United States. *Global Change Biology, 22*, 2329–2352, https://doi.org/10.1111/gcb.13160

Coelho, M. T. P., Barreto, E., Rangel, T. F., Diniz-Filho, J. A. F., Wüest, R. O., Bach, W., & Graham, C. H. (2023). The geography of climate and the global patterns of species diversity. *Nature, 622*(7983), 537–544.

Cohen, S., & Miller, K. 2001. North America. In J. J. McCarthy, O. F. Canziani, N. A. Leary, D. J. Dokken, & K. S. White (Eds.), *Climate change 2001: Impacts, adaptation, and vulnerability* (pp. 735–800). Cambridge University Press.

Colwell, R. K., Brehm, G., Cardelu, C. L., Gilman, A. C., & Longino, J. T. (2008). Global warming, elevational range shifts, and lowland biotic attrition in the wet tropics. *Science*, *322*, 258–261.

Cosofret, C., & Bouriaud, L. (2019). Which silvicultural measures are recommended to adapt forests to climate change? A literature review. *Bulletin of the Transilvania University of Brasov, Series II: Forestry, Wood Industry, Agricultural Food Engineering*, *12*, 13–34.

Dainese, M., Kuhn, I., & Bragazza, L. (2014). Alien plant species distribution in the European Alps: Influence of species' climatic requirements. *Biological Invasions*, *16*, 815–831.

Dale, V. H., et al. (2001). Climate change and forest disturbances. *BioScience*, *52*, 723–234.

Dennison, P. E., Brewer, S. C., Arnold, J. D., & Moritz, M. A. (2014). Large wildfire trends in the western United States, 1984–2011. *Geophysical Research Letters*, *41*, 2928–2933. https://doi.org/10.1002/2014GL059576

Doelman, J. C., Stehfest, E., van Vuuren, D. P., Tabeau, A., Hof, A. F., Braakhekke, M. C., Gernaat, D. E. H. J., Berg, M. v. D., van Zeist, W., & Daioglou, V., et al. (2019). Afforestation for climate change mitigation: Potentials, risks and trade-offs. *Global Change Biology*, *26*, 1576–1591.

Domke, G., Williams, C. A., Birdsey, R., Coulston, J., Finzi, A., Gough, C., Haight, B., Hicke, J., Janowiak, M., Jong, B., deKurz, W. A., Lucash, M., Ogle, S., Olguín-Álvarez, M., Pan, Y., Skutsch, M., Smyth, C., Swanston, C., Templer, P., Wear, D., & Woodall, C. W.(2018). Forests. In N. Cavallaro, G. Shrestha, R. Birdsey, M. A. Mayes, R. G. Najjar, S. C. Reed, P. Romero-Lankao, & Z. Zhu (Eds.), *Second state of the carbon cycle report (SOCCR2)*. U.S. Global Change Research Program.

Donovan, L. A., Rosenthal, D. M., Sanchez-velenosi, M., Rieseberg, L. H., & Ludwig, F. (2010). Are hybrid species more fit than ancestral parent species in the current hybrid species habitats? *Journal of Evolutionary Biology*, *23*, 805–816.

Dusenge, M. E., Duarte, A. G., & Way, D. A. (2019). Plant carbon metabolism and climate change: Elevated CO2 and temperature impacts on photosynthesis, photorespiration and respiration. *New Phytologist*, *221*, 32–49.

Elsen, P. R., Saxon, E. C., Simmons, B. A., Ward, M., Williams, B. A., Grantham, H. S., …, & Watson, J. E. (2022). Accelerated shifts in terrestrial life zones under rapid climate change. *Global Change Biology*, *28*(3), 918–935.

Eriksson, E., Gillespie, A. R., Gustavsson, L., Langvall, O., Olsson, M., Sathre, R., & Stendahl, J. (2007). Integrated carbon analysis of forest management practices and wood substitution. *Canadian Journal of Forest Research*, *37*(3), 671–681.

Fensham, R. J., & Fairfax, R. J. (2005). Preliminary assessment of gidgee (*Acacia cambagei*) woodland thickening in the Longreach district, Queensland. *The Rangeland Journal*, *27*, 159–168.

Fensham, R. J., & Fairfax, R. J. (2007). Drought-related tree death of savanna eucalypts: Species susceptibility, soil conditions and root architecture. *Journal of Vegetation Science*, *18*, 71–80.

Fitter, A. H., & Fitter, R. S. (2002). Rapid changes in flowering time in British plants. *Science*, *296*, 1689–1691.

Fitzgerald, G. (2000). Triage. In P. Cameron, G. Jelinke, A. M. Kelly, L. Murray, & J. Heyworth (Eds.), *Textbook of adult emergency medicine* (pp. 584–588). Churchill Livingston.

Fleming, R. A., Candau, J.-N., & McAlpine, R. S. (2002). Landscape-scale analysis of interactions between insect defoliation and forest fire in central Canada. *Climatic Change*, *55*, 251–272.

Food and Agriculture Organization of the United Nations (FAO). (2020). *Global Forest Resources Assessment 2020: Main Report*. FAO.

Forest, C. E. (1999). Paleoaltimetry incorporating atmospheric physics and botanical estimates of paleoclimate. *Geological Society of American Bulletin*, *111*(4), 497–511.

Forrester, D. I., & Bauhus, J. (2016). A review of processes behind diversity—Productivity relationships in forests. *Current Forestry Reports*, *2*, 45–61.

Franco-Cisterna, M., Ramos-Jiliberto, R., & de Espanés, P. M., et al. (2020). *Phenological shifts drive biodiversity loss in plant-pollinator networks*. https://doi.org/10.1101/2020.04.03.023457

Gitay, H., Brown, S., Easterling, W., & JaUow, B.(2001). Ecosystems and their goods and services.In J. J. McCarthy, O. F. Canziani, N. A. Leary, D. J. Dokken, & K. S. White (Eds.) *Climate change 2001: Impacts, adaptation, and vulnerability* (pp. 235–342). Contribution of Working Group I1 to the Third Assessment Report of the Intergovernmental Panel on Climate Change. New York Cambridge University Press.

Gustafson, E. J., Kern, C. C., Miranda, B. R., Sturtevant, B. R., Bronson, D. R., & Kabrick, J. M.(2020). Climate adaptive silviculture strategies: How do they impact growth, yield, diversity and value in forested landscapes? *Forest Ecology and Management*, *470*, 118208.

Haberstroh, S., & Werner, C. (2022). The role of species interactions for forest resilience to drought. *Plant Biology*, *24*, 1098–1107.

Hansen, A. J., Neilson, R. P., Dale, V. H., Flather, C. H., Iverson, L. R., Currie, D. J., Shafer, S., Cook, R., & Bartlein, P. J. (2001). Global change in forests: Response of species, communities, and biomes. *BioScience*, *51*, 765–779.

Hao, Y., Widagdo, F. R. A., Liu, X., Quan, Y., Dong, L., & Li, F.(2020). Individual tree diameter estimation in small-scale forest inventory using UAV laser scanning. *Remote Sensing*, *13*(1), 24. https://doi.org/10.3390/rs13010024.

Harris, J. A., Hobbs, R. J., Higgs, E., & Aronson, J. (2006). Ecological restoration and global climate change. *Restoration Ecology*, *14*, 170–176.

Hart, R., Salick, J., Ranjitkar, S., et al. (2014). Herbarium specimens show contrasting phenological responses to Himalayan climate. *Proceedings of the National Academy of Sciences of the USA*, *111*, 10615–10619.

He, C., Kim, H., Hashizume, M., Lee, W., Honda, Y., Kim, S. E., & Kan, H. (2022). The effects of night-time warming on mortality burden under future climate change scenarios: A modelling study. *The Lancet Planetary Health*, *6*(8), e648–e657.

Hember, R. A., Kurz, W. A., & Coops, N. C. (2017). Relationships between individual-tree mortality and water-balance variables indicate positive trends in water stress-induced tree mortality across North America. *Global Change Biology*, *23*, 1691–1710. https://doi.org/10.1111/gcb.13428

Hiremath, A. J., & Sundaram, B. (2005). The fire-lantana cycle hypothesis in Indian forests. *Conservation and Society*, *3*(1), 26–42.

Hobday, A. J., Oliver, E. C. J., Gupta, A. S., Benthuysen, J. A., Burrows, M. T., Donat, M. G., Holbrook, N. J., Moore, P. J., Thomsen, M. S., Wernberg, T., & Smale, D. A. (2018). Categorizing and naming marine heatwaves. *Oceanography*, *31*, 162–173. https://doi.org/10.5670/oceanog.2018.205

Hoffmann, A. A., & Sgro, C. M. (2011). Climate change and evolutionary adaptation: A review. *Nature*, *470*, 479–485.

Hogg, E. H., Brandt, J. P., & Michaellian, M. (2008). Impacts of a regional drought on the productivity, dieback, and biomass of western Canadian aspen forests. *Canadian Journal of Forest Research-Revue Canadienne De Recherche Forestiere*, *38*, 1373–1384.

Howard, J., Sutton-Grier, A., Herr, D., Kleypas, J., Landis, E., Mcleod, E., et al. (2017). Clarifying the role of coastal and marine systems in climate mitigation. *Frontiers in Ecology and the Environment*, *15*, 42–50. https://doi.org/10.1002/fee.1451

IPCC. (2007). Summary for policymakers. Summary for Policymakers. In: Climate Change 2007: Impacts, Adaptation and Vulnerability. Contribution of Working Group II to the Fourth Assessment Report of the Intergovernmental Panel on Climate Change, M.L. Parry, O.F. Canziani, J.P. Palutikof, P.J. van der Linden and C.E. Hanson, Eds., Cambridge University Press, Cambridge, UK, 7–22.

IPCC. (2022). Summary for Policymakers [H. O. Pörtner, D. C. Roberts, E. S. Poloczanska, K. Mintenbeck, M. Tignor, A. Alegría, M. Craig, S. Langsdorf, S. Löschke, V. Möller, A. Okem (eds.)]. In: Climate Change 2022: Impacts, Adaptation and Vulnerability. Contribution of Working Group II to the Sixth Assessment Report of the Intergovernmental Panel on Climate Change [H.-O. Pörtner, D.C. Roberts, M. Tignor, E.S. Poloczanska, K. Mintenbeck, A. Alegría, M. Craig, S. Langsdorf, S. Löschke, V. Möller, A. Okem, B. Rama (eds.)]. Cambridge University Press, Cambridge, UK and New York, NY, USA, pp. 3–33, doi:10.1017/9781009325844.001.

Jactel, H., Bauhus, J., Boberg, J., Bonal, D., Castagneyrol, B., Gardiner, B., Gonzalez-Olabarria, J. R., Koricheva, J., Meurisse, N., & Brockerhoff, E. G. (2017). Tree diversitydrivesforeststandresistance to naturaldisturbances. *Current Forestry Reports*, *3*, 223–243.

Jay, A., Reidmiller, D. R., Avery, C. W., Barrie, D., DeAngelo, B. J., Dave, A., Dzaugis, M., Kolian, M., Lewis, K. L. M., Reeves, K., & Winner, D.(2018). Overview. In D. R. Reidmiller, C. W. Avery, D. R. Easterling, K. E. Kunkel, K. L. M. Lewis, T. K. Maycock, & B. C. Stewart (Eds.), *Impacts, risks, and adaptation in the United States: Fourth national climate assessment. II* (pp. 33–71). U.S. Global Change Research Program.

Johnson, E. (2009). Goodbye to carbon neutral: Getting biomass footprints right. *Environmental Impact Assessment Review*, *29*(3), 165–168.

Kauhanen, H., Wallenius, T., Kuuluvainen, T., Aakala, T., & Mikkola, K.(2008). Extensive mortality of spruce forests in Arkhangelsk Region: satellite image analysis. In: Poster Presentation At: International Conference "Adaptation of Forests and Forest Management to Changing Climate with Emphasis on Forest Health: A Review of Science, Policies, and Practices", Umeaer Sweden, FAO/IUFRO, 25–28 August 2008.

Keatley, M. R., Fletcher, T. D., Hudson, I. L., & Ades, P. K. (2002). Phenological studies in Australia: Potential application in historical and future climate analysis. *International Journal of Climatology*, *22*, 1769–1780.

Keith, H., Mackey, B. G., & Lindenmayer, D. B. (2009). Re-evaluation of forest biomass carbon stocks and lessons from the world's most carbon-dense forests. *Proceedings of the National Academy of Sciences*, *106*(28), 11635–11640. https://doi.org/10.1073/pnas.0901970106

Kelly, A. E., &Goulden, M. L. (2008). Rapid shifts in plant distribution with recent climate change. *Proceedings of the National Academy of Sciences*, *105*, 11823–11826. https://doi.org/10.1073/pnas.0802891105

Khan, J. A., Rodgers, W. A., Johnsingh, A. J. T., & Mathur, P. K. (1994). Tree and shrub mortality and debarking by sambar Cervus-Unicolor (Kerr) in Gir after a drought in Gujarat, India. *Biological Conservation*, *68*, 149–154.

Korner, C., Sarris, D., & Christodoulakis, D. (2005). Long-term increase in climatic dryness in the East-Mediterranean as evidenced for the island of Samos. *Regional Environmental Change Journal*, *5*, 27–36.

Krankina, O. N., & Harmon, M. E. 2006. Forest management strategies for carbon storage. In M. Cloughesy (Ed.), *Forests, carbon, and climate change: A synthesis of science findings* (pp. 79–92). Oregon Forest Resources Institute.

Kurz, W. A., Dymond, C. C., Stinson, G., Rampley, G. J., Neilson, E. T., Carroll, A. L., Ebata, T., & Safranyik, L. (2008). Mountain pine beetle and forest carbon feedback to climate change. *Nature*, *452*, 987–990.

Lande, R. (2009). Adaptation to an extraordinary environment by evolution of phenotypic plasticity and genetic assimilation. *Journal of Evolutionary Biology*, *22*, 1435–1446.

Lim, J. H., Chun, J. H., Woo, S. Y., & Kim, Y. K., 2008. Increased declines of Korean fir forest caused by climate change in Mountain Halla, Korea. In: Oral Presentation At: International Conference "Adaptation of Forests and Forest Management to Changing Climate with Emphasis on Forest Health: A Review of Science, Policies, and Practices", Umea P Sweden, FAO/IUFRO, 25–28 August 2008.

Lindner, M., Lasch, P., & Erhard, M. (2000). Alternative forest management strategies under climate change: Prospects for gap model applications in risk analysis. *Silva Fennica, 34*, 101–111.

Lipton, D., Rubenstein, M. A., Weiskopf, S. R., Carter, S., Peterson, J., Crozier, L., Fogarty, M., Gaichas, S., Hyde, K. J. W., Morelli, T. L., Morisette, J., Moustahfid, H., Muñoz, R., Poudel, R., Staudinger, M. D., Stock, C., Thompson, L., Waples, R., & Weltzin, J. F.(2018). Ecosystems, ecosystem services, and biodiversity. In: D. R. Reidmiller, C. W. Avery, D. R. Easterling, K. E. Kunkel, K. L. M. Lewis, T. K. Maycock, &B. C. Stewart, (Eds.),*Impacts, risks, and adaptation in the United States: Fourth national climate assessment, volume II*. U.S. Global Change Research Program.

Lloyd, A. H., & Bunn, A. G.(2007). Responses of the circumpolar boreal forest to 20th century climate variability. *Environmental Research Letters, 2*, 045013, https://doi.org/10:1088/1748-9326/2/4/045013

Loehman, R. A., Clark, J. A., & Keane, R. E. (2011). Modeling effects of climate change and fire management on Western whitepine (*Pinusmonticola*) in the Northern Rocky Mountains, USA, *Forests, 2*, 832–860, https://doi.org/10.3390/f2040832

Loehman, R. A., Keane, R. E., Holsinger, L. M., & Wu, Z. (2017). Interactions of landscape disturbances and climate change dictate ecological pattern and process: Spatial modeling of wildfire, insect, and disease dynamics under future climates. *Landscape Ecology, 32*, 1447–1459. https://doi.org/10.1007/s10980-016-0414-6

Lwanga, J. S. (2003). Localized tree mortality following the drought of 1999 at Ngogo, Kibale National Park, Uganda. *African Journal of Ecology, 41*, 194–196.

Malmsheimer, R. W., Heffernan, P., Brink, S., Crandall, D., Deneke, F., Galik, C., Gee, E., Helms, J. A., McClure, N., Mortimer, M., Ruddell, S., Smith, M., & Stewart, J. (2008). Forest management solutions for mitigating climate change in the United States. *Journal of Forestry, 106*(3), 115–117.

Martin-R, M. H., Cox, J. R., & Ibarraf, F. (1995). Climatic effects on buffelgrass productivity in the Sonoran Desert. *Journal of Range Management, 48*, 60–63.

McLachlan, J. S., Hellmann, J., & Schwartz, M. (2007). A framework for debate of assisted migration in an era of climate change. *Conservation Biology, 21*(2), 297–302.

Melillo, J. M., Frey, S. D., DeAngelis, K. M., Werner, W. J., Bernard, M. J., Bowles, F. P., Pold, G., Knorr, M. A., & Grandy, A. S. (2017). Long-term pattern and magnitude of soil carbonfeedback to the climate system in a warming world. *Science (80-.), 358*, 101–105. https://doi.org/10.1126/science.aan2874

Menzel, A., Sparks, T. H., & Estrella, N. (2006). European phonological response to climate change matches the warming pattern. *Global Change Biology, 12*, 1969–1976.

Millar, C. I. (1998). Reconsidering the conservation of Monterey pine. *Fremontia, 26*, 12–16.

Millar, C. I., & Brubaker, L. B. 2006. Climate change and paleoecology: New contexts for restoration ecology. In M. Palmer, D. Falk, & J. Zedler (Eds.), *Restoration science* (pp. 315–340). Island Press.

Millar, C. I., Skog, K. E., McKinley, D. C. C., Birdsey, R. A., Swanston, C. W., Hines, S. J., Woodall, C. W., Reinhardt, E. D., Peterson, D. L., & Vose, J. M.(2012). Adaptation and mitigation. In: J. M. Vose, D. L. Peterson,&T. Patel-Weynand (Eds.),*Effects of climatic variability and change on forest ecosystems: A comprehensive science synthesis for the U.S. Forest sector* (pp. 125–192). U.S. Department of Agriculture, Forest Service, Pacific Northwest Research Station.

Mina, M., Messier, C., Duveneck, M. J., Fortin, M. J., & Aquilué, N. (2022). Managing for the unexpected: Building resilient forest landscapes to cope with global change. *Global Change Biology, 28*(14), 4323–4341.

Mina, U., Dimri, A. P., & Farswan, S. 2023. Forest fires and climate attributes interact in central Himalayas: An overview and assessment. *Fire Ecology, 19*(14). https://doi.org/10.1186/s42408-023-00177-4

Morichetti, M., Vangi, E., & Collalti, A. 2024. Predicted future changes in the mean seasonal carbon cycle due to climate change. *Forests*. mdpi.com

Namkoong, G. (1984). Strategies for gene conservation. In C. W. Yeatman, D. Kafton, & G. Wilkes (Eds.), *Plant gene resources: A conservation imperative* (pp. 79–89). American Association for the Advancement of Science, Selected Symposium 87.

NAS (2001). *Climate change science: An analysis of some key questions*. National Academy of Sciences (NAS), National Academy Press.

Neilson, R. F., Prentice, I. C., Smith, B., Kittel, T. G. F., & Viner, D.(1998). Simulated changes in vegetation distribution under global warming. In R. T. Watson, M. C. Zinyowera, R. H. Moss, D. J. Dokken (Eds.), *The regional impacts of climate change: An assessment of vulnerability* (p. 43M56). Cambridge University Press

Norby, R. J., Warren, J. M., Iversen, C. M., Medlyn, B. E., & McMurtrie, R. E. (2010). CO_2 enhancement of forest productivity constrained by limited nitrogen availability. *Proceedings of the National Academy of Sciences of the United States of America*, *107*, 19368–19373, https://doi.org/10.1073/pnas.1006463107

Noss, R. F. (2001). Beyond Kyoto: Forest management in a time of rapid climate change. *Conservation Biology*, *15*, 578–590.

Ollinger, S. V., Goodale, C. L., Hayhoe, K., & Jenkins, J. P. (2008). Effects of predicted changes in climate and atmospheric composition on ecosystem processes in northeastern U.S. Forests. *Mitigation and Adaptation Strategies for Global Change*, *13*, 467–485.

Oppenheimer, C. (2003). Climatic, environmental and human consequences of the largest known historic eruption: Tambora volcano (Indonesia) 1815. *Progress in Physical Geography*, *27*(2), 230–259.

Outhwaite, C. L., McCann, P., & Newbold, T. 2022. Agriculture and climate change are reshaping insect biodiversity worldwide. *Nature*. UCL.ac.uk.

Papadopol, C. S. (2000). Impacts of climate warming on forests in Ontario: Options for adaptation and mitigation. *Forestry Chronicle*, *76*, 139–149.

Parker, W. C., Colombo, S. J., Cherry, M. L., Flannigan, M. D., Greifenhagen, S., McAlpine, R. S., Papadopol, C., & Scarr, T. (2000). Third millennium forestry: What climate change might mean to forests and forest management in Ontario. *Forestry Chronicle*, *76*, 445–463.

Pathak, R., Negi, V. S., Rawal, R. S., & Bhatt, I. D. (2019). Alien plant invasion in the Indian Himalayan Region: State of knowledge and research priorities. *Biodiversity and Conservation*, *28*(12), 3073–3102. https://doi.org/10.1007/s10531-019-01829-1

Patricola, C. M., & Wehner, M. F. (2018). Anthropogenic influences on major tropical cyclone events. *Nature*, *563*, 339–346, https://doi.org/10.1038/s41586-018-0673-2

Pillay, R., Venter, M., Aragon Osejo, J., González del Pliego, P., Hansen, A. J., Watson, J. E., & Venter, O. (2022). Tropical forests are home to over half of the world's vertebrate species. *Frontiers in Ecology and the Environment*, *20*(1), 10–15.

Pimentel, D., Lach, L., Zuniga, R., & Morrison, D. (2000). Environmental and economic costs of nonindigenous species in the United States. *Bioscience*, *50*, 53–65.

Pucttmann, K. J. (2011). Silvicultural challenges and options in the context of global change: "simple" fixes and opportunities for new management approaches. *Journal of Forestry*, *31*, 321–331.

Pugh, T. A. M., Lindeskog, M., Smith, B., Poulter, B., Arneth, A., Haverd, V., & Calle, L. (2019). Role of forest regrowth in global carbon sink dynamics. *Proceedings of the National Academy of Sciences of the United States of America*, *116*, 4382–4387. [Google Scholar] [CrossRef] [PubMed]

Rafferty, N. E., Diez, J. M., & Bertelsen, C. D. (2020). Changing climate drives divergent and nonlinear shifts in flowering phenology across elevations. *Current Biology*, *30*, 432–441.

Rana, N., Manish, K., & Pandit, M. K.. 2024. Effect of climate change on the flowering phenology of *Rhododendron arboretum* Sm. in the Western Himalaya. *Journal of Asia-Pacific Biodiversity*. https://doi.org/10.1016/j.japb.2024.10.004

Regniere, J. (2009). Predicting insect continental distributions from species physiology. *Unasylva*, *60*, 37–42.

Reyes-Palomeque, G., Dupuy, J. M., Portillo-Quintero, C. A., Andrade, J. L., Tun-Dzul, F. J., & Hernández-Stefanoni, J. L. (2021). Mapping forest age and characterizing vegetation structure and species composition in tropical dry forests. *Ecological Indicators*, *120*, 106955.

Rogers, C. E., & McCarty, J. P. (2000). Climate change and ecosystems of the mid-Atlantic region. *Climate Research*, *14*, 235–244.

Sagar, N., Suresh, K. P., Naveesh, Y. B., Archana, C. A., Hemadri, D., Patil, S. S., Archana, V. P., Raaga, R., Nandan, A. S., & Chethan, A. J. (2024). Forest fire dynamics in India (2005–2022): Unveiling climatic impacts, spatial patterns, and interface with anthrax incidence. *Ecological Indicators*, *166*, 112454. https://doi.org/10.1016/j.ecolind.2024.112454

SCBD [Secretariat of the Convention on Biological Diversity]. (2001). Impacts of human-caused fires on biodiversity and ecosystem functioning, and their causes in tropical, temperate, and boreal forest biomes. CBD Technical Series No. 5. Montreal, Quebec, Canada: Secretariat of the Convention on Biological Diversity. Available online at: https://www.cbd.int/doc/publications/cbd-ts-05.pdf

Schepaschenko, D., Moltchanova, E., Fedorov, S., Karminov, V., Ontikov, P., Santoro, M., …, & Kraxner, F. (2021). Russian forest sequesters substantially more carbon than previously reported. *Scientific Reports*, *11*(1), 12825. nature.com

Schwartz, M. D., & Reiter, B. E. (2000). Changes in North American spring. *International Journal of Climatology*, *20*, 929–932.

Siwecki, R., & Ufnalksi, K. (1998). Review of oak stand decline with special reference to the role of drought in Poland. *European Journal of Forest Pathology*, *28*, 99–112.

Solberg, S. (2004). Summer drought: A driver for crown condition and mortality of Norway spruce in Norway. *Forest Pathology*, *34*, 93–107.

Song, Z., Du, Y., Primack, R. B., et al. (2021). Surprising roles of climate in regulating flowering phenology in a subtropical ecosystem. *Ecography*, *44*, 1379–1390.

Spittlehouse, D. L., & Stewart, R. B. (2003). Adaptation to climate change in forest management. *BC Journal of Ecosystems and Management*, *4*(1), 1–11.

Stephens, S. L., & Moghaddas, J. J. (2005). Experimental fuel treatment impacts on forest structure, potential fire behavior, and predicted tree mortality in a mixed conifer forest. *Forest Ecology and Management*, *215*, 21–36.

Sykes, M. T. (2009). *Climate change impacts: Vegetation. In encyclopedia of life sciences (ELS)*. John Wiley & Sons, Ltd.

Szmyt, J., & Dering, M. (2024). Adaptive silviculture and climate change—A forced marriage of the 21st century? *Sustainability*, *16*, 2703.

Tafangenyasha, C. (2001). Decline of the mountain acacia, *Brachystegia glaucescens* in Gonarezhou National Park, southeast Zimbabwe. *Journal of Environmental Management*, *63*, 37–50.

Teo, H. C., Zeng, Y., Sarira, T. V., Fung, T. K., Zheng, Q., Song, X. P., Chong, K. Y., & Koh, L. P. (2021). Global urban reforestation can be an important natural climate solution. *Environmental Research Letters*, *16*, 034059.

Ter-Mikaelian, M. T., Colombo, S. J., & Chen, J. (2015). The burning question: Does forest bioenergy reduce carbon emissions? A review of common misconceptions about forest carbon accounting. *Journal of Forestry*, *113*(1), 57–68.

Thomas, C. D., et al. (2004). Extinction risk from climate change. *Nature*, *427*, 145–148.

Trottier, N., Groeneveld, E., & Lavoie, C. (2017). Giant hogweed at its northern distribution limit in North America: Experiments for a better understanding of its dispersal dynamics along rivers. *River Research and Applications*, *33*, 1098–1106. https://doi.org/10.1002/rra.3149.

Tschora, H., & Cherubini, F. (2020). Co-benefits and trade-offs of agroforestry for climate change mitigation and other sustainability goals in West Africa. *Global Ecology and Conservation*, *22*, e00919, https://doi.org/10.1016/j.gecco.2020.e00919

Tsopelas, P., Angelopoulos, A., Economou, A., & Soulioti, N. (2004). Mistletoe (*Viscumalbum*) in the fir forest of Mount Parnis, Greece. *Forest Ecology and Management, 202*, 59–65.

Vacek, Z., Vacek, S., & Cukor, J. 2023. European forests under global climate change: Review of tree growth processes, crises and management strategies. *Journal of Environmental Management.*

van Nieuwstadt, M. G. L., & Sheil, D. (2005). Drought, fire and tree survival in a Borneo rain forest, East Kalimantan, Indonesia. *Journal of Ecology, 93*, 191–201.

Viljoen, A. J. (1995). The influence of the 1991/92 drought on the woody vegetation of the Kruger National Park. *Koedoe, 32*, 85–97.

Wade, C. M., Baker, J. S., Jones, J. P., Austin, K. G., Cai, Y., de Hernandez, A. B., … & McCarl, B. (2022). Projecting the impact of socioeconomic and policy factors on greenhouse gas emissions and carbon sequestration in US Forestry and Agriculture. *Journal of Forest Economics, 37*(1), 127–131. Nowpublishers.com

Walker, W. S., Gorelik, S. R., Cook-Patton, S. C., Baccini, A., Farina, M. K., Solvik, K. K., … & Griscom, B. W. (2022). The global potential for increased storage of carbon on land. *Proceedings of the National Academy of Sciences, 119*(23), e2111312119. pnas.org

Wang, J., Taylor, A. R., & D'Orangeville, L. (2023). Warming-induced tree growth may help offset increasing disturbance across the Canadian boreal forest. *Proceedings of the National Academy of Sciences, 120*(2), e2212780120.

Wang, Y., Yang, X. D., Ali, A., et al. (2020). Flowering phenology shifts in response to functional traits, growth form, and phylogeny of woody species in a desert area. *Frontiers in Plant Science, 11*, 536.

Watson, R. T., Noble, I. R., Bolin, B., Ravindranath, N. H., Verardo, D. J., & Dokken, D. J. (2000). *IPCC special report on land-use, land-use change and forestry*. Cambridge University Press.

Wheaton, E. (2001): Changing fire risk in a changing climate: a literature review and assessment; Saskatchewan Research Council, Publication No. 11341-2E01, prepared for Climate Change Action Fund (CCAF).

White, A., Cannel, M. G. R., & Friend, A. D. (1999). Climate change impacts on ecosystems and the terrestrial carbon sink: A new assessment. *Global Environmental Change*, 1999, *9*, S3–S19.

Williams, S. E., Shoo, L. P., Isaac, J. L., Hoffmann, A. A., & Langham, G. (2008). Towards an integrated framework for assessing the vulnerability of species to climate change. *PLoS Biology, 6*, 2621–2626.

WRI. (2024). *Global forest review: Key findings*. World Resources Institute. https://www.wri.org/research/global-forest-review-key-findings

Yadav, V., Singh, N. B., Singh, H., Singh, A., & Hussain, I. (2016). Allelopathic invasion of alien plant species in India and their management strategies: A review. *Tropical Plant Research, 3*(1), 87–101.

Zu, K., Chen, F., Li, Y., Shrestha, N., Fang, X., Ahmad, S., Nabi, G., & Wang, Z. (2023). Climate change impacts flowering phenology in Gongga Mountains, Southwest China. *Plant Diversity*. https://doi.org/10.1016/j.pld.2023.07.007

3 Interplay of Biotic and Abiotic Components for Forest Functioning and Productivity

Khoisnam Vramari Devi, Salam Suresh Singh, Keshav Kumar Upadhyay, Rajdeep Chanda, Ningthoujam Ranjana Devi, Ngangbam Somen Singh, David C. Vanlalfakawma, and Shri Kant Tripathi

3.1 INTRODUCTION

Forests are intricate, self-regulating systems in which the interaction of biotic and abiotic factors determines their structure, functioning, and long-term sustainability (Cudlín et al., 2013; Kharmawlong et al., 2023a). Such interactions drive essential ecological processes like nutrient cycling, energy flow, carbon sequestration, and regeneration to collectively support biodiversity and the health of the ecosystem (Feng et al., 2024; Kharmawlong et al., 2023b). Roles of biotic and abiotic factors are critical for effective management of forests, especially in the wake of global environmental challenges that include deforestation, climate change, and habitat degradation (Thompson et al., 2009; Upadhyay & Tripathi, 2019).

All biotic factors are living components of the ecosystem, from the trees and understory to the animals, fungi, and microorganisms, and their interaction influences crucial ecosystem processes (Liebhold et al., 2017; Huidrom et al., 2024). For instance, plants take a leading role in photosynthesis in converting solar energy into biomass, which is the foundational input to the forest food web (Huryna & Pokorný, 2016). Animals are involved in activities like seed dispersal and pollination, thereby enhancing genetic variation and facilitating forest regeneration (Neuschulz et al., 2016). Fungi and bacteria, in decomposing organic matter (OM) and recycling nutrients, facilitate soil fertility that, in turn, sustains plant growth and enhances productivity in a healthy ecosystem (Biswas & Kole, 2017). Every species, ranging from the microscopic microbe to the massive predator, has its own role to play in this sensitive balance of life in a forest ecosystem (Perry et al., 2008).

Abiotic factors, which include all non-living environmental elements, including climate, soil type, water, light, pollutants and air, constitute the physical context in which biotic interactions take place (Jain, 2021; Singh et al., 2024). These factors directly influence the distribution, growth, and survival of organisms in the forest.

 DOI: 10.1201/9781003588580-3

For example, temperature and precipitation regimes determine the rate of plant growth and species composition, while soil properties such as texture, pH, and nutrient content influence root development and water retention (Babushkina et al., 2022; Chanda et al., 2024). Water availability regulates tree physiology, influencing transpiration rates and carbon sequestration capacity (Upadhyay et al., 2021). Sunlight penetration through the canopy impacts understory plant diversity, further influencing the ecosystem's overall productivity and resilience (Teshome et al., 2020).

Interplay among biotic and abiotic factors is highly dynamic as well as varies among diverse forest types and climatic zones (Klanderud et al., 2015). Abundant rainfall and stable temperatures usually support high biodiversity and robust biotic interactions (Blois et al., 2013). However, in boreal forests, abiotic factors take precedence over biotic factors. The extreme cold and nutrient-poor soil conditions limit species diversity, making abiotic factors the primary drivers of ecosystem processes in these regions. This delicate balance is further affected by natural disturbances such as wildfires, storms, and droughts, which can modify both biotic and abiotic components, bringing alterations to forest structure and function (Roy et al., 2020).

The various human activities such as logging, land conversion, pollution, and climate change have imposed immense pressures on forest ecosystems (Singh et al., 2024). These disturbances interrupt the natural balance between biotic and abiotic factors, leading to a reduction in biodiversity, changes in nutrient cycles, and weakened ecosystem resilience (Roy et al., 2022). For instance, the process of deforestation not only eliminates the vegetation but also interrupts the soil composition, hydrological cycles, and microclimatic conditions, leading to long-term impacts on the ecosystem functions (Chhabra et al., 2006). Climate change exacerbates these impacts by changing temperature and precipitation regimes, increasing the frequency of extreme weather events, and redistributing species, thus testing the limits of the potential for forest ecosystems to adapt and thrive (Trenberth, 2018).

This chapter discusses the interlinked roles of biotic and abiotic factors in regulating key forest ecosystem processes, such as nutrient cycling, regeneration, carbon sequestration, and productivity. The contribution of these factors, both individually and in their dynamic interactions, gives us a better understanding of the mechanisms that sustain forest ecosystems. This knowledge is essential for informing strategies for sustainable forest management and conservation, ensuring the long-term health and functionality of these vital ecosystems.

3.1.1 Nutrient Cycling

Nutrient cycling is a complex process that includes the transfer of essential nutrients between living and non-living organisms in various ecosystems. This process ensures the continuous recycling and availability of various essential nutrients, such as carbon, nitrogen, phosphorus, and water, to organisms for growth, energy, and reproduction (Singh et al., 2025a). The continuous cycling of nutrients in forests is driven by various biotic and abiotic factors. Among biotic factors, plant litter type and litter production rate, type of decomposers and litter decomposition rate, and plant root depth play an important role in nutrient cycling in various ecosystems (Cornwell et al., 2008; Giweta, 2020). Similarly, nutrient cycling is also affected by abiotic factors such as temperature, rainfall, soil texture, soil organic content, topography, and soil moisture content. Although leaf litter contains only a small proportion of nutrients in above-ground forest biomass (typically <10%), it dominates the intra-annual cycle of nutrients through rapid turnover (Landuyt et al., 2020; Tripathi et al., 2025). The quantity and quality of litter determine the functioning of forest ecosystems (Patil et al., 2020) and are essential for balanced ecosystem processes. Furthermore, litter production patterns among ecosystems vary depending on elevation, latitude, soil fertility, stand structure, climate, tree species composition, and season (Becker et al., 2015; Parsons et al., 2014). In most tropical forests, the amount of litter in the soil varies depending on the season (Zhang et al., 2015). Although ground vegetation in sparse, open stands can significantly contribute to the overall litterfall, tree litter typically serves as the primary natural source of organic material and nutrients entering the forest floor (Giweta, 2020). Furthermore, the canopy characteristics play a crucial role in determining the quantity and composition of leaf litter, which significantly impacts the recycling of nutrients, the structure of soil microbial and faunal communities, and the overall availability of nutrients in the ecosystem (Ita et al., 2022).

3.1.2 Factors Affecting Litter Production

Litter production is a fundamental process that supports nutrient cycling, sustains soil organisms, and maintains ecosystem health and productivity. It plays a central role in ecosystem functioning, intricately linked to primary productivity and global biogeochemical cycles. Litter contributes to carbon and nutrient dynamics while forming a protective organic layer that regulates soil-atmosphere interactions (Chen & Chen, 2018). The forest canopy, as the primary source of leaf litter, creates stratified layers that interconnect with other ecosystem components. Litterfall, comprising senescent plant material, contributes significantly to plant litter productivity and serves as a key nutrient source for forest ecosystems (Nakamura et al., 2017). Tree growth and ecosystem productivity are influenced by critical abiotic factors, including water, nutrients, and light, which exert a more direct impact on vegetation structure and function compared to biotic influences (van der Sande et al., 2017). Soil fertility and other abiotic elements drive variations in tree species diversity, growth, and biomass productivity, underscoring the interconnected roles of biotic and abiotic factors. Neglecting the influence of abiotic factors on tree diversity and productivity risks overlooking their reciprocal impact on ecosystem functionality (Guo et al., 2020).

The production of litter is closely tied to the productivity of the plant community at a given site. Biotic factors such as tree species diversity and stand structural attributes, including stem diameter, tree density, and variation in tree size, positively influence aboveground biomass (AGB) (Poorter et al., 2017). The studies reveal a positive correlation between AGB and annual litter production, with litter yield increasing in tandem with vegetation structural complexity (Costa et al., 2020). Higher AGB is often associated with increased species diversity and structural variation. Tree species with high relative growth rates show greater productivity, leading to faster growth and higher litter production. Tree species diversity enhances productivity through mechanisms such as the niche complementarity hypothesis, which

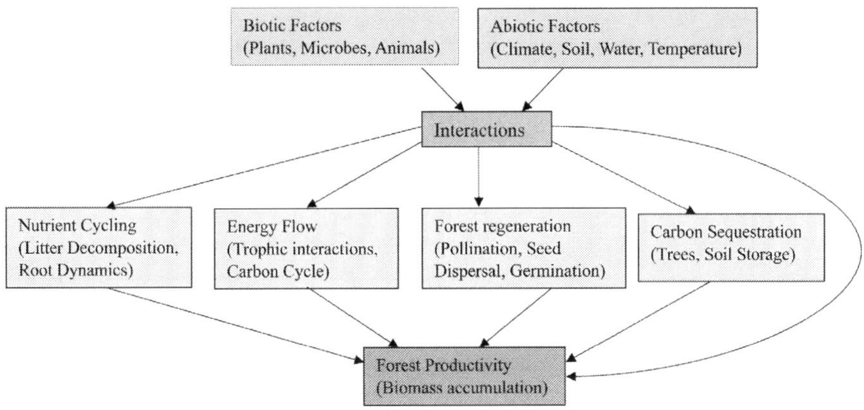

FIGURE 3.1 The figure illustrates the interactions between biotic and abiotic factors across various ecosystem functions, highlighting their collective impact on forest productivity.

posits that species or functional diversity improves biomass productivity by enhancing resource-use efficiency through niche partitioning and facilitation among species (Kunz et al., 2019). The effects of tree diversity on annual litter production are influenced by abiotic factors such as canopy light availability, rainfall, soil moisture, soil nutrients, and geomorphology, all of which affect tree size (Zheng et al., 2019).

Diverse stand structures, characterized by variations in tree size, enhance light capture and utilization across different canopy heights. Larger trees, which often dominate upper canopy strata (Figure 3.1), exhibit faster growth due to their ability to access greater light availability (Hildebrand et al., 2021). Litter generation is positively correlated with factors such as basal area, volume, and dominance of the upper canopy, as observed in subtropical forest fragments (Capellesso et al., 2016). Stand age and basal area also influence litter production, with older stands typically generating more litterfall (Bičkovskis et al., 2023). The type and origin of forests can impact the relationship between vegetation structure and litter generation. For instance, natural forests generally exhibit stronger correlations between structural attributes and litterfall compared to planted forests. Consequently, stand structure affects annual litter production both directly and indirectly (Guo et al., 2019). Litter biomass accumulation varies across forest types, with mixed forests generally producing the highest levels, followed by coniferous and broad-leaved forests. Deciduous broad-leaved forests (DBF) exhibit a seasonal pattern, producing significant litterfall in autumn, while monsoon evergreen broad-leaved forests show irregular patterns of litterfall throughout the year due to gradual leaf turnover (Lili et al., 2024).

Forest ecosystems rely on complex interactions among their various components. However, anthropogenic disturbances reduce structural complexity, leading to diminished species diversity, stability, and ecosystem functioning compared to structurally complex stands (Caviedes & Ibarra, 2017). Activities such as firewood collection, logging, and road construction degrade structural attributes and eliminate understory vegetation, resulting in reduced litter production (Mishra et al., 2004; Wiebe, 2014). Undisturbed forests typically produce more annual litterfall than disturbed ones. For example, virgin forests in eastern Nepal generate 54.2% more litterfall than disturbed areas, with leaves constituting 69–76% of the total litterfall (Gautam & Mandal, 2016).

Abiotic factors, including temperature, precipitation, radiation, and seasonal events, significantly influence litter production by affecting plant physiological processes such as phenology, evapotranspiration, and photosynthesis. Climate change can alter plant phenology, shifting litterfall patterns and affecting soil fertility and ecosystem productivity (Andivia et al., 2018). Rainfall demonstrates a complex relationship with litterfall, as it can trigger leaf shedding and influence the timing of peak litterfall (Lopes et al., 2015). Extreme climatic events, such as droughts, can drastically reduce litter production, as observed in Brazilian mangrove forests (de Oliveira Gomes et al., 2021). Understanding these interactions between biotic and abiotic factors is essential for sustainable forest ecosystem management.

3.1.3 Factors Affecting the Rate of Litter Decomposition

The decomposition of leaf litter is a crucial component of nutrient cycling and the food web, playing a significant role in biochemical processes (Baker et al., 2001). Litter decomposition involves the breakdown of OM into CO_2 and the mineralization of nutrients through physical, biological, and chemical pathways. This process provides a major source of nutrients and energy for heterotrophic organisms (Jia et al., 2018; Singh et al., 2022). During decomposition, carbon is released as CO_2 into the atmosphere through heterotrophic respiration by soil microorganisms and fauna. Decomposition rates can vary, with slower rates leading to the accumulation of OM and increased nutrient stocks in the soil, while faster rates facilitate the availability of nutrients required by plants (Isaac & Nair, 2005). The rate of litter decomposition is generally influenced by three main factors including litter quality (physical structure and chemical composition), the decomposer community (microorganisms and soil fauna), and the abiotic environment (e.g., temperature, soil moisture, and soil pH) (Chakravarty et al., 2020).

Litter quality, assessed by the chemical composition and nutrient concentrations of leaf litter, plays a significant role in determining decomposition rates (Hoeber et al., 2020; Rahman et al., 2013). Two primary physical traits of litter that influence decomposition are leaf toughness and leaf mass per unit area. Leaf toughness, which acts as a defense mechanism against herbivores, slows down decomposition and nutrient release (de la Riva et al., 2019). Additionally, tough leaves with lower nitrogen concentrations decompose more slowly than softer leaves. Conversely, leaves with lower mass per unit area and higher nutrient content tend to decompose more quickly, accelerating nutrient cycling and releasing nutrients into the environment (Li et al., 2023). Herb litter also significantly affects tree litter decomposition and soil fauna composition in forest ecosystems. Nutrient-rich summer herb litter can speed up the decomposition of recalcitrant tree litter by increasing nutrient availability and supporting soil fauna populations (Wang, 2021).

Freshly senesced leaf litter and green leaves (newly matured, sun-exposed leaves) differ substantially in chemical composition (Rosenfield et al., 2020). Green leaves, rich in nutrients such as nitrogen (N), phosphorus (P), calcium (Ca), and magnesium (Mg), decompose more rapidly due to their high palatability to decomposers. In contrast, senesced litter typically contains high concentrations of structural compounds like lignin and cellulose, as well as refractory substances such as tannins and polyphenols, which can inhibit microbial colonization and slow decomposition (Sun et al., 2022). Lignin, in particular, strengthens leaves and slows their decomposition, requiring specialized microorganisms to break it down, as seen in species like oak. Litter with higher nitrogen concentrations decomposes more rapidly than litter with low nitrogen levels. Deciduous species, which tend to have papery leaves with higher nutrient concentrations and lower levels of lignin and phenolic compounds, generally decompose faster than evergreen species, which produce more resistant litter (Rawat et al., 2020). Cellulose and hemicellulose are key components of nutrient cycling in forest ecosystems. Their decomposition provides carbon and energy essential for soil microorganisms. In temperate forests, cellulose breaks down relatively slowly, with only about 40% of the cellulose being lost over two years (Grossman et al., 2020). In contrast, phenolic compounds and condensed tannins influence litter quality and decomposition by binding to proteins, inhibiting microbial activity, and exerting toxic effects on soil fauna, all of which slow down decomposition (Lei et al., 2024).

Although litter chemistry influences the potential rate of decomposition, the actual breakdown largely depends on the activity of decomposer communities residing on the forest floor. A diverse community of algae, actinomycetes, bacteria, fungi, and soil fauna drives the sequential stages of litter decomposition. As decomposition progresses, small molecules like amino acids and sugars in the cell walls are readily broken down by fast-growing microorganisms within the decomposing community (Adair et al., 2008; Fioretto et al., 2005). The rate of litter decomposition, nutrient mineralization, and soil fertility are closely linked to the functional roles and metabolic activities of these decomposer communities. However, their specific contributions and mechanisms in the decomposition process vary.

Ectomycorrhizal and ericoid mycorrhizal fungi produce enzymes that break down OM. These fungi colonize freshly fallen litter, facilitating the transfer of nitrogen (N) and carbon (C) between litter layers by colonizing newly fallen leaves (Nicolás et al., 2019). In contrast, bacteria rely on the flow of substrate into their cells for decomposition (Laganière et al., 2010; Vivanco & Austin, 2008;). While both fungi and bacteria play important roles in decomposition, fungi are responsible for the majority of litter and wood breakdown due to their greater ability to decompose wood and leaf polymers and their superior capacity to colonize plant tissues (Baldrian, 2017; Purahong et al., 2016). Bacteria, by comparison, have a more limited ability to decompose complex organic materials like wood and leaf polymers (Boer et al., 2005).

In addition to biotic factors, climatic conditions play a significant role in litter decomposition by directly influencing the activity of decomposers. Climate change can alter soil fauna activity by modifying soil microclimates, such as temperature and humidity within the litter layer, which impacts ecosystem functions and decomposition rates (Singh et al., 2025b). Among these climatic factors, temperature and precipitation are particularly important in affecting decomposer metabolism. Temperature directly influences the activity of decomposer organisms, thereby affecting the rate of litter decomposition (Esquivel et al., 2020; Lalthakimi et al., 2023b). In general, warm and moist conditions, such as those found in tropical forests, promote decomposer activity and lead to faster decomposition. High temperatures and increased precipitation can enhance soil acidification, which may increase the fungi-to-bacteria ratio. Rising temperatures can also reduce soil microbial carbon-use efficiency (Bölscher et al., 2020). Litter properties, such as lignin content, also determine decomposition rates, which tend to be faster under warmer temperatures and lower pH levels (Cai et al., 2020). Studies have shown that the decomposition of low-quality, recalcitrant litter is more sensitive to temperature than that of high-quality, labile litter (Cai et al., 2021). Additionally, decreases in rainfall can suppress decomposition rates (Zhou et al., 2018). This is because higher precipitation increases litter moisture content, accelerating decomposition through leaching (Kutlu et al., 2018).

Photodegradation, the breakdown of OM by solar radiation, can also accelerate decomposition. Solar radiation enhances the decomposition of SOM and litter through photochemical reactions (photopriming) or by producing labile materials that facilitate microbial activity (Lin et al., 2017). UV-B wavelengths (280–320 nm) are primarily responsible for photodegradation, with lignin being the most affected compound in litter. Studies have shown that solar radiation reduces lignin content, thereby increasing litter's degradability and promoting microbial decomposition (Austin et al., 2016). Photodegradation occurs in the presence of lignin but not in pure cellulose substrates.

Soil properties also influence plant litter decomposition, with decomposition rates typically being slower in sandy soils compared to clay soils (Frøseth & Bleken, 2015). However, soil properties generally have minimal effects on the overall kinetics of litter decomposition (Cai et al., 2020; Gregorich et al., 2017). Soil texture plays a key role, as it influences nutrient and water dynamics, porosity, permeability, and surface area, all of which impact decomposition rates. Soil temperature significantly affects the population and activity of soil biota involved in decomposition (Salinas et al., 2011). Soil moisture is also crucial for microbial activity, as optimal moisture levels accelerate litter decomposition, whereas waterlogged conditions slow the process by limiting oxygen availability to decomposers (Peña-Peña & Irmler, 2016). Additionally, OM in the soil can increase the population density of macroorganisms, which helps in litter mixing and decomposition (Akpor et al., 2006). Soil nutrients also play an important role in microbial activity (Ge et al., 2013). Fresh litter often immobilizes nutrients, particularly nitrogen and phosphorus, early in the decomposition process, indicating that newly fallen litter may lack sufficient nutrients for decomposers (Hobbie & Vitousek, 2000). In nutrient-rich forests, decomposition tends to occur more rapidly, with nutrients retained more easily during the early stages of decomposition. Conversely, in nutrient-poor soils, litter decomposition is slower due to the high C/N ratio and the presence of more resistant compounds (Shaolin & Qiang, 2002).

3.1.4 Factors Affecting the Root Mediated Nutrient Dynamics

Rhizospheric nutrient dynamics are complex processes influenced by various factors, including root morphology, soil characteristics, and environmental conditions. Nutrient absorption by plants involves the movement of ions from the soil solution to the root surface, followed by their uptake by the plant (Ishaya & Rasmussen, 2024; Kiri & Hausa, 2023). Understanding this process is essential for improving nutrient efficiency in crops, as it is critical for plant growth and development. The root system plays a vital role in nutrient uptake, and different root types (e.g., primary, seminal, crown, and brace roots), having distinct nutrient uptake capabilities, contribute differently to this process (Ishaya & Rasmussen, 2024). Root length density, surface area, and root radius are the key factors that significantly influence nutrient uptake. The researchers have documented that root length density varies with soil depth and growth stage, with around 90% of the total root length found in the 0–20 cm soil depth throughout the growing season (Teo et al., 1995).

Nutrient flow into the roots also changes as the plant grows. The studies have demonstrated the highest nutrient flow in the early growth stages, which declines rapidly with increasing plant age (Mengel & Barber, 1974). This highlights the importance of considering plant age when studying nutrient uptake kinetics. Mathematical models can also describe nutrient intake processes. These models predict nutrient uptake by considering variables such as the initial nutrient concentration in the solution, buffer power, effective diffusion coefficient, and root system properties (Cushman, 1984; Oates & Barber, 1987). Advanced models can incorporate time-varying root characteristics, such as maximum nutrient input rate, root outflow, and absorption power (Cushman, 1984). Such models are useful tools for evaluating soil and plant properties and predicting nitrogen uptake under various conditions.

Biotic factors, such as microbes, insects, and weeds, also significantly influence root development and nutrient dynamics (Kumari et al., 2021). Arbuscular mycorrhizal fungi (AMF) are an example of beneficial microorganisms that interact with plant roots to improve nutrient uptake and plant health. Higher levels of AMF are associated with higher-quality sites, which enhance soil biota diversity and nutrient cycling (Souza et al., 2022). Seasonal fluctuations also play a significant role in root nutrient dynamics. In colder climates with slower nutrient cycling, dead fine root biomass tends to peak in the fall, influencing nutrient release and cycling in the soil (Fortier et al., 2019).

Additionally, root chemistry changes across seasons, genotypes, and environmental conditions, further affecting nutrient dynamics throughout the year (Fortier et al., 2019). A particularly interesting aspect of root nutrient dynamics is the impact of night-time nutrient uptake. Global estimates indicate that 43% and 45% of the annual phosphorus and nitrogen uptake, respectively, occur during the night when photosynthesis does not take place (Riley et al., 2018). This night-time uptake has significant implications for nutrient cycling and competition for nutrients between plants, microorganisms, and abiotic processes. The formation of organic horizons in the soil profile has been shown to significantly influence root nutrient dynamics. High-quality sites that promote the development of these organic horizons facilitate nutrient release and enhance plant nutrient content. This, in turn, stimulates root growth in the surface layers and creates a positive plant-soil feedback loop that supports primary production and nutrient cycling, ultimately preserving soil quality (Souza et al., 2022).

Environmental factors, particularly temperature, can significantly affect nutrient uptake. Global warming and increased heat stress, especially at temperatures between 40°C and 42°C, may impair root nutrient uptake. Heat stress can reduce crop yield and nutritional quality by lowering total protein concentrations, nitrogen absorption, and protein assimilation in roots (Giri et al., 2017). Soil fertility is another crucial factor in root nutrient dynamics. Recent studies have shown a negative correlation between fine root biomass and soil fertility, suggesting that plants may allocate more resources to root growth in nutrient-poor soils, enabling them to explore a larger soil volume in search of limited nutrients (Fortier et al., 2019).

Soil type, season, and depth also have significant effects on the dynamics of phosphorus (P) in soil. Non-labile phosphorus forms (e.g., NaOH-P, HCl-P, Mix-P, and Residual-P) dominate many soil systems, while labile phosphorus forms (e.g., H_2O-P and $NaHCO_3$-P) are typically less abundant. Biotic factors, particularly fungi, have been found to have a greater impact on phosphorus fractions than abiotic factors, with fungi influencing refractory phosphorus more than bacteria (Luo et al., 2022). The relationship between below-ground and above-ground biomass production is another key factor in root nutrient dynamics.

3.1.5 Energy Flow in Forest Ecosystems

Energy flow is a fundamental component of forest ecosystems, driving their dynamics, function, and structure. Energy is transferred through various trophic levels, starting with primary producers (plants), followed by consumers and decomposers, which collectively sustain the ecosystem's functioning. The primary energy source in these ecosystems is solar radiation, which is captured by plants and converted into chemical energy through photosynthesis. This chemical energy is stored as organic compounds, forming the foundation of the forest food web (Sato et al., 2011; Waring & Running, 2007). Trees, as the dominant vegetation in forest ecosystems, play a central role in energy capture and storage, contributing the bulk of primary production and biomass generation. Below the canopy, understory vegetation such as herbs and shrubs captures a smaller share of energy. Energy stored in plant tissues is accessed by herbivores, decomposers, and symbiotic organisms through consumption, decomposition, and nutrient cycling (Temple, 1999).

Energy flow in forests is intricately linked with the biogeochemical cycling of carbon. Through photosynthesis, trees and other plants sequester atmospheric carbon, storing it in their biomass and helping regulate carbon balances across ecosystems. Carbon stored in biomass is eventually released back into the atmosphere via ecological processes such as respiration and decomposition or through anthropogenic disturbances like fire and logging (Korhonen et al., 2001; Lalthakimi et al., 2023b; Väisänen et al., 1994). Global climate change exerts a significant influence on energy flow and carbon dynamics in forests. Simulations for southern Finland, for example, suggest that a 5°C increase in temperature over a century could boost photosynthesis rates by 6–8% under current CO_2 levels and by 8–10% at elevated CO_2 levels. However, increased respiration due to higher temperatures and CO_2 concentrations offsets this gain, leading to only a modest rise in total stem production (Väisänen et al., 1994). These findings underscore the complex interactions between climate change and energy dynamics in forest ecosystems.

The study of energy flow and material cycling in natural forests has informed practices in industrial ecology. In Finland, for instance, industries have adopted sustainable models that emphasize recycling matter, nutrients, and energy. These approaches minimize the use of virgin materials and energy inputs while reducing waste and emissions (Korhonen et al., 2001). Understanding energy flow in forest ecosystems is crucial for developing effective conservation and management strategies. Insights into energy dynamics help assess ecosystem resilience, productivity, and responses to disturbances. Moreover, this knowledge is vital for predicting the impacts of climate change on forests and formulating appropriate adaptation measures (Lamsal et al., 2018; Väisänen et al., 1994).

3.1.6 Influence of Abiotic and Biotic Factors on Ecosystem Energy Flow

Both abiotic and biotic factors have a significant impact on energy flow in various ecosystems. They have a consequence on the ecosystem's overall energy capture, transfer, utilization, and dissipation. Each component contributes in the following manner:

Abiotic factors play a pivotal role in shaping energy flow within forest ecosystems by directly influencing primary production and indirectly affecting ecological interactions and processes. Environmental elements such as light, temperature, humidity, and soil characteristics are fundamental in determining energy distribution and ecosystem functionality. These abiotic factors significantly interact with biotic components, underscoring their importance in understanding forest ecological dynamics (Schwarz et al., 2014). Light availability and temperature are key determinants of primary production—the conversion of CO_2, water, and solar energy into biomass by plants. This process forms the foundation of the forest food web and dictates energy availability for supporting various organisms within the ecosystem (Schwarz et al., 2014).

Additionally, abiotic factors influence energy flow through their impact on biogeochemical processes. For instance, grazing by cattle in birch forest patches alters soil abiotic properties, such as reducing the carbon-to-nitrogen ratio while increasing phosphorus content, pH, and bulk density (Proesmans et al., 2022). These changes cascade through microbial and invertebrate communities, affecting decomposition rates and nutrient cycling, which ultimately regulate energy flow within the ecosystem. The effects of abiotic factors become particularly evident during extreme events. Disturbances like storms, floods, and fires can disrupt photosynthesis, reduce primary production, and weaken tree resilience, thereby diminishing energy availability in forest ecosystems (Sierota et al., 2022). Such disruptions can lead to long-lasting impacts on forest health and energy dynamics.

The energy flow in forest ecosystems is intricately linked to biotic factors, which profoundly influence the carbon cycle, nutrient dynamics, and overall ecosystem functioning. Research indicates that species diversity and density play critical roles in enhancing forest growth rates and carbon dynamics, particularly in terms of carbon accumulation across various forest types (Li et al., 2022). For instance, DBF and coniferous broad-leaved mixed forests (CBMF) exhibit the highest growth and mortality rates, highlighting their significant contributions to carbon cycling (Li et al., 2022). The relationship between biotic factors and energy flow is complex and context-dependent. While tree species richness often enhances ecosystem functions, in certain ecosystems, it can reduce the spatial stability of soil microbial biomass, underscoring the nuanced and variable nature of biotic interactions (Gottschall et al.,

2022). These interactions illustrate the multifaceted ways in which living organisms influence energy flow within forest ecosystems.

A comprehensive understanding of the roles of both abiotic and biotic factors is vital for elucidating their regulatory effects on energy flow and carbon cycling. Such knowledge is essential for predicting forest responses to environmental changes and for formulating effective management strategies to ensure the sustainability of forest ecosystems (Gottschall et al., 2022; Li et al., 2022).

3.1.7 Forest Regeneration Process

Plant regeneration is a fundamental process that underpins forest biodiversity and ecosystem functionality. It involves multiple interconnected stages, including flowering, pollination, seed production, seed dispersal, germination, and the survival and establishment of seedlings. This comprehensive cycle ensures the replacement of mature plants with new recruits, maintaining population dynamics and supporting the long-term sustainability of forest ecosystems (Díaz et al., 2021). Each stage plays a critical role in facilitating the growth of new individuals and preserving the ecological balance within forest habitats.

3.1.8 Impact of Biotic and Abiotic Factors on Flowering of Plants

Flowering in plants is one of the critical stages in the life cycle of forest species that has a direct bearing on their reproductive success and the functioning of forest ecosystems. Biotic and abiotic factors that influence flowering are known to be interlinked in intricate ways, and understanding them helps one to understand how the functioning of forest ecosystems happens and how populations of plants exist in space and time.

Biotic factors also play an important role in the regulation of flowering. One of the most important biotic factors is competition for resources, which can affect flowering. Plants growing in crowded environments, where light, water, and nutrients are limited, may experience delayed or reduced flowering. Herbivory by insects or mammals can also impair flowering, especially if they browse or graze and damage flowers or inhibit their development altogether (Haas & Lortie, 2020). Plant diseases and pests can further impact the health of flowers and decrease the chances of successful pollination and seed production (Balla et al., 2021). Other plant species also may influence flowering in mixed-species forests. Some plants flower earlier or later depending on neighboring species to avoid competition for pollinators or exploit certain niches in pollinators (Howlett et al., 2021). The mutualistic relationships between plants and their biotic partners also regulate flowering. For example, some tree species depend on particular mycorrhizal fungi or nitrogen-fixing bacteria to enhance nutrient uptake, which impacts their flowering success (Zhao & Riaz, 2024). Similarly, in forests where certain species have evolved specific interactions with animals, such as seed dispersers or herbivores, these relationships can influence the timing and success of flowering events.

Similarly, abiotic factors that include light, temperature, humidity, soil composition, and availability of water play a huge role in the timing and outcome of flowering in plants. Light is one of the most important environmental cues as it governs photoperiod-sensitive plants that flower according to changes in day length (Li et al., 2016). For instance, some species in temperate forests, for example, trees and shrubs, have their flowering process initiated by certain conditions of light. Apart from light, temperature is the next influential factor in the process of flower development. The process of flowering is often initiated at certain temperatures by most plant species. Some species of plants require a period of exposure to cold, known as vernalization, to be able to flower (Amasino, 2005). Moreover, water availability affects flowering by influencing plant physiological processes. Insufficient water during critical flowering periods can lead to reduced flower production or even flower abortion (Bykova et al., 2019). Soil characteristics, such as nutrient availability and pH, can also impact flowering. Poor soil fertility or imbalanced nutrient levels can stress plants, potentially delaying or reducing flower formation (Copeland et al., 2012). Similarly, the level of humidity in an environment affects the rate of transpiration and water uptake, which then influences flowering timing and success. Extreme fluctuations in abiotic factors, such as temperature and moisture stress, may also cause plants to delay flowering or alter the duration of flowering periods, which can affect pollination success and overall reproductive output (Piao et al., 2019).

3.1.9 Role of Biotic and Abiotic Factors in Pollination Process

The plant-pollinator relationship is a vital mutualistic interaction where both parties benefit, playing a key role in ecosystem function. Flower morphology plays a crucial role in predicting pollination modes across the biotic to abiotic continuum. Wind-pollinated species tend to have longer reproductive structures, whereas insect-pollinated species tend to have shorter reproductive structures (Martínez-Gómez et al., 2023). The structure, composition, and age of forests play pivotal roles in shaping insect pollinator communities (Ulyshen et al., 2023). Open forest canopies enhance floral abundance and diversity in the understory, fostering microclimates with favourable conditions such as optimal temperature and soil moisture, which are crucial for pollinator nesting and foraging (Hyjazie & Sargent, 2022; Sobek et al., 2009). The composition of tree species within a forest also significantly impacts pollinator diversity. Broadleaf forests typically support greater pollinator diversity than conifer-dominated forests. This discrepancy is likely due to the higher nutritional value and palatability of pollen from broadleaf trees, which are generally preferred by pollinators over conifers (Ulyshen et al., 2024). Forest age further influences pollinator diversity, revealing distinct patterns between coniferous and broadleaf ecosystems. In conifer forests, pollinator diversity declines as canopy closure progresses, reducing light availability on the forest floor and thereby limiting floral resource production. Conversely, aging broadleaf canopies often enhance pollinator abundance and diversity. This is because broadleaf trees provide increasing floral resources as they mature, including tree flowers that attract forest specialist pollinators (Ulyshen et al., 2024).

Climate change is projected to have profound effects on plant-pollinator interactions, particularly in sensitive alpine ecosystems (Trunschke et al., 2024). Rising global temperatures can cause phenological shifts in both plants and pollinators, potentially leading to temporal mismatches between these mutualistic partners. Such mismatches disrupt the availability of floral resources, hindering pollination success and destabilizing plant-pollinator networks. Warming also impacts key plant traits, including flower, nectar, and pollen production, while influencing pollinator foraging behavior, body size, and lifespan. These combined effects ultimately reduce pollen flow and pollination efficiency (Inouye, 2020). For example, heat waves have been shown to exert species-specific effects on insect pollinators, particularly cold-adapted species such as bumblebees. Heat stress in these species negatively affects male reproductive traits, including sperm viability, DNA integrity, and pheromone production, which are critical for successful reproduction (Martinet et al., 2021).

3.1.10 Role of Biotic and Abiotic Affecting Seed Production

The quantity and quality of seeds produced by a plant are critical determinants of successful tree establishment. Higher seed production increases the likelihood of colonizing new areas, enhancing the survival and spread of tree species (Bruun & Poschlod, 2006). Additionally, high-quality seeds are more likely to survive and establish under stressful environmental conditions (Moles & Westoby, 2002; Walters & Reich, 2000). Conversely, low seed production, both in quantity and quality, can hinder these processes.

Seed production is influenced by multiple factors, including the microclimate, physiology of flowering and fruiting, and the allocation of resources. These factors are further modulated by pollen availability, genetic traits, hormonal regulation, and the tree's position in the forest stand (Vacchiano et al., 2018). Resource availability is a key determinant of seed production. Ample resources promote the development of high-quality flowers, which are more likely to produce viable seeds. Limited resources, however, can result in poor seed quality and reduced output (Bogdziewicz et al., 2020; Garcia-Barreda et al., 2021). A well-documented pattern in seed production is "mast seeding," characterized by years of high yields followed by years of low or negligible output. This cyclical phenomenon is partly due to resource depletion, as reproductive processes demand substantial energy and nutrient investment. Trees require time to accumulate sufficient resources before subsequent high-yield events (Bogdziewicz et al., 2020; Garcia-Barreda et al., 2021).

Biological and environmental factors also significantly affect seed production. Phenological synchrony, temperature, rainfall, wind, light availability, and drought can limit pollen production and subsequent seed development. Such limitations can drastically reduce fruit yield (Koenig et al., 2015; Misson et al., 2011). For example, high precipitation can wash away pollen, impeding pollination, while frost, drought, and extreme temperatures can lead to flower abortion and reduced seed output (Bogdziewicz et al., 2020; Pulido & Díaz, 2005). The social position of a tree within the canopy plays a pivotal role in determining seed production. Trees with expansive crowns and upper-canopy positions receive more light and often produce higher yields of flowers and seeds than those in lower canopy layers (Kremer & Bauhus, 2020). Since flowering and fruiting predominantly occur in the outer crown, larger crown dimensions are associated with increased seed yields (Gonçalves & Pommerening, 2012; Gonçalves et al., 2017).

Abiotic stressors such as frost, drought, and high temperatures further impact seed yield. Frost and low temperatures can impair pollination processes, while drought can limit flower development and reduce nectar availability, thereby hindering pollinator activity. High temperatures can induce pollen sterility by preventing anther dehiscence and pollen release, ultimately reducing fertilization and seed production (De Storme & Geelen, 2014; Lora et al., 2012).

3.1.11 Role of Biotic and Abiotic Factors in Seed Dispersal

Seed dispersal is the movement of seeds, spores, or propagules away from the parent plant and is essential for the survival, spread, and regeneration of plant populations. Seed dispersal occurs primarily through abiotic (ballistic, gravity, wind, water dispersal) and biotic (animal-mediated) dispersal. While abiotic dispersal is significant, particularly in open habitats, biotic dispersal, especially by frugivorous animals, is dominant in woody plants and forest ecosystems (Calama et al., 2017; Kozlowski, 2002). Gymnosperms like *Pinus*, *Abies*, and *Picea* produce winged seeds dispersed by wind (Salazar-Tortosa et al., 2019). Similarly, angiosperms such as *Acer*, *Fraxinus*, and *Ulmus* produce samaras for aerial dispersal. Other species, like *Populus*, *Salix*, and *Catalpa*, release lightweight seeds aided by a pappus, enabling long-distance wind transport (Landhäusser et al., 2019; Kim et al., 2022). Seed mass influences dispersal range; lighter seeds travel farther, while heavier seeds tend to drop near the parent plant. Some seeds may also undergo secondary dispersal after landing, through wind, gravity, or water (Zhu et al., 2019).

The effectiveness of animal-mediated dispersal depends on animal behaviour, dietary preferences, movement patterns, and habitat choices. For successful dispersal, the animal must encounter and consume the fruit. The seed must survive handling, and the animal must be transported far enough to deposit it away from the parent plant (Morales & Morán López, 2022). When encountering fruits, frugivores must decide whether to forage or move on, based on factors such as size, ripeness, and pest infestation (Roger et al., 2017). Fruit selection depends on the nutritional needs of frugivores and fruit traits such as size and chemical content. For instance, frugivorous birds adjust their preferences seasonally using more accessible or nutritious fruits (Albrecht et al., 2018). Behaviorally, bolder individuals are more likely to explore and locate visible or abundant fruits, while shy ones remain in safer zones and depend on cues from others (Sih & Del Giudice, 2012). While handling fruits, frugivores may either crush seeds ("mashers") or swallow them ("gulpers"). For example, primates either swallow seeds for distant dispersal or spit them out, usually near the parent plant (Naniwadekar et al., 2019). Fruit handling significantly affects seed dispersal effectiveness by influencing seed quantity, viability, and deposition patterns. Spitting and dropping reduce dispersed seed numbers, whereas handling affects seed survival and post-dispersal success (Fricke et al., 2019).

Land use changes, habitat degradation/disturbances, climate, and extreme weather conditions have a significant influence on the seed dispersal mechanisms of various groups of plants in the ecosystem. (Tucker et al., 2021). Habitat degradation reduces habitat quality and diversity, significantly affecting seed deposition by decreasing the vegetation cover and fruit availability for bird dispersers. Global habitat modifications due to human expansion have reduced animal movement by an average of 30%, which likely disrupts seed dispersal and related ecological functions (Tucker et al., 2018). Changes in wind speed and frequency, particularly extreme winds, can impact wind-dispersed plants by limiting the seed dispersal distance. When wind speed is insufficient for long-distance dispersal, seeds remain close to the parent plants, leading to higher seedling density, increased competition, and greater susceptibility to pathogens and predators. In fragmented landscapes, strong winds can disperse seeds into inhospitable areas (e.g., roads, crops, and urban zones), reducing seed survival and the establishment of native species (Novaes et al., 2020). Environmental variations such as wind and moisture affect leaf sprouting and seed dispersal. During the dry season, reduced water availability and strong winds trigger leaf shedding and fruit ripening. Leaf fall can also facilitate seed dispersal by opening the canopy and allowing more wind to pass through. Lower rainfall, humidity, and higher temperatures further promote abscission and reduce propagule mass, thereby enhancing dispersal potential (Hernandez et al., 2023).

3.1.12 Role of Biotic and Abiotic Factors on Seed Germination

Germination, a critical sub-process of natural regeneration, begins with seed imbibition and culminates in the full elongation of the radicle (Calama Sainz et al., 2017). Similarly, physiologists define germination as the emergence of the radicle through the seed coat (Pardos et al., 2005).. Biotic factors like species traits, particularly seed dormancy, play a key role in germination outcomes. Certain seeds, such as those of *Quercus* species, *Castanea sativa*, and *Pinus pinaster*, germinate soon after ripening under favorable conditions. Conversely, other species exhibit dormancy, delaying germination and hindering population regeneration in unsuitable environments (Alvarado & Bradford, 2002). Seed sizes also significantly affect germination success; larger seeds, such as those of *Avicennia officinalis*, are generally more successful in establishing roots and proliferating (Chowdhury et al., 2023). Seasonal patterns in flowering and fruiting further influence seed characteristics. Larger and heavier seeds, like those of *Terminalia bellirica*, *Terminalia chebula*, and *Artocarpus chama*, demonstrate better germination and establishment outcomes (Chowdhury et al., 2023). However, Hu et al. (2024) found that seed mass and shape correlate with germination thresholds, where larger, rounder seeds often have lower base water potentials for germination.

Temperature and water availability are critical environmental factors governing seed germination and seedling establishment in natural settings. Temperature variations significantly influence seedgermination process across various species. For instance, low temperatures inhibit enzyme activity in seeds, thereby slowing physiological processes and reducing germination efficiency. Studies on *Betula platyphylla* show that germination rates rise with increasing temperature until an optimal threshold, beyond which they decline (Luo et al., 2022). Water stress exacerbates germination challenges, with seeds becoming more sensitive to drought under lower temperatures (Luo et al., 2022). Therefore, climate change is expected to disrupt germination and seedling establishment, especially in tropical dry forests, where reduced rainfall is predicted to be a greater limiting factor than increased soil temperature by 2055

(Dantas et al., 2020). Light intensity also plays a significant role in seed germination and early seedling growth. Many small-seeded species require light for germination, preventing them from sprouting in deeply shaded areas or beneath thick leaf litter during forest succession (Jankowska-Blaszczuk & Daws, 2007). Light requirements vary across species; for example, *Cunninghamia lanceolata* germinates better under lower light intensity, while *Schima superba* shows optimal germination at intermediate light levels (Liu et al., 2019). In contrast, certain species from jarrah forests, including *Eucalyptus marginata* and *Eucalyptus calophylla*, exhibit reduced germination rates under high light intensities compared to darkness (Rokich & Bell, 1995). Soil properties exert a profound influence on seed germination and early seedling development. Factors such as soil porosity and moisture content are particularly significant. Clayey soils often promote higher germination rates due to superior water retention (Bocchese et al., 2008). Conversely, soil compaction restricts root and microorganism access to essential resources like water, oxygen, and nutrients (Picchio et al., 2020). Additionally, land use changes, such as shifting cultivation, can alter soil properties and negatively impact germination capacity (Lalthakimi et al., 2023a).

3.1.13 Role of Biotic and Abiotic Factors on Seedling Survival and Establishment

Establishment, survival, and early seedling growth are considered critical stages during the early life stages of plants (Kozlowski, 2002). For tree communities, the transition from seedling to sapling has the highest mortality rate during natural regeneration, affecting tree establishment. Juveniles suffer most from both biotic and abiotic constraints than mature plants (Wright, 2002). Seedling survival is determined by a seed's resources or a seedling's vigor, seed predation, and intraspecific or interspecific competition.

Understory species generally produce fewer seeds than canopy species, resulting in fewer recruits than canopy trees (Terborgh et al., 2014). Fewer recruits reduce conspecific negative density dependence (CNDD) and associated repercussions (Lu et al., 2015). Seed predators, pathogens, and herbivores reduce the survival and recruitment of seedlings when they occur in localized, conspecific density (LCD) areas. In addition, an asymmetric resource competition with conspecific nearby adults may result in high seedling mortality (Abiem et al., 2023). Seedling survival may also be influenced by the timing of recruitment, especially in seasonal forests (Lin et al., 2017). Seedlings that establish in the wet season may have a higher survival rate than those that establish in the dry season because they are likely to grow faster, thus acquiring an initial size advantage (Bai et al., 2012).

Abiotic factors are essential for seedling recruitment (Martini et al., 2019). For instance, light is one of the most important abiotic factors limiting seedlings in tropical forests (Comita et al., 2009). Under proper light conditions, the growth and survival rate of seedlings is increased (Lu et al., 2018). Similarly, under an open canopy environment, the survival rate of seedlings was increased under optimum light conditions for both shade-tolerant and shade-intolerant species; however, under an extreme-light environment, the survival rate increased only for shade-intolerant species and decreased for shade-tolerant species. Oshima et al. (2015) in their study of *Dipterocarps* in South-East Asia reported that canopy openness was a major factor affecting seedling survival. Among topographic factors, slopes and aspect affect seedling survival, with steep slopes having the lowest survival rates due to poor water retention, increased runoff, and soil erosion, hindering seedling establishment. While aspect also influences seedling survival and species diversity, which are significantly higher on south- and east-facing slopes compared to north- and west-facing ones (Abiem et al., 2023).

3.1.14 Carbon Sequestration in Forest Ecosystems

Forest ecosystems play an essential role in sequestering atmospheric carbon through photosynthesis and tree growth, storing it in both aboveground and belowground biomass (Liu et al., 2016). Trees, as the primary components of forests, absorb significant amounts of CO_2 via photosynthesis and release it back into the atmosphere through autotrophic and heterotrophic respiration. However, a portion of the carbon remains stored in vegetation, litter, and soil over time (Schulze et al., 2019). Terrestrial and soil carbon serve as major carbon sinks, reducing atmospheric CO_2 concentration (Schmidt et al., 2011). These sinks contain over 80% of all terrestrial aboveground carbon (AGC) and more than 70% of all soil organic carbon (SOC) (Santoro et al., 2021). In forests, most carbon is held in intermediate pools such as wood, litter, and partially decomposed OM (Smith et al., 2004). The three primary carbon pools in forest ecosystems are (i) AGB (ii) belowground biomass, and (iii) SOC.

Carbon fluxes in forests are largely governed by the balance between CO_2 uptake through photosynthesis and its release through respiration and decomposition (Mukul et al., 2020). All organisms in the ecosystem play a role in regulating this carbon exchange. During photosynthesis, trees assimilate atmospheric CO_2 and distribute the fixed carbon to different tissues, including trunks, branches, leaves, and roots (Kishwan et al., 2012). The stem wood contains the largest proportion of AGC, serving as a major long-term storage pool.

Belowground carbon is found in a complex mixture of roots, soil organisms, and decaying organic matter. Dead organic material, such as litter and microbial residues, is decomposed and incorporated into the soil through the action of soil fauna, microbes, and root exudates (Perez-Quezada et al., 2021; Singh et al., 2025b). This decomposition process includes mineralization, converting organic carbon into forms usable by plants and humification, which contributes to the formation and stabilization of soil organic matter (Jimenez & Lal, 2006).

The rate and capacity of carbon sequestration in forests are influenced by multiple factors, including climate, soil characteristics, tree species, management practices, and the chemical composition of the forest floor (Lal, 2005).

3.1.15 Role of Biotic and abiotic Factors in Carbon Sequestration

Biota plays a crucial role in terrestrial carbon cycling by transforming CO_2 into OM through assimilation by primary producers (plants, algae, autotrophic prokaryotes) and releasing CO_2 (or CH_4) via respiration by consumers (animals), microbial decomposers (fungi, bacteria, archaea), and primary producers (Lorenz et al., 2018).

The interaction between heterotrophic microorganisms and plant root systems represents a vital energy source for soil and rhizosphere fauna, such as nematodes and collembolans, influencing OM production, decomposition, and soil carbon turnover through top-down control of microbial communities and herbivory on plant roots (Gan & Wicklings, 2020). Plants depend on heterotrophic soil organisms for their nutrient supply and can stimulate the heterotrophic decomposition of dead OM by providing decomposers with energy-rich substrates (i.e., priming). In turn, heterotrophs require plant-derived organic compounds for energy production and enhance plant productivity by making nutrients available for uptake. Soong et al. (2020) show that microbial communities in soil are primarily limited by organic carbon or energy, while plant productivity is nutrient-limited, requiring the integration of these components to understand ecosystem carbon cycling.

Changes in species composition can radically alter ecosystem carbon (C) dynamics through changes in the functional types of plants, soil fauna (e.g., earthworms), and soil microbes (De Deyn et al. 2008). Slow-growing plants decompose slowly, supporting C sequestration through several mechanisms (De Deyn et al., 2008). Encroachment of N-fixing shrubs frequently increases ecosystem C turnover and sequestration. Conversely, disease-mediated decline in native N fixers can reduce productivity and induce disequilibrium in the C cycling processes (Ruess et al., 2009). Fungi favour C sequestration through their higher growth efficiency than bacteria and through the recalcitrance of the compounds they produce (De Deyn et al., 2008).

Tree density and species richness were more strongly and positively correlated with growth than with mortality, thereby generating an average increase in net tree C stocks. Tree density, in addition to tree species diversity, is important for forest productivity. Higher stand densities increase forest C storage and wood production through greater canopy cover, which intercepts more light for photosynthesis (Ouyang et al., 2019).

Recent studies have highlighted the complex interactions among tree functional traits, forest biomass, species diversity, and site properties that drive forest soil organic carbon (SOC) storage. Tree functional traits and forest biomass explain the variability in SOC, which is influenced by climatic and soil conditions (Augusto & Boča, 2022). Increased tree species richness can enhance SOC storage, particularly when biomass overyielding occurs in mixed forests (Augusto & Boča, 2022). Tree productivity and specific functional traits, such as root diameter and litter C: N ratio, influence soil carbon concentrations and microbial biomass (Beugnon et al., 2023).

Abiotic variables are essential for carbon sequestration in several ecosystems. Soil moisture and temperature have major effects on carbon sequestration under desert conditions. Low temperatures can improve carbon sequestration in shifting sand. However, high soil moisture can promote rapid CO_2 release by boosting microbial activity and OM transport (Yang et al., 2023). The two main factors influencing carbon sequestration in forest ecosystems are soil characteristics and climate. Multifaceted ecosystem functioning, including carbon storage, is inversely correlated with soil texture, particularly sand content (Wang et al., 2023). Elevation also has an indirect impact on carbon sequestration by affecting essential tree characteristics, including height, diameter at breast height, and crown area (Ali et al., 2023). It's interesting to keep in mind that by influencing both above- and below-ground biodiversity, abiotic factors might have a greater direct and indirect impact on ecosystem multifunctionality than biotic factors

(Wang et al., 2023). The inhibitory effects of phenolic chemicals produce an "enzymatic latch" on breakdown in peatlands, which contributes to their remarkable carbon sequestration capacity due to their high carbon storage densities (Freeman et al., 2012). The potential carbon sequestration is also affected by climate change. Statistical tools like the SSP pathway demonstrated the best rate of carbon sequestration in forest vegetation among several Shared Socioeconomic Pathways (Guan et al., 2023). These results emphasized the importance of considering abiotic factors when developing carbon sequestration plans and attempting to mitigate climate change.

3.1.16 Forest Productivity

The pace at which biomass accumulates in a forest over time is commonly referred to as forest productivity and is an essential indicator of ecosystem health and function. It is commonly measured in cubic meters per hectare per year (m^3 ha^{-1} $year^{-1}$) and is reported as the increase in AGB (Diodato & Bellocchi, 2020; Ren et al., 2021). This idea includes the capacity of the carbon stored by forests to generate timber and other ecosystem services.

Forest productivity is an intricate concept that integrates various ecosystem functions and services, and is influenced by biodiversity, stand structures, environmental conditions, and landscape patterns. For sustainable forest management, it is essential to understand and forecast forest production, particularly considering the growing demand for forest goods and services. The use of bio-stimulants, balanced nutrition additions, and modification of soil-microbe-plant connections are some of the novel approaches to increasing forest production that are now being investigated within environmental constraints (Clinton, 2018).

Our level of understanding, based on forest productivity, has expanded through many studies that have included below-ground components. Many studies have indicated that mixed-species forests have a positive impact on fine root length, improve water and nutrient uptake, and promote increased forest productivity (Huang et al., 2023). This emphasizes the importance of preserving a variety of tree species to increase forest production and sequester carbon. Large-scale forest productivity is estimated using a variety of techniques, including (i) remotely sensed data for continuous net primary productivity estimates, (ii) flux tower observations documenting plant-atmosphere gas exchange, and (iii) forest growth data from research plots and inventory sampling points based on repeated tree observations (Hasenauer et al., 2012). By combining these techniques, the productivity evaluation in sizable wooded areas can be enhanced.

3.1.16.1 Factor Affecting Forest Productivity

Forest productivity is influenced by the intricate relationships between biotic and abiotic elements. Abiotic factors include weather patterns, topography, and soil characteristics, whereas biotic factors include species variety, structural diversity, tree density, and functional quality. Approximately 31% of the variation in productivity in species-rich broad-leaved forests can be attributed to species diversity, which has been shown to have a major impact on production (Bayat et al., 2021). However, the relationship between productivity and variety remains unclear. According to Yue et al. (2020), species variety may not always directly affect productivity. This was observed in temperate forest plots.

Productivity can be affected both favorably and unfavorably by structural diversity, which is the variation in tree size and spatial arrangement within a forest. In certain instances, structural diversity reduces the annual AGB increments of surviving trees and the entire tree community while increasing the AGB increments of recruits (Yue et al., 2020). This implies that asymmetric competition shapes forest productivity dynamics. Forest production has frequently been found to be significantly affected by tree density. In temperate forests, density variables have the greatest impact on the annual AGB increments of surviving trees as well as the overall tree community (Yue et al., 2020). Reduced inter-annual variance in canopy complexity has been linked to high stem density, which can increase production (Fotis et al., 2018).

Forest productivity is greatly influenced by the functional properties of the trees, such as the features of their leaves and branches. These characteristics are extremely flexible in response to variations in the climate and nutrient availability (Zarzosa et al., 2021). In certain species, such as *Quercus ilex*, intraspecific variability in functional attributes has been demonstrated to have a substantial impact on plant biomass and growth, which ultimately explains forest productivity (Zarzosa et al., 2021).

Climate variables are among the most significant abiotic factors that affect forest productivity. Primary productivity is a climate regulation service that is significantly affected by the interactions between temperature and precipitation (Serna-Chavez et al., 2017). Functional features and, in turn, forest productivity are affected by the aridity gradient, which is represented by mean annual precipitation (Zarzosa et al., 2021). Forest productivity is also greatly influenced by topography and soil characteristics. In certain forest ecosystems, solar radiation and topographic wetness index are significant determinants of productivity (Bayat et al., 2021). Forest productivity can be affected by the availability of soil resources, such as nutrients and texture (Zarzosa et al., 2021).

Furthermore, the type of forest and the surrounding environment might affect the relative significance of biotic and abiotic elements. Abiotic factors may not always affect productivity as biotic factors do. For example, tree size and local forest structure were found to be more significant predictors of tree growth in Qinghai Province's plantation forests than geography or availability of soil resources (Lin et al., 2021). Both biotic and abiotic factors affect the relationship between the biomass and production. The scaling exponent of the power function that links productivity to biomass and the slope of the linear regression model are influenced by tree age, density, height, and diameter at breast height (DBH) (Hui et al., 2012). When forecasting tree productivity based on biomass, it is crucial to consider both biotic and abiotic factors because this relationship varies with elevation, latitude, and longitude.

3.2 CONCLUSION

Forest ecosystems are complex and dynamic systems determined by the complex interactions of biotic and abiotic factors. These interactions govern some of the most essential processes in an ecosystem, like nutrient cycling, regeneration, carbon sequestration, and productivity. The improved understanding of these interdependencies is essential for sustainable forest management and addressing climate change challenges. The biotic factors including plant diversity, microbial communities, and soil fauna underpin forest functions. Species-specific traits, such as seed size, dormancy, and carbon allocation patterns, have a large impact on forest regeneration and nutrient cycling. Microbial decomposers and soil fauna play important roles in breaking down OM, releasing nutrients needed for plant growth, and maintaining soil health. Tree species diversity and functional traits, such as root architecture and litter quality, enhance SOC storage and microbial biomass, underlining the importance of biodiversity for resilience and productivity. Similarly, the abiotic factors such as topography, temperature, precipitation, and soil properties are equally important. Soil texture, moisture, and nutrient availability influence seed germination, root growth, and microbial activity, whereas climatic factors control the growth and productivity of the forest. Temperature and precipitation control carbon cycling through their influence on photosynthesis, respiration, and decomposition rates. Topography, including elevation and slope, modifies microclimates and resource distribution, influencing both biotic communities and ecosystem processes.

Carbon sequestration, a critical ecosystem service, underscores the synergy between biotic and abiotic factors. Tree density, species richness, and functional traits determine carbon storage, while soil moisture and temperature regulate microbial decomposition and carbon fluxes. Biotic interactions, such as plant-microbe relationships, further shape carbon dynamics. Forest productivity, an indicator of ecosystem health, benefits from higher species richness and structural diversity, which improve resource efficiency and resilience. However, abiotic factors like nutrient scarcity or arid conditions can limit productivity. Therefore, a holistic integration of biotic and abiotic factors will be essential in sustaining forest functions and ecosystem services. Adaptive management strategies that prioritize biodiversity, soil health, and conservation of functional traits will ensure sustainable forest productivity, effective carbon sequestration, and resilience to climate change and habitat degradation.

REFERENCES

Abiem, I., Kenfack, D., & Chapman, H. M. (2023). Assessing the impact of abiotic and biotic factors on seedling survival in an African montane forest. *Frontiers in Forests and Global Change*, *6*, 1108257.

Adair, E. C., Parton, W. J., Del Grosso, S. J., Silver, W. L., Harmon, M. E., Hall, S. A., ..., & Hart, S. C. (2008). Simple three-pool model accurately describes patterns of long-term litter decomposition in diverse climates. *Global Change Biology*, *14*(11), 2636–2660.

Akpor, O. B., Okoh, A. I., & Babalola, G. O. (2006). *Culturable microbial population dynamics during decomposition of Theobroma cacao leaf litters in a tropical soil setting.*

Albrecht, J., Hagge, J., Schabo, D. G., Schaefer, H. M., & Farwig, N. (2018). Reward regulation in plant–frugivore networks requires only weak cues. *Nature Communications*, *9*(1), 4838.

Ali, S., Khan, S. M., Ahmad, Z., Siddiq, Z., Ullah, A., Yoo, S., ..., & Raposo, A. (2023). Carbon sequestration potential of different forest types in Pakistan and its role in regulating services for public health. *Frontiers in Public Health*, *10*, 1064586.

Alvarado, V., & Bradford, K. J. (2002). A hydrothermal time model explains the cardinal temperatures for seed germination. *Plant, Cell & Environment*, *25*(8), 1061–1069.

Amasino, R. M. (2005). Vernalization and flowering time. *Current Opinion in Biotechnology*, *16*(2), 154–158.

Andivia, E., Bou, J., Fernandez, M., Caritat, A., Alejano, R., Vilar, L., & Vazquez-Pique, J. (2018). Assessing the relative role of climate on litterfall in Mediterranean cork oak forests. *iForest-Biogeosciences and Forestry*, *11*(6), 786.

Augusto, L., & Boča, A. (2022). Tree functional traits, forest biomass, and tree species diversity interact with site properties to drive forest soil carbon. *Nature Communications*, *13*(1), 1097.

Austin, A. T., Méndez, M. S., & Ballaré, C. L. (2016). Photodegradation alleviates the lignin bottleneck for carbon turnover in terrestrial ecosystems. *Proceedings of the National Academy of Sciences*, *113*(16), 4392–4397.

Babushkina, E. A., Sitnikov, G. A., Upadhyay, K. K., Zhirnova, D. F., Zelenov, G. K., Vaganov, E. A., & Belokopytova, L. V. (2022). Seasonal growth of pine tree rings: Comparison of direct observations and simulation. *Forests*, *13*(12), 1978.

Bai, X., Queenborough, S. A., Wang, X., Zhang, J., Li, B., Yuan, Z., …, & Hao, Z. (2012). Effects of local biotic neighbors and habitat heterogeneity on tree and shrub seedling survival in an old-growth temperate forest. *Oecologia*, *170*, 755–765.

Baker, T. T. III, Lockaby, B. G., Conner, W. H., Meier, C. E., Stanturf, J. A., & Burke, M. K.. (2001). Leaf litter decomposition and nutrient dynamics in four southern forested floodplain communities. *Soil Science Society of America Journal*, *65*(4), 1334–1347.

Baldrian, P. (2017). Forest microbiome: Diversity, complexity and dynamics. *FEMS Microbiology Reviews*, *41*(2), 109–130.

Balla, A., Silini, A., Cherif-Silini, H., Chenari Bouket, A., Moser, W. K., Nowakowska, J. A., …, & Belbahri, L. (2021). The threat of pests and pathogens and the potential for biological control in forest ecosystems. *Forests*, *12*(11), 1579.

Bayat, M., Bettinger, P., Heidari, S., Hamidi, S. K., & Jaafari, A. (2021). A combination of biotic and abiotic factors and diversity determine productivity in natural deciduous forests. *Forests*, *12*(11), 1450.

Becker, J., Pabst, H., Mnyonga, J., & Kuzyakov, Y. (2015). Annual litterfall dynamics and nutrient deposition depending on elevation and land use at Mt. Kilimanjaro. *Biogeosciences*, *12*(19), 5635–5646.

Beugnon, R., Bu, W., Bruelheide, H., Davrinche, A., Du, J., Haider, S., …, & Eisenhauer, N. (2023). Abiotic and biotic drivers of tree trait effects on soil microbial biomass and soil carbon concentration. *Ecological Monographs*, *93*(2), e1563.

Bičkovskis, K., Samariks, V., & Jansons, Ā (2023). Tree litter production in grey alder (*Alnus incana*) stands on dry and wet mineral soils. *International Multidisciplinary Scientific GeoConference: SGEM*, *23*(3.2), 343–350.

Biswas, T., & Kole, S. C. (2017). Soil organic matter and microbial role in plant productivity and soil fertility. *Advances in soil microbiology: Recent trends and future prospects: Volume 2: Soil-Microbe-Plant Interaction* (pp. 219–238).

Blois, J. L., Zarnetske, P. L., Fitzpatrick, M. C., & Finnegan, S. (2013). Climate change and the past, present, and future of biotic interactions. *Science*, *341*(6145), 499–504.

Bocchese, R. A., Oliveira, A. D., Favero, S., Garnés, S. D. S., & Laura, V. A. (2008). Seed rain and seedling establishment from the use of isolated trees and artificial perches by seed-dispersing birds in a Cerrado area, Mato Grosso do Sul, Brazil. *Revista Brasileira de Ornitologia*, *16*(3), 207–213.

Boer, W. D., Folman, L. B., Summerbell, R. C., & Boddy, L. (2005). Living in a fungal world: Impact of fungi on soil bacterial niche development. *FEMS Microbiology Reviews*, *29*(4), 795–811.

Bogdziewicz, M., Ascoli, D., Hacket-Pain, A., Koenig, W. D., Pearse, I., Pesendorfer, M., …, & Tanentzap, A. (2020). From theory to experiments for testing the proximate mechanisms of mast seeding: An agenda for an experimental ecology. *Ecology Letters*, *23*(2), 210–220.

Bölscher, T., Ågren, G. I., & Herrmann, A. M. (2020). Land-use alters the temperature response of microbial carbon-use efficiency in soils–a consumption-based approach. *Soil Biology and Biochemistry*, *140*, 107639.

Bruun, H. H., & Poschlod, P. (2006). Why are small seeds dispersed through animal guts: Large numbers or seed size per se? *Oikos*, *113*, 402–411.

Bykova, O., Chuine, I., & Morin, X. (2019). Highlighting the importance of water availability in reproductive processes to understand climate change impacts on plant biodiversity. *Perspectives in Plant Ecology, Evolution and Systematics*, *37*, 20–25.

Cai, A., Chang, N., Zhang, W., Liang, G., Zhang, X., Hou, E., … & Luo, Y. (2020). The spatial patterns of litter turnover time in Chinese terrestrial ecosystems. *European Journal of Soil Science*, *71*(5), 856–867.

Cai, A., Liang, G., Yang, W., Zhu, J., Han, T., Zhang, W., & Xu, M. (2021). Patterns and driving factors of litter decomposition across Chinese terrestrial ecosystems. *Journal of Cleaner Production*, *278*, 123964.

Calama Sainz, R., Manso González, R., Lucas Borja, M. E., Espelta Morral, J. M., Piqué Nicolau, M., Bravo Oviedo, F., …, & Pardos Mínguez, M. (2017). *Natural regeneration in Iberian pines: A review of dynamic processes and proposals for management.*

Capellesso, E. S., Scrovonski, K. L., Zanin, E. M., Hepp, L. U., Bayer, C., & Sausen, T. L. (2016). Effects of forest structure on litter production, soil chemical composition and litter-soil interactions. *Acta Botanica Brasilica*, *30*, 329–335.

Caviedes, J., & Ibarra, J. T. (2017). Influence of anthropogenic disturbances on stand structural complexity in Andean temperate forests: Implications for managing key habitat for biodiversity. *PLoS One*, *12*(1), e0169450.

Chakravarty, S., Rai, P., Pala, N. A., & Shukla, G. (2020). Litter production and decomposition in tropical forest. In *Handbook of research on the conservation and restoration of tropical dry forests* (pp. 193–212). IGI Global.

Chanda, R., Singh, S. S., Singh, N. S., Upadhyay, K. K., & Tripathi, S. K. (2024). Two-decadal climate impacts on growth of major forest types of Eastern Himalaya. *Trees, Forests and People*, *15*, 100491.

Chen, X., & Chen, H. Y. (2018). Global effects of plant litter alterations on soil CO_2 to the atmosphere. *Global Change Biology*, *24*(8), 3462–3471.

Chhabra, A., Geist, H., Houghton, R. A., Haberl, H., Braimoh, A. K., Vlek, P. L., …, & Lambin, E. F. (2006). Multiple impacts of land-use/cover change. In *Land-use and land-cover change: Local processes and global impacts* (pp. 71–116). Springer Berlin Heidelberg.

Chowdhury, M. I. H., Rakib, M. H., & Das, C. (2023). Intra and interspecific relation of tree species aspects of seed biometry, phenology, seed dispersion, and germination. *Journal of Agriculture Sustainability and Environment ISSN*, *2997*, 271X.

Clinton, P. W. (2018). Future expectations of forest soils: Increasing productivity within environmental limits using new knowledge. *New Zealand Journal of Agricultural Research*, *61*(3), 389–401.

Comita, L. S., Uriarte, M., Thompson, J., Jonckheere, I., Canham, C. D., & Zimmerman, J. K. (2009). Abiotic and biotic drivers of seedling survival in a hurricane-impacted tropical forest. *Journal of Ecology*, *97*(6), 1346–1359.

Copeland, S. M., Bruna, E. M., Silva, L. V. B., Mack, M. C., & Vasconcelos, H. L. (2012). Short-term effects of elevated precipitation and nitrogen on soil fertility and plant growth in a Neotropical savanna. *Ecosphere*, *3*(4), 1–20.

Cornwell, W. K., Cornelissen, J. H., Amatangelo, K., Dorrepaal, E., Eviner, V. T., Godoy, O., …, & Westoby, M. (2008). Plant species traits are the predominant control on litter decomposition rates within biomes worldwide. *Ecology Letters*, *11*(10), 1065–1071.

Costa, A. N., Souza, J. R., Alves, K. M., Penna-Oliveira, A., Paula-Silva, G., Becker, I. S., … & Vieira-Neto, E. H. (2020). Linking the spatiotemporal variation of litterfall to standing vegetation biomass in Brazilian savannas. *Journal of Plant Ecology*, *13*(5), 517–524.

Cudlín, P., Seják, J., Pokorný, J., Albrechtová, J., Bastian, O., & Marek, M. (2013). Forest ecosystem services under climate change and air pollution. *Developments in Environmental Science*, *13*, 521–546.

Cushman, J. H. (1984). Numerical study of some age-dependent parameters in root nutrient uptake. *Plant and Soil*, *79*, 123–141.

Dantas, B. F., Moura, M. S., Pelacani, C. R., Angelotti, F., Taura, T. A., Oliveira, G. M., ..., & Seal, C. E. (2020). Rainfall, not soil temperature, will limit the seed germination of dry forest species with climate change. *Oecologia*, *192*(2), 529–541.

De Deyn, G. B., Cornelissen, J. H., & Bardgett, R. D. (2008). Plant functional traits and soil carbon sequestration in contrasting biomes. *Ecology Letters*, *11*(5), 516–531.

de la Riva, E. G., Prieto, I., & Villar, R. (2019). The leaf economic spectrum drives leaf litter decomposition in Mediterranean forests. *Plant and Soil*, *435*, 353–366.

de Oliveira Gomes, L. E., Vescovi, L. C., & Bernardino, A. F. (2021). The collapse of mangrove litterfall production following a climate-related forest loss in Brazil. *Marine Pollution Bulletin*, *162*, 111910.

De Storme, N., & Geelen, D. (2014). The impact of environmental stress on male reproductive development in plants: Biological processes and molecular mechanisms. *Plant, Cell & Environment*, *37*(1), 1–18.

Díaz, M., Sánchez-Mejía, T., & Morán-López, T. (2021). Long-term tree regeneration of fragmented agroforestry systems under varying climatic conditions. *Frontiers in Ecology and Evolution*, *9*, 640143.

Diodato, N., & Bellocchi, G. (2020). Spatial probability modelling of forest productivity indicator in Italy. *Ecological Indicators*, *108*, 105721.

Esquivel, J., Park, B. B., Casanoves, F., Delgado, D., Park, G. E., & Finegan, B. (2020). Altitude and species identity drive leaf litter decomposition rates of ten species on a 2950 m altitudinal gradient in Neotropical rain forests. *Biotropica*, *52*(1), 11–21.

Feng, Q., Yang, H., Liu, Y., Liu, Z., Xia, S., Wu, Z., & Zhang, Y. (2024). Interdisciplinary perspectives on forest ecosystems and climate interplay: A review. *Environmental Reviews*, *33*, 1–21.

Fioretto, A., Di Nardo, C., Papa, S., & Fuggi, A. (2005). Lignin and cellulose degradation and nitrogen dynamics during decomposition of three leaf litter species in a Mediterranean ecosystem. *Soil Biology and Biochemistry*, *37*(6), 1083–1091.

Fortier, J., Truax, B., Gagnon, D., & Lambert, F. (2019). Abiotic and biotic factors controlling fine root biomass, carbon and nutrients in closed-canopy hybrid poplar stands on post-agricultural land. *Scientific Reports*, *9*(1), 6296.

Fotis, A. T., Morin, T. H., Fahey, R. T., Hardiman, B. S., Bohrer, G., & Curtis, P. S. (2018). Forest structure in space and time: Biotic and abiotic determinants of canopy complexity and their effects on net primary productivity. *Agricultural and Forest Meteorology*, *250*, 181–191.

Freeman, C., Fenner, N., & Shirsat, A. H. (2012). Peatland geoengineering: An alternative approach to terrestrial carbon sequestration. *Philosophical Transactions of the Royal Society A: Mathematical, Physical and Engineering Sciences*, *370*(1974), 4404–4421.

Fricke, E. C., Bender, J., Rehm, E. M., & Rogers, H. S. (2019). Functional outcomes of mutualistic network interactions: A community-scale study of frugivore gut passage on germination. *Journal of Ecology*, *107*(2), 757–767.

Frøseth, R. B., & Bleken, M. A. (2015). Effect of low temperature and soil type on the decomposition rate of soil organic carbon and clover leaves, and related priming effect. *Soil Biology and Biochemistry*, *80*, 156–166.

Gan, H., & Wickings, K. (2020). Root herbivory and soil carbon cycling: Shedding "green" light onto a "brown" world. *Soil Biology and Biochemistry*, *150*, 107972.

Garcia-Barreda, S., Sangüesa-Barreda, G., Madrigal-González, J., Seijo, F., de Andrés, E. G., & Camarero, J. J. (2021). Reproductive phenology determines the linkages between radial growth, fruit production and climate in four Mediterranean tree species. *Agricultural and Forest Meteorology*, *307*, 108493.

Gautam, T. P., & Mandal, T. N. (2016). Effect of disturbance on litter dynamics in moist tropical forest of eastern Nepal. *Our Nature*, *14*(1).

Ge, X., Zeng, L., Xiao, W., Huang, Z., Geng, X., & Tan, B. (2013). Effect of litter substrate quality and soil nutrients on forest litter decomposition: A review. *Acta Ecologica Sinica*, *33*(2), 102–108.

Giri, A., Heckathorn, S., Mishra, S., & Krause, C. (2017). Heat stress decreases levels of nutrient-uptake and-assimilation proteins in tomato roots. *Plants*, *6*(1), 6.

Giweta, M. (2020). Role of litter production and its decomposition, and factors affecting the processes in a tropical forest ecosystem: A review. *Journal of Ecology and Environment*, *44*(1), 11.

Gonçalves, A. C., Afonso, A., Pereira, D. G., & Pinheiro, A. (2017). Influence of umbrella pine (*Pinus pinea* L.) stand type and tree characteristics on cone production. *Agroforestry Systems*, *91*, 1019–1030.

Gonçalves, A. C., & Pommerening, A. (2012). Spatial dynamics of cone production in Mediterranean climates: A case study of *Pinus pinea* L. in Portugal. *Forest Ecology and Management*, *266*, 83–93.

Gottschall, F., Cesarz, S., Auge, H., Kovach, K. R., Mori, A. S., Nock, C. A., & Eisenhauer, N. (2022). Spatiotemporal dynamics of abiotic and biotic properties explain biodiversity–ecosystem-functioning relationships. *Ecological Monographs*, *92*(1), e01490.

Gregorich, E. G., Janzen, H., Ellert, B. H., Helgason, B. L., Qian, B., Zebarth, B. J., …, & Dyck, M. F. (2017). Litter decay controlled by temperature, not soil properties, affecting future soil carbon. *Global Change Biology*, *23*(4), 1725–1734.

Grossman, J. J., Cavender-Bares, J., & Hobbie, S. E. (2020). Functional diversity of leaf litter mixtures slows decomposition of labile but not recalcitrant carbon over two years. *Ecological Monographs*, *90*(3), e01407.

Guan, D., Nie, J., Zhou, L., Chang, Q., & Cao, J. (2023). How to simulate carbon sequestration potential of forest vegetation? A forest carbon sequestration model across a typical mountain city in China. *Remote Sensing*, *15*(21), 5096.

Guo, Y., Chen, H. Y., Mallik, A. U., Wang, B., Li, D., Xiang, W., & Li, X. (2019). Predominance of abiotic drivers in the relationship between species diversity and litterfall production in a tropical karst seasonal rainforest. *Forest Ecology and Management*, *449*, 117452.

Guo, Y., Wang, B., Li, D., Xiang, W., Ding, T., Huang, F., & Li, X. (2020). Multivariate relationships between litter productivity and its drivers in a tropical karst seasonal rainforest. *Flora*, *273*, 151728.

Haas, S. M., & Lortie, C. J. (2020). A systematic review of the direct and indirect effects of herbivory on plant reproduction mediated by pollination. *PeerJ*, *8*, e9049.

Hasenauer, H., Petritsch, R., Zhao, M., Boisvenue, C., & Running, S. W. (2012). Reconciling satellite with ground data to estimate forest productivity at national scales. *Forest Ecology and Management*, *276*, 196–208.

Hernandez, J. O., Naeem, M., & Zaman, W. (2023). How does changing environment influence plant seed movements as populations of dispersal vectors decline? *Plants*, *12*(7), 1462.

Hildebrand, M., Perles-Garcia, M. D., Kunz, M., Härdtle, W., von Oheimb, G., & Fichtner, A. (2021). Tree-tree interactions and crown complementarity: The role of functional diversity and branch traits for canopy packing. *Basic and Applied Ecology*, *50*, 217–227.

Hobbie, S. E., & Vitousek, P. M. (2000). Nutrient limitation of decomposition in Hawaiian forests. *Ecology*, *81*(7), 1867–1877.

Hoeber, S., Fransson, P., Weih, M., & Manzoni, S. (2020). Leaf litter quality coupled to Salix variety drives litter decomposition more than stand diversity or climate. *Plant and Soil*, *453*, 313–328.

Howlett, B. G., Todd, J. H., Willcox, B. K., Rader, R., Nelson, W. R., Gee, M., & Davidson, M. M.. (2021). Using non-bee and bee pollinator-plant species interactions to design diverse plantings benefiting crop pollination services. In *Advances in ecological research* (Vol. 64, pp. 45–103). Academic Press.

Hu, G., He, X., Wang, N., Liu, J. E., & Zhou, Z. (2024). The impact of water potential and temperature on native species' capability for seed germination in the Loess Plateau Region, China. *Plants*, *13*(5), 693.

Huang, C., Chen, H. Y., Chang, S. X., Cahill, J. F. Jr, & Ma, Z. (2023). Species mixtures increase fine root length to support greater stand productivity in a natural boreal forest. *Journal of Ecology*, *111*(5), 1139–1150.

Hui, D., Wang, J., Le, X., Shen, W., & Ren, H. (2012). Influences of biotic and abiotic factors on the relationship between tree productivity and biomass in China. *Forest Ecology and Management*, *264*, 72–80.

Huidrom, M., Singh, S. S., & Meitei, M. D. (2024). Role of Naga Nullah tributary as a major polluter of Loktak: A Ramsar site from Manipur, India with 3 decades in the Montreux records of the wetland's convention. *Sustainable Water Resources Management*, *10*(2), 56.

Huryna, H., & Pokorný, J. (2016). The role of water and vegetation in the distribution of solar energy and local climate: A review. *Folia Geobotanica*, *51*, 191–208.

Hyjazie, B. F., & Sargent, R. D. (2022). Floral resources predict the local bee community: Implications for conservation. *Biological Conservation*, *273*, 109679.

Inouye, D. W. (2020). Effects of climate change on alpine plants and their pollinators. *Annals of the New York Academy of Sciences*, *1469*(1), 26–37.

Isaac, S. R., & Nair, M. A. (2005). Biodegradation of leaf litter in the warm humid tropics of Kerala, India. *Soil Biology and Biochemistry*, *37*(9), 1656–1664.

Ishaya, F. D., & Rasmussen, A. (2024). Methods for measuring nutrient uptake in maize using nitrogen stable isotopes. *Cold Spring Harbor Protocols*. *2025*(3).

Ita, R. E., Ogbemudia, F. O., & Udo, E. D. (2022). Canopy types on nutrient availability in soil and litter pools of a forest ecosystem. *Acta Scientifica Malaysia (ASM)*, *6*(2), 43–47.

Jain, S. V. (2021). *Applied ecology and sustainable environment*. BFC Publications.

Jankowska-Blaszczuk, M., & Daws, M. I. (2007). Impact of red: Far red ratios on germination of temperate forest herbs in relation to shade tolerance, seed mass and persistence in the soil. *Functional Ecology*, *21*(6), 1055–1062.

Jia, B., Xu, Z., Zhou, G., & Yin, X. (2018). Statistical characteristics of forest litterfall in China. *Science China Life Sciences*, *61*(3), 358–360.

Jimenez, J. J., & Lal, R. (2006). Mechanisms of C sequestration in soils of Latin America. *Critical Reviews in Plant Sciences*, *25*(4), 337–365.

Kharmawlong, R., Upadhyay, K. K., & Tripathi, S. K. (2023a). Detecting the legacies of climatic extremes through radial growth releases in *Pinus kesiya* Royle ex Gordon in Shillong, Meghalaya, Northeast India. *Environment and Ecology*, *41*(4B), 2766–2774.

Kharmawlong, R., Upadhyay, K. K., & Tripathi, S. K. (2023b). Tree ring analysis of Mexican weeping pine (*Pinus patula* Schiede ex Schltdl. and Cham.) from Nagaland, Northeast India. *Environment and Ecology*, *41*(4A), 2537–2546.

Kim, M., Lee, S., Lee, S., Yi, K., Kim, H. S., Chung, S., …, & Yoon, T. K. (2022). Seed dispersal models for natural regeneration: A review and prospects. *Forests*, *13*(5), 659.

Kiri, I., & Hausa, K. (2023). Mechanisms of nutrient uptake and assimilation processes in some plants: A review. *Dutse Journal of Pure and Applied Sciences*, *9*, 223–237.

Kishwan, J., Pandey, R., & Dadhwal, V. K. (2012). Emission removal capability of India's forest and tree cover. *Small-Scale Forestry*, *11*, 61–72.

Klanderud, K., Vandvik, V., & Goldberg, D. (2015). The importance of biotic vs. abiotic drivers of local plant community composition along regional bioclimatic gradients. *PLoS One*, *10*(6), e0130205.

Koenig, W. D., Knops, J. M., Carmen, W. J., & Pearse, I. S. (2015). What drives masting? The phenological synchrony hypothesis. *Ecology*, *96*(1), 184–192.

Korhonen, J., Wihersaari, M., & Savolainen, I. (2001). Industrial ecosystem in the Finnish forest industry: Using the material and energy flow model of a forest ecosystem in a forest industry system. *Ecological Economics*, *39*(1), 145–161.

Kozlowski, T. T. (2002). Physiological ecology of natural regeneration of harvested and disturbed forest stands: Implications for forest management. *Forest Ecology and Management, 158*(1–3), 195–221.

Kremer, K. N., & Bauhus, J. (2020). Drivers of native species regeneration in the process of restoring natural forests from mono-specific, even-aged tree plantations: A quantitative review. *Restoration Ecology, 28*(5), 1074–1086.

Kumari, S., Patra, D., Mishra, N., & Panigrahi, K. C. (2021). Impact of abiotic stress on the root growth and development. *Microbial Management of Plant Stresses*, 235–250.

Kunz, M., Fichtner, A., Härdtle, W., Raumonen, P., Bruelheide, H., & von Oheimb, G. (2019). Neighbour species richness and local structural variability modulate aboveground allocation patterns and crown morphology of individual trees. *Ecology Letters, 22*(12), 2130–2140.

Kutlu, T., Guber, A. K., Rivers, M. L., & Kravchenko, A. N. (2018). Moisture absorption by plant residue in soil. *Geoderma, 316*, 47–55.

Laganière, J., Pare, D., & Bradley, R. L. (2010). How does a tree species influence litter decomposition? Separating the relative contribution of litter quality, litter mixing, and forest floor conditions. *Canadian Journal of Forest Research, 40*(3), 465–475.

Lal, R. (2005). Forest soils and carbon sequestration. *Forest Ecology and Management, 220*(1–3), 242–258.

Lalthakimi, C., Joshi, M., Monsang, N. P., Singh, N. S., Vanlalfakawma, D. C., Upadhyay, K. K., & Tripathi, S. K. (2023b). Ecological analysis of climate change on soil characteristics under shifting cultivation practice in tropical moist regions. *Environment and Ecology, 41*(4B), 2701–2712.

Lalthakimi, C., Singh, N. S., Vanlalfakawma, D. C., Upadhyay, K. K., & Tripathi, S. K. (2023a). Land use change effects on soil physical and biochemical properties during wet and dry season in forest and shifting cultivation (Jhum) sites in Northeast India. *Environment and Ecology, 41*(4A), 2584–2594.

Lamsal, P., Kumar, L., Atreya, K., & Pant, K. P. (2018). Forest ecosystem services in Nepal: A retrospective synthesis, research gaps and implications in the context of climate change. *International Forestry Review, 20*(4), 506–537.

Landhäusser, S. M., Pinno, B. D., & Mock, K. E. (2019). Tamm review: Seedling-based ecology, management, and restoration in aspen (*Populus tremuloides*). *Forest Ecology and Management, 432*, 231–245.

Landuyt, D., Ampoorter, E., Bastias, C. C., Benavides, R., Müller, S., Scherer-Lorenzen, M., …, & Verheyen, K. (2020). Importance of overstorey attributes for understorey litter production and nutrient cycling in European forests. *Forest Ecosystems, 7*, 1–11.

Lei, L., Zeng, J., Liu, Q., Luo, L., Ma, Z., Chen, Y., & Liu, Y. (2024). Effects of soil fauna on the home-field advantage of litter total phenol and condensed tannin decomposition. *Forests, 15*(2), 389.

Li, L., Li, X., Liu, Y., & Liu, H. (2016). Flowering responses to light and temperature. *Science China Life Sciences, 59*, 403–408.

Li, T., Liu, Y., Wang, Q., Lai, C., Qiu, Y., Tissue, D. T., … & Peng, L. (2022). The impact of abiotic and biotic factors on growth, mortality and net tree C stock in mountain forest ecosystems in southwest China. *Environmental Research Letters, 17*(12), 124037.

Li, W., Liu, Z., Zhao, J., Ma, L., Wu, J., Qi, J., & Wang, H. (2023). Leaf mechanical properties as potential predictors of leaf-litter decomposability. *Forestry Research, 3*(1).

Liebhold, A. M., Brockerhoff, E. G., Kalisz, S., Nuñez, M. A., Wardle, D. A., & Wingfield, M. J. (2017). Biological invasions in forest ecosystems. *Biological Invasions, 19*(11), 3437–3458.

Lili, Z., Vivek, Y., Hengchen, Z., Yaya, S., Xiangxiang, H., Xiaoxia, W., …, & Yongxiang, K. (2024). Climate drivers of litterfall biomass dynamics in three types of forest stands on the Loess Plateau. *Ecological Indicators, 163*, 112088.

Lin, S., Li, Y., Li, Y., Chen, Q., Wang, Q., & He, K. (2021). Influence of tree size, local forest structure, topography, and soil resource availability on plantation growth in Qinghai Province, China. *Ecological Indicators*, *120*, 106957.

Lin, Y. C., Comita, L. S., Johnson, D. J., Chen, M. R., & Wu, S. H. (2017). Biotic vs abiotic drivers of seedling persistence in a tropical karst forest. *Journal of Vegetation Science*, *28*(1), 206–217.

Liu, Q. Q., Huang, Z. J., Guo, S., Wang, D. Y., Wang, C. H., Wang, Z. N., ..., & Liu, B. (2019). Responses of seed germination and seedling growth of *Cunninghamia lanceolata* and *Schima superba* to different light intensities. *Ying Yong Sheng tai xue bao= The Journal of Applied Ecology*, *30*(9), 2955–2963.

Liu, Y., Li, S., Sun, X., & Yu, X. (2016). Variations of forest soil organic carbon and its influencing factors in east China. *Annals of Forest Science*, *73*, 501–511.

Lopes, M. C. A., Araújo, V. F. P., & Vasconcellos, A. (2015). The effects of rainfall and vegetation on litterfall production in the semiarid region of northeastern Brazil. *Brazilian Journal of Biology*, *75*, 703–708.

Lora, J., Herrero, M., & Hormaza, J. I. (2012). Pollen performance, cell number, and physiological state in the early-divergent angiosperm *Annona cherimola* Mill. (Annonaceae) are related to environmental conditions during the final stages of pollen development. *Sexual Plant Reproduction*, *25*, 157–167.

Lorenz, K., & Lal, R. (2018). Carbon sequestration in wetland soils. In R. Lal (Ed.), *Carbon sequestration in agricultural ecosystems* (pp. 211–234). Springer International Publishing.

Lu, D., Wang, G. G., Yu, L., Zhang, T., & Zhu, J. (2018). Seedling survival within forest gaps: The effects of gap size, within-gap position and forest type on species of contrasting shade-tolerance in Northeast China. *Forestry: An International Journal of Forest Research*, *91*(4), 470–479.

Lu, J., Johnson, D. J., Qiao, X., Lu, Z., Wang, Q., & Jiang, M. (2015). Density dependence and habitat preference shape seedling survival in a subtropical forest in central China. *Journal of Plant Ecology*, *8*(6), 568–577.

Luo, Y., Cheng, J., Yan, X., Zhang, J., & Zhang, J. (2022). Germination of seeds subjected to temperature and water availability: Implications for ecological restoration. *Forests*, *13*(11), 1854.

Martinet, B., Zambra, E., Przybyla, K., Lecocq, T., Anselmo, A., Nonclercq, D., ..., & Hennebert, E. (2021). Mating under climate change: Impact of simulated heatwaves on the reproduction of model pollinators. *Functional Ecology*, *35*(3), 739–752.

Martínez-Gómez, J., Park, S., Hartogs, S. R., Soza, V. L., Park, S. J., & Di Stilio, V. S. (2023). Flower morphology as a predictor of pollination mode in a biotic to abiotic pollination continuum. *Annals of Botany*, *132*(1), 61–76.

Martini, F., Zou, C., & Goodale, U. M. (2019). Intrinsic biotic factors and microsite conditions drive seedling survival in a species with masting reproduction. *Ecology and Evolution*, *9*, 14261–14272. doi:10.1002/ece3.5861

Mengel, D. B., & Barber, S. A. (1974). Development and distribution of the corn root system under field conditions 1. *Agronomy Journal*, *66*(3), 341–344.

Mishra, B. P., Tripathi, O. P., Tripathi, R. S., & Pandey, H. N. (2004). Effects of anthropogenic disturbance on plant diversity and community structure of a sacred grove in Meghalaya, northeast India. *Biodiversity & Conservation*, *13*, 421–436.

Misson, L., Degueldre, D., Collin, C., Rodriguez, R., Rocheteau, A., Ourcival, J. M., & Rambal, S. (2011). Phenological responses to extreme droughts in a Mediterranean forest. *Global Change Biology*, *17*(2), 1036–1048.

Moles, A. T., & Westoby, M. (2002). Seed addition experiments are more likely to increase recruitment in larger-seeded species. *Oikos*, *99*(2), 241–248.

Morales, J. M., & Morán López, T. (2022). Mechanistic models of seed dispersal by animals. *Oikos*, *2022*(2).

Mukul, S. A., Halim, M. A., & Herbohn, J. (2020). Forest carbon stock and fluxes: Distribution, biogeochemical cycles, and measurement techniques. In W. Leal Filho, A. M. Azul, L. Brandli, P. G. Özuyar, & T. Wall (Eds.), *Life on land, encyclopedia of the UN sustainable development goals* (pp. 365–380). Springer.

Nakamura, A., Kitching, R. L., Cao, M., Creedy, T. J., Fayle, T. M., Freiberg, M., …, & Ashton, L. A. (2017). Forests and their canopies: Achievements and horizons in canopy science. *Trends in Ecology & Evolution*, *32*(6), 438–451.

Naniwadekar, R., Chaplod, S., Datta, A., Rathore, A., & Sridhar, H. (2019). Large frugivores matter: Insights from network and seed dispersal effectiveness approaches. *Journal of Animal Ecology*, *88*(8), 1250–1262.

Neuschulz, E. L., Mueller, T., Schleuning, M., & Böhning-Gaese, K. (2016). Pollination and seed dispersal are the most threatened processes of plant regeneration. *Scientific Reports*, *6*(1), 29839.

Nicolás, C., Martin-Bertelsen, T., Floudas, D., Bentzer, J., Smits, M., Johansson, T., & Tunlid, A. (2019). The soil organic matter decomposition mechanisms in ectomycorrhizal fungi are tuned for liberating soil organic nitrogen. *The ISME Journal*, *13*(4), 977–988.

Novaes, L. R., Calixto, E. S., de Oliveira, M. L., Alves-de-Lima, L., de Almeida, O., & Torezan-Silingardi, H. M. (2020). Environmental variables drive phenological events of anemocoric plants and enhance diaspore dispersal potential: A new wind-based approach. *Science of the Total Environment*, *730*, 139039.

Oates, K., & Barber, S. A. (1987). Nutrient uptake: A microcomputer program to predict nutrient absorption from soil by roots. *Journal of Agronomic Education*, *16*(2), 65–68.

Oshima, C., Tokumoto, Y., & Nakagawa, M. (2015). Biotic and abiotic drivers of dipterocarp seedling survival following mast fruiting in Malaysian Borneo. *Journal of Tropical Ecology*, *31*(2), 129–137.

Ouyang, S., Xiang, W., Wang, X., Xiao, W., Chen, L., Li, S., … & Peng, C. (2019). Effects of stand age, richness and density on productivity in subtropical forests in China. *Journal of Ecology*, *107*(5), 2266–2277.

Pardos, M., Del Castillo, J. R., Cañellas, I., & Montero, G. (2005). Ecophysiology of natural regeneration of forest stands in Spain. *Forest Systems*, *14*(3), 434–445.

Parsons, S. A., Valdez-Ramirez, V., Congdon, R. A., & Williams, S. E. (2014). Contrasting patterns of litterfall seasonality and seasonal changes in litter decomposability in a tropical rainforest region. *Biogeosciences*, *11*(18), 5047–5056.

Patil, M., Kumar, A., Kumar, P., & Singh, A. (2020). Litter decomposability traits and their linkage with the cycling of nutrients in the forest ecosystems under the Lens of climate change: A conceptual overview. *Advances in Zoology and Botany*, *8*, 269–277.

Peña-Peña, K., & Irmler, U. (2016). Moisture seasonality, soil fauna, litter quality and land use as drivers of decomposition in Cerrado soils in SE-Mato Grosso, Brazil. *Applied Soil Ecology*, *107*, 124–133.

Perez-Quezada, J. F., Pérez, C. A., Brito, C. E., Fuentes, J. P., Gaxiola, A., Aguilera-Riquelme, D., & Lopatin, J. (2021). Biotic and abiotic drivers of carbon, nitrogen and phosphorus stocks in a temperate rainforest. *Forest Ecology and Management*, *494*, 119341.

Perry, D. A., Oren, R., & Hart, S. C. (2008). *Forest ecosystems*. JHU Press.

Piao, S., Liu, Q., Chen, A., Janssens, I. A., Fu, Y., Dai, J., …, & Zhu, X. (2019). Plant phenology and global climate change: Current progresses and challenges. *Global Change Biology*, *25*(6), 1922–1940.

Picchio, R., Mederski, P. S., & Tavankar, F. (2020). How and how much, do harvesting activities affect forest soil, regeneration and stands? *Current Forestry Reports*, *6*(2), 115–128.

Poorter, L., van der Sande, M. T., Arets, E. J., Ascarrunz, N., Enquist, B. J., Finegan, B., …, & Peña-Claros, M. (2017). Biodiversity and climate determine the functioning of Neotropical forests. *Global Ecology and Biogeography*, *26*(12), 1423–1434.

Proesmans, W., Andrews, C., Gray, A., Griffiths, R., Keith, A., Nielsen, U. N., …, & Vanbergen, A. J. (2022). Long-term cattle grazing shifts the ecological state of forest soils. *Ecology and Evolution*, *12*(4), e8786.

Pulido, F. J., & Díaz, M. (2005). Regeneration of a Mediterranean oak: A whole-cycle approach. *Ecoscience*, *12*(1), 92–102.

Purahong, W., Wubet, T., Lentendu, G., Schloter, M., Pecyna, M. J., Kapturska, D., …, & Buscot, F. (2016). Life in leaf litter: Novel insights into community dynamics of bacteria and fungi during litter decomposition. *Molecular Ecology*, *25*(16), 4059–4074.

Rahman, M. M., Tsukamoto, J., Tokumoto, Y., & Shuvo, M. A. R. (2013). The role of quantitative traits of leaf litter on decomposition and nutrient cycling of the forest ecosystems. *Journal of Forest and Environmental Science*, *29*(1), 38–48.

Rawat, M., Arunachalam, K., Arunachalam, A., Alatalo, J. M., & Pandey, R. (2020). Predicting litter decomposition rate for temperate forest tree species by the relative contribution of green leaf and litter traits in the Indian Himalayas region. *Ecological Indicators*, *119*, 106827.

Ren, S., Yang, Q., Liu, H., Shen, G., Zheng, Z., Zhou, S., …, & Wang, X. (2021). The driving factors of subtropical mature forest productivity: Stand structure matters. *Forests*, *12*(8), 998.

Riley, W. J., Zhu, Q., & Tang, J. Y. (2018). Weaker land–climate feedbacks from nutrient uptake during photosynthesis-inactive periods. *Nature Climate Change*, *8*(11), 1002–1006.

Rogers, H. S., Buhle, E. R., HilleRisLambers, J., Fricke, E. C., Miller, R. H., & Tewksbury, J. J. (2017). Effects of an invasive predator cascade to plants via mutualism disruption. *Nature communications*, *8*(1), 14557.

Rokich, D. P., & Bell, D. T. (1995). Light quality and intensity effects on the germination of species from the jarrah (*Eucalyptus marginata*) forest of Western Australia. *Australian Journal of Botany*, *43*(2), 169–179.

Rosenfield, M. V., Keller, J. K., Clausen, C., Cyphers, K., & Funk, J. L. (2020). Leaf traits can be used to predict rates of litter decomposition. *Oikos*, *129*(10), 1589–1596.

Roy, N., Roychoudhury, S., Nautiyal, S., Agarwal, S. K., & Baksi, S. (2020). Socio-economic and eco-biological dimensions in resource use and conservation: Epilogue. In S. K. Agarwal, S. Nautiyal, N. Roy, S. Roychoudhury, & S. Baksi (Eds.), *Socio-economic and eco-biological dimensions in resource use and conservation: Strategies for sustainability* (pp. 547–553). Springer International Publishing.

Roy, P. S., Ramachandran, R. M., Paul, O., Thakur, P. K., Ravan, S., Behera, M. D., & Kanawade, V. P. (2022). Anthropogenic land use and land cover changes—A review on its environmental consequences and climate change. *Journal of the Indian Society of Remote Sensing*, *50*(8), 1615–1640.

Ruess, R. W., McFarland, J. M., Trummer, L. M., & Rohrs-Richey, J. K. (2009). Disease-mediated declines in n-fixation inputs by *Alnus tenuifolia* to early-successional floodplains in interior and South-Central Alaska. *Ecosystems*, *12*, 489–502.

Salazar-Tortosa, D., Saladin, B., Zimmermann, N. E., Castro, J., & de Casas, R. R. (2019). The evolution of seed dispersal is associated with environmental heterogeneity in Pinus. *Perspectives in Plant Ecology, Evolution and Systematics*, *41*, 125464.

Salinas, N., Malhi, Y., Meir, P., Silman, M., Roman Cuesta, R., Huaman, J., …, & Farfan, F. (2011). The sensitivity of tropical leaf litter decomposition to temperature: Results from a large-scale leaf translocation experiment along an elevation gradient in Peruvian forests. *New Phytologist*, *189*(4), 967–977.

Santoro, M., Cartus, O., Carvalhais, N., Rozendaal, D. M., Avitabile, V., Araza, A., …, & Willcock, S. (2021). The global forest above-ground biomass pool for 2010 estimated from high-resolution satellite observations. *Earth System Science Data*, *13*(8), 3927–3950.

Sato, T., Watanabe, K., Kanaiwa, M., Niizuma, Y., Harada, Y., & Lafferty, K. D. (2011). Nematomorph parasites drive energy flow through a riparian ecosystem. *Ecology*, *92*(1), 201–207.

Schmidt, M. W., Torn, M. S., Abiven, S., Dittmar, T., Guggenberger, G., Janssens, I. A., …, & Trumbore, S. E. (2011). Persistence of soil organic matter as an ecosystem property. *Nature*, *478*(7367), 49–56.

Schulze, ED., Beck, E., Buchmann, N., Clemens, S., Müller-Hohenstein, K., Scherer-Lorenzen, M. (2018). Interactions Between Plants, Plant Communities and the Abiotic and Biotic Environment. In: *Plant Ecology*. Springer, Berlin, Heidelberg.

Schwarz, K., Herrmann, D. L., & McHale, M. R. (2014). Abiotic drivers of ecological structure and function in urban systems. In Scheible, R. M., & McHale, M. R. (Eds.), *Urban wildlife conservation: Theory and practice* (pp. 55–74). Springer.

Serna-Chavez, H. M., Swenson, N. G., Weiser, M. D., Van Loon, E. E., Bouten, W., Davidson, M. D., & Van Bodegom, P. M. (2017). Strong biotic influences on regional patterns of climate regulation services. *Global Biogeochemical Cycles*, *31*(5), 787–803.

Shaolin, P., & Qiang, L. I. U. (2002). The dynamics of forest litter and its responses to global warming. *Acta Ecologica Sinica*, *22*(9), 1534–1544.

Sierota, Z., Kwaśna, H., Grodzki, W., & Tarwacki, G. (2022). Abiotic factors affecting forest tree health. In *Forest microbiology* (pp. 77–97). Academic Press.

Sih, A., & Del Giudice, M. (2012). Linking behavioural syndromes and cognition: A behavioural ecology perspective. *Philosophical Transactions of the Royal Society B: Biological Sciences*, *367*(1603), 2762–2772.

Singh, N. S., Upadhyay, K. K., & Tripathi, S. K. (2022). Leaf litter decomposition of *Melocanna baccifera* (Roxb.) Kurz under field and laboratory microcosm in Northeast India. *Indian Journal of Ecology*, *49*(1), 114–118.

Singh, S. S., Meitei, M. D., Upadhyay, K. K., Chanda, R., Mawi, R., Singh, N. S., Brearly, F. Q & Tripathi, S. K. (2024). Bioremediation of toxic metals using aquatic macrophytes: Challenges and opportunities. *CLEAN–Soil, Air, Water*, *52*(11), e202400273.

Singh, S. S., Lamare, E., Chopade, B. A., Chanda, R., Devi, N. R., Upadhyay, K. K., ... & Tripathi, S. K. (2025a). The hidden players: Soil microbes in ecosystems sustainability. In *Climate Change and soil microorganisms for environmental sustainability* (pp. 295–320). Springer Nature Singapore.

Singh, S. S., Singh, N. S., Lamare, E., Devi, N. R., Devi, S. A., Kaguijenlui, R., Takum, B., Upadhyay, K. K., & Tripathi, S. K. (2025b). Drought mitigation in plants through root exudate-mediated rhizosphere interactions: Opportunities for future research. *Current Plant Biology*, 100504.

Smith, J. E., Heath, L. S., & Woodbury, P. B. (2004). How to estimate forest carbon for large areas from inventory data. *Journal of Forestry*, *102*(5), 25–31.

Sobek, S., Tscharntke, T., Scherber, C., Schiele, S., & Steffan-Dewenter, I. (2009). Canopy vs. understory: Does tree diversity affect bee and wasp communities and their natural enemies across forest strata?. *Forest Ecology and Management*, *258*(5), 609–615.

Soong, J. L., Fuchslueger, L., Marañon-Jimenez, S., Torn, M. S., Janssens, I. A., Penuelas, J., & Richter, A. (2020). Microbial carbon limitation: The need for integrating microorganisms into our understanding of ecosystem carbon cycling. *Global Change Biology*, *26*(4), 1953–1961.

Souza, T., Dobner, M. Jr, Schmitt, D. E., da Silva, L. J. R., & Schneider, K. (2022). Soil biotic and abiotic traits as driven factors for site quality of *Araucaria angustifolia* plantations. *Biologia*, *77*(5), 1219–1230.

Sun, Z., Tian, P., Zhao, X., Wang, Y., Wang, S., Fang, X., …, & Liu, S. (2022). Temporal shifts in the explanatory power and relative importance of litter traits in regulating litter decomposition. *Forest Ecosystems*, *9*, 100072.

Temple, P. J. (1999). Effects of ozone on understory vegetation in the mixed conifer forest. In *Oxidant air pollution impacts in the Montane forests of Southern California: A case study of the San Bernardino mountains* (pp. 208–222). Springer New York.

Teo, Y. H., Beyrouty, C. A., Norman, R. J., & Gbur, E. E. (1995). Nutrient uptake relationship to root characteristics of rice. *Plant and Soil, 171*, 297–302.

Terborgh, J., Zhu, K., Álvarez-Loayza, P., & Cornejo Valverde, F. (2014). How many seeds does it take to make a sapling? *Ecology, 95*(4), 991–999.

Teshome, D. T., Zharare, G. E., & Naidoo, S. (2020). The threat of the combined effect of biotic and abiotic stress factors in forestry under a changing climate. *Frontiers in Plant Science, 11*, 601009.

Thompson, I., Mackey, B., McNulty, S., & Mosseler, A. (2009). Forest resilience, biodiversity, and climate change. In *Secretariat of the convention on biological diversity, Montreal. Technical Series no. 43* (Vol. 43, pp. 1–67).

Trenberth, K. E. (2018). Climate change caused by human activities is happening and it already has major consequences. *Journal of Energy & Natural Resources Law, 36*(4), 463–481.

Tripathi, S. K. et al. (2025). The Role of Plant–Microbe Partnerships in Ecosystem Functionality. In: Saraf, M., Goswami, D., Maheshwari, D. K. (eds) *Climate Change and Soil Microorganisms for Environmental Sustainability*. Microorganisms for Sustainability, vol 28. Springer, Singapore. https://doi.org/10.1007/978-981-96-3425-5_6

Trunschke, J., Junker, R. R., Kudo, G., Alexander, J. M., Richman, S. K., & Till-Bottraud, I. (2024). Effects of climate change on plant-pollinator interactions and its multitrophic consequences. *Alpine Botany*, 134(2), 115–121.

Tucker, M. A., Böhning-Gaese, K., Fagan, W. F., Fryxell, J. M., Van Moorter, B., Alberts, S. C., …, & Mueller, T. (2018). Moving in the Anthropocene: Global reductions in terrestrial mammalian movements. *Science, 359*(6374), 466–469.

Tucker, M. A., Busana, M., Huijbregts, M. A., & Ford, A. T. (2021). Human-induced reduction in mammalian movements impacts seed dispersal in the tropics. *Ecography, 44*(6), 897–906.

Ulyshen, M., Adams, C., Adams, J., Adams, S. B., Bland, M., Bragg, D. C., …, & Young, A. D.. (2024). Spatiotemporal patterns of forest pollinator diversity across the southeastern United States. *Diversity and Distributions, 30*(8), e13869.

Ulyshen, M., Urban-Mead, K. R., Dorey, J. B., & Rivers, J. W. (2023). Forests are critically important to global pollinator diversity and enhance pollination in adjacent crops. *Biological Reviews, 98*(4), 1118–1141.

Upadhyay, K. K., Shah, S. K., Roy, A., & Tripathi, S. K. (2021). Dendroclimatology of teak indicates prevailing climatic conditions of tropical moist forests in India. *Ecological Indicators, 129*, 107888.

Upadhyay, K. K., & Tripathi, S. K. (2019). Sustainable forest management under climate change: A dendrochronological approach. *Environment and Ecology, 37*, 998–1006.

Vacchiano, G., Ascoli, D., Berzaghi, F., Lucas-Borja, M. E., Caignard, T., Collalti, A., Mairota, P., Palaghianu, C., Reyer, C. P. O., Sanders, T. G. M., Schermer, E., Wohlgemuth, T., & Hacket-Pain, A. (2018). Reproducing reproduction: How to simulate mast seeding in forest models. *Ecological modelling, 376*, 40–53.

Väisänen, H., Strandman, H., & Kellomäki, S. (1994). A model for simulating the effects of changing climate on the functioning and structure of the boreal forest ecosystem: an approach based on object-oriented design. *Tree Physiology, 14*(7-8-9), 1081–1095.

van der Sande, M. T., Peña-Claros, M., Ascarrunz, N., Arets, E. J., Licona, J. C., Toledo, M., & Poorter, L. (2017). Abiotic and biotic drivers of biomass change in a Neotropical forest. *Journal of Ecology, 105*(5), 1223–1234.

Vivanco, L., & Austin, A. T. (2008). Tree species identity alters forest litter decomposition through long-term plant and soil interactions in Patagonia, Argentina. *Journal of Ecology*, *96*(4), 727–736.

Walters, M. B., & Reich, P. B. (2000). Seed size, nitrogen supply, and growth rate affect tree seedling survival in deep shade. *Ecology*, *81*(7), 1887–1901.

Wang, B. (2021). *The functional role of understorey herbs for litter decomposition in temperate forests* (Doctoral dissertation, Ghent University).

Wang, C., Yu, W., Ma, L., Ye, X., Erdenebileg, E., Wang, R., ... & Liu, G. (2023). Biotic and abiotic drivers of ecosystem multifunctionality: Evidence from the semi-arid grasslands of northern China. *Science of the Total Environment*, *887*, 164158.

Waring, R. H., & Running, S. W. (2007). *Forest Ecosystems*: Analysis at multiple scales. Academic Press.

Wiebe, K. L. (2014). Responses of cavity-nesting birds to fire: testing a general model with data from the Northern Flicker. *Ecology*, *95*(9), 2537–2547.

Wright, J. S. (2002). Plant diversity in tropical forests: a review of mechanisms of species coexistence. *Oecologia*, *130(1)*, 1–14.

Yang, F., Huang, J., Zhou, C., Yang, X., Mamtimin, A., Zheng, X., Wang, Y., Liang, Y., Zhang, W., Wang, X., Wang, H., & Zhu, C. (2023). Desert abiotic carbon sequestration weakening by precipitation. *Environmental Science & Technology*, *57*(18), 7174–7184.

Yue, Q., Hao, M., Li, X., Zhang, C., von Gadow, K., & Zhao, X. (2020). Assessing biotic and abiotic effects on forest productivity in three temperate forests. *Ecology and evolution*, *10*(14), 7887–7900.

Zarzosa, P. S., Herraiz, A. D., Olmo, M., Ruiz-Benito, P., Barrón, V., Bastias, C. C., Castro, J., Coello, J., Gámez, L., Gimeno, T., Morales, F., Prieto-Pérez, P., Quero, J. L., Viana, H., & Villar, R. (2021). Linking functional traits with tree growth and forest productivity in Quercus ilex forests along a climatic gradient. *Science of The Total Environment*, *786*, 147468.

Zhang, Z., Yan, C., Krebs, C. J., & Stenseth, N. C. (2015). Ecological non-monotonicity and its effects on complexity and stability of populations, communities and ecosystems. *Ecological Modelling*, *312*, 374–384.

Zheng, L. T., Chen, H. Y., & Yan, E. R. (2019). Tree species diversity promotes litterfall productivity through crown complementarity in subtropical forests. *Journal of Ecology*, *107*(4), 1852–1861.

Zhou, S., Huang, C., Xiang, Y., Tie, L., Han, B., & Scheu, S. (2018). Effects of reduced precipitation on litter decomposition in an evergreen broad-leaved forest in western China. *Forest Ecology and Management*, *430*, 219–227.

Zhao, S., Riaz, M. (2024). Plant–Soil Interactions and Nutrient Cycling Dynamics in Tropical Rainforests. In: Fahad, S., Saud, S., Nawaz, T., Gu, L., Ahmad, M., Zhou, R. (eds) *Environment, Climate, Plant and Vegetation Growth*. Springer, Cham. https://doi.org/10.1007/978-3-031-69417-2_8

Zhu, J., Liu, M., Xin, Z., Liu, Z., & Schurr, F. M. (2019). A trade-off between primary and secondary seed dispersal by wind. *Plant Ecology*, *220*, 541–552.

4 Forest Community Structure and Species Migration in the Changing Climatic Conditions

Ruchi Thakur, Dhirender Kumar, Alisha Keprate, C. L. Thakur, D. R. Bhardwaj, and Gurwinder Singh

4.1 INTRODUCTION

Climate plays a major role in defining Earth's major biomes, such as forests, grasslands, and other critical ecosystems. An increase in greenhouse gas (GHG) concentration results in predictable alterations in the global climate, which are expected to profoundly affect these biomes, particularly forests, by changing tree growth, establishment, and reproductive patterns (Gebeyehu & Hirpo, 2019). These alterations result in changes in the community composition, structure, species distribution, and overall function of the forests. Although the biosphere is in a stable state of adaptation and rarely achieves complete equilibrium with the climate, the rapid rate of climate change results in an imbalance, increasing the likelihood of nonlinear shifts and complex feedback mechanisms between biospheric and atmospheric systems (Scheffer & Carpenter, 2003). As climatic conditions diverge from previous norms, plant species may shift their ranges from lower altitudes to higher altitudes in search of suitable habitats. These climate-related shifts in plant distribution adjust forest ecosystem characteristics and change their structure and ecological functions (Grimm et al., 2013; Neilson et al., 2005). Migration among plant species not only results in reorganization within forest communities but also disrupts the long-established interactions among species, including mutualistic, competitive, and other ecological relationships. As a result, the risk of extinction of some species increases if they cannot adapt or relocate themselves, potentially decreasing the functional resilience and stability of forest ecosystems (Neilson et al., 2005; Thomas et al., 2004).

The response of forest ecosystems to climate change is closely related to alterations in community structure, particularly through their influence on forest growth and establishment (Khaine & Woo, 2015). Changes in climatic variables, including temperature, precipitation, and drought, affect the growth rate, establishment, and reproduction of tree species. Favorable conditions such as short and moderate dry periods, adequate water availability, and rising temperatures may increase forest

DOI: 10.1201/9781003588580-4

growth (Toledo et al., 2011). However, these conditions may not always be beneficial, as the susceptibility of tree species to strong climate variation significantly influences their ability to establish and reproduce. Changing climatic patterns, especially increasing temperatures and inconsistent water supply, significantly affect the regeneration of species. Seedlings, which are highly susceptible to climatic fluctuations, encounter increased hazards, and even slight alterations in temperature and precipitation patterns can significantly diminish their survival and establishment. Projections indicate that climate change will substantially affect tree regeneration in the *Eucalyptus* forests of Australia over the next few decades, potentially altering the structure of this vital ecosystem (Khaine & Woo, 2015).

Forest community structure is affected by climate change. Changing species distribution, growth patterns, and disrupted regeneration processes will result in changes in species composition and forest dynamics. Hence, this chapter focuses on the effects of climate change on species distribution, forest ecosystem function, and development, thus offering an in-depth overview of the challenges confronting forest ecosystems and the possible adaptation strategies in a period of accelerated climatic change.

4.1.1 Forest Community Structure and Climate Change

Forest ecosystems are considered complex systems in which various biotic components, including plants, animals, and microorganisms, interact with abiotic components, including light, heat, water, air, and soil. Trees play an important role in preserving ecosystem equilibrium because of these interactions, which accelerate the nutrient cycle and energy flow. Forest ecosystems provide several ecosystem services that are essential for environmental sustainability (Brockerhoff et al., 2017), including minimizing deforestation and forest degradation, encouraging forest conservation and better management techniques, and mitigating climate change through carbon sequestration (Chen et al., 2022). Additionally, these systems serve as "green water reservoirs," playing a major role in water conservation. They also serve as a "green gene pool" that promotes the release of oxygen and carbon sequestration. In addition, forests also act as green carbon pools that promote biodiversity conservation and as a "green oxygen bar," which helps to purify the atmosphere (Guo et al., 2023).

A forest community is defined as a group of animals, plants, fungi, and microorganisms that live together and interact within a forest ecosystem consisting of layers of herbs, shrubs, trees, and mosses, depending on the geographical characteristics and forest types (Agarwal et al., 2022). For instance, forests in central China are well distinguished by their physical characteristics, tree density, and niche makeup. Complex interactions between biotic and abiotic factors determine the organization of forest communities. These factors influence species abundance and distribution. Biotic factors, including mutualism and competition among plants, animals, and microbes, influence species abundance and interactions, whereas abiotic factors, including light, temperature, precipitation, and soil type, influence species distribution and vertical stacking, resulting in different forest types. Additionally, anthropogenic factors, such as pollution, deforestation, and climate change, interfere with natural dynamics, which further results in habitat fragmentation, changes

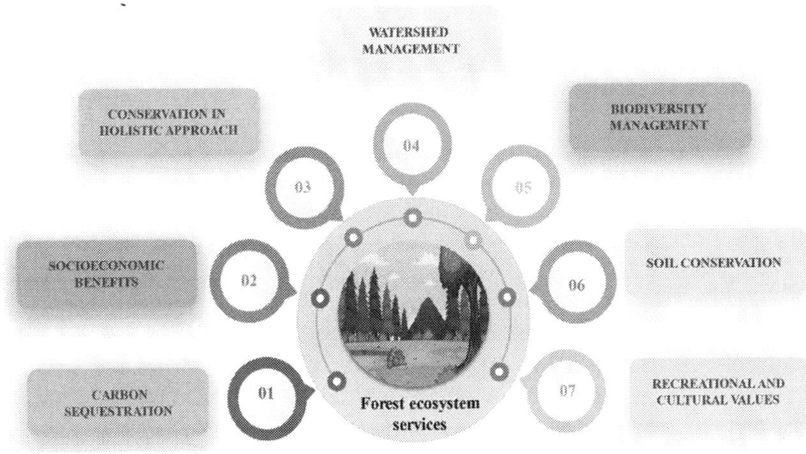

FIGURE 4.1 Major forest ecosystem services.

in biodiversity, ecosystem services, and their functions (Figure 4.1). Interactions between biotic and abiotic factors also impact the general structure, resilience, and biodiversity of forest ecosystems across temporal and geographical scales. Tropical forests are recognized for their rich biodiversity, whereas boreal forests are known for their ability to store carbon (Rao et al., 2007). Various forest types, including tropical, temperate, boreal, and subtropical, support species richness that is appropriate for their specific environment. In the Himalayas, altitudinal variations affect forest community composition as different tree species dominate at different elevational ranges (Ao et al., 2024). In forest ecosystems, various associations or interactions are established among plant species, such as symbiotic associations between fungi and trees for exchanging nutrients. Forest communities are dynamic and susceptible to succession and disturbances, such as storms and fire-initiated species replacement, which promotes biodiversity enhancement and ecosystem rejuvenation. However, human-induced activities such as deforestation, pollution, and climate change pose serious challenges to these systems, resulting in the degradation and loss of biodiversity. Climate change is an emerging global threat to biodiversity and forest ecosystems. Alterations in ecosystem structure and function, along with the resources and services provided by this natural system to society, are derived from their impact on individual communities and their interactions with other organisms and environments. It is now well acknowledged that developing countries are more susceptible to the impact of climate change than developed countries. This heightened susceptibility arises from their complex ecological systems, increased threats to economic structures, and low-income levels of the majority of the population, which diminish their ability to adjust. Continuous urbanization, changed river patterns, evolving agricultural practices, severe erosion, and enhanced desertification are significantly transforming land uses, contributing roughly 4.3–5.5 $GtCO_2e$/year in global emissions (Sannigrahi et al., 2020). China and India significantly contribute to global carbon dioxide (CO_2) and GHG emissions, primarily because of their growing population and rapid industrialization (Pathak, 2023). India witnessed

573 disasters from 1970 to 2021 that were associated with climate change, extreme weather, and water events, leading to 138,377 fatalities, as reported by the Indian Meteorological Department (IMD Annual Report, 2021). India has just emerged as the world's most populous nation, with 1.4 billion people, accounting for 17.7% of the global population (Salunke, 2022). This expansion is anticipated to result in an increasing carbon footprint.

Various recent changes in global and regional climates, including an increase in average annual temperatures and more frequent extreme events such as heatwaves and droughts, are well recognized. Various events regarding impact of climate change activities during different years has been presented in Table 4.1. The increasing emissions of GHGs, particularly carbon dioxide (CO_2), are considered a principal catalyst of contemporary global warming and climate change, and are widely recognized by the scientific community (Hui et al., 2017). Deforestation and the combustion of fossil fuels have raised atmospheric carbon dioxide (CO_2) levels from 280 ppm prior to the Industrial Revolution to 394 ppm now (Penuelas et al., 2013). As a result, the mean global surface temperature has increased at an unprecedented rate. Changes in global surface temperature, precipitation patterns, and nitrogen deposition significantly affect tree development, forest dynamics, and carbon sequestration in forests (Zuidema et al., 2013). Understanding the collective effects of these climate changes, such as global warming, elevated CO_2 levels, changes in precipitation, and nitrogen deposition, is essential for predicting carbon dynamics in forest ecosystems.

Forest ecosystems serve as both carbon sinks and sources of carbon emissions and play an important role in the global climate system. Forests play an important role in carbon sequestration and cover approximately 30% of the Earth's land surface (Thornton et al., 2002). Tropical forests absorb approximately 1.8 Gt of atmospheric carbon dioxide each year, highlighting their importance in climate change mitigation (Zhao et al., 2019) Nevertheless, human activities, such as deforestation, agricultural expansion, and land-use changes, accelerate the release of stored carbon from these ecosystems, contributing to approximately 25% of all anthropogenic GHG emissions (Wang, 2024). A thorough understanding of the complex interactions between forests and climate change is important for formulating strategies to address global warming and promote the sustainable management of forest ecosystems.

Climate change has both positive and negative effects on forest ecosystems. Climate-related events, including fluctuations in temperature and precipitation patterns, create stress on vegetation and affect soil health and overall ecosystem functioning. In contrast, forests can adapt to changing climatic conditions and improve ecosystem dynamics to mitigate negative climatic effects.

4.1.1.1 Negative Impacts

4.1.1.1.1 Forest Disturbances

Climate change can alter disturbance regimes, thereby influencing the structure and function of forest ecosystems (Grimm et al., 2013). Forest structure and function can be greatly disrupted when disturbances (Figure 4.2) exceed natural variability (Battles et al., 2008; Lasco et al., 2008). Natural disturbances, such as fires, insect outbreaks, and windthrows, play an important role in forest ecosystem dynamics worldwide. A steady increase in temperature will have an impact on these ecosystems as well as

TABLE 4.1
Impact of Climate Change Activities during Different Years

Events	Description	Reference
Flood	In December 2015, Chennai faced a significant one-day precipitation event that resulted in extensive floods. The probable cause was determined to be the interaction between aerosols and greenhouse gases.	Van Oldenborgh et al. (2016)
	In June 2013, climate change was identified as the cause of the total rainfall in northern India.	Singh et al. (2014)
	According to CMIP5 simulations, thc warm mcan surfacc temperatures observed in 2015 were a result of anthropogenic warming.	Kam ct al. (2016), Nanditha et al. (2020)
	Between July and September, Bihar witnessed the most extensive floodwater inundation, affecting 20,837 km^2, followed by West Bengal with 17,307.1 km^2 and Assam with 13,460.1 km^2. This occurrence was associated with global warming, increasing sea levels, and the accelerated melting of glaciers as a consequence of climate change.	Pandey et al. (2022)
Storm and cyclone	In recent decades, storms and cyclones have become significantly more frequent compared to historical records.	Subramanian et al. (2023)
	Dust storm occurrences lead to elevated levels of fine particulate matter in the atmosphere, which are associated with detrimental health outcomes, including premature mortality, cardiovascular complications, respiratory disorders, lung cancer, and severe respiratory tract infections.	Maji and Sonwani (2022)
Forest fire	Climate change has significantly increased the incidence and frequency of forest fires in central India. This increase is evident not only during the typical fire-prone season but also throughout the remainder of the year.	Jain et al. (2021)
	The incidence, frequency, and severity of forest fires are strongly correlated with changing climate conditions, including rising temperatures, decreased precipitation, more dry days, and El Nino-Southern Oscillation (ENSO) events, all of which contribute to an increase in fire occurrences.	Nurdiati et al. (2022)
Drought	Climate-change-induced variation in precipitation, soil moisture, and surface water levels. These variables when coupled with excessive groundwater extraction lead to the development of drought conditions.	Mukherjee et al. (2018)
	Around 21.06% of India's land area was affected by drought-like conditions.	Roy et al. (2022)
Heat waves and cold waves	CMIP projections indicated the occurrence of severe heat waves in South Indian cities.	Nandi and Swain (2022)
	Ahmedabad experienced extreme heat events (EHE) in 2010 and 2016, with severe heat waves recorded temperatures up to 50°C.	Patel et al. (2022)
	The El Niño event in India disrupts cold wave activities.	Pai and Nair (2022)

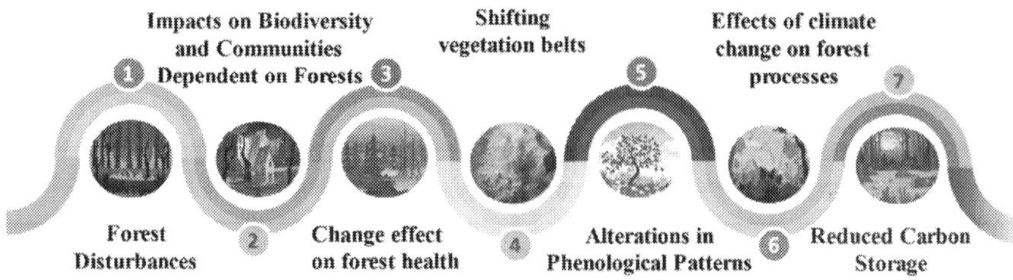

FIGURE 4.2 Negative impacts of climate change on the forest ecosystem.

on species development (Hellmann et al., 2008). During the seedling stage, factors such as higher temperatures, low soil moisture, and competition among species also impede tree regeneration. Some climate change effects on forests may be temporary, while in other cases, the effects can be long-term. Certain forests may benefit from specific climate-related changes; for example, warmer temperatures may encourage increased tree and plant development in areas where cold weather normally shortens the growing season. In some forests, rising temperatures allow invasive species to thrive (Hellmann et al., 2008; Rustad et al., 2012). Furthermore, warmer temperatures may promote the survival and growth of insects. Disturbances can interact, increasing the danger to forests. For example, dryness can damage trees, making them more vulnerable to wildfires or insect infestations. Wildfires may also leave woodlands that are more susceptible to pest infestation. Similarly, climate change can be combined with other stresses, such as land development, reducing a forest's ability to adapt.

4.1.1.1.2 Impacts on Biodiversity and Communities Dependent on Forests

Climate change results in biodiversity loss by altering habitat, causing species to migrate from their native ranges, and changing natural cycles, such as migration and breeding. Extreme weather events, including droughts, floods, wildfires, and ocean acidification, as well as the expansion of invasive species or insect and pest attacks, further disturb ecosystems and reduce species abundance. Many villagers rely heavily on forest resources, as India dedicates more than one-fifth of its land to forests (Gopalakrishnan et al., 2011). The consequences of climate change differ by geographical area and directly affect the lives of communities reliant on forests (Hussain et al., 2024). Forests are considered the primary source of income for approximately 1 billion individuals globally, of whom approximately 300 million people show relatively higher reliance. For instance, about two-thirds of the population in Africa is dependent on forest resources for their basic requirements, including fuel and food (Forner et al., 2018). The livelihood of these forest-based communities is unacquainted with the concept of climate change, although they are adversely impacted by its consequences and worsening their situations (Rahman & Alam, 2016)

4.1.1.1.3 Change Effect on Forest Health

Climate change is expected to increase pests and pathogens in forest ecosystems, which may be due to direct effects on their number or distribution, as well as indirect effects, such as increased water stress or wind damage, which cause trees to be more vulnerable to attack (Linnakoski et al., 2019). Climate change may change the disturbance dynamics of native forest insects and diseases, while simultaneously promoting the establishment and spread of non-native species (Marshet & Fekadu, 2019). Pathogens can exploit the changing climate, potentially leading to significant effects on tree health and survival. Forest diseases exhibit different infection and transmission rates depending on the temperature and moisture levels within the forest environment, and fungal pathogens, in particular, can live and be infectious at a wide variety of temperatures. However, the ideal temperature range for epidemic growth in most fungal diseases is limited to a few degrees Celsius for most tree species (Gebeyehu & Hirpo, 2019).

4.1.1.1.4 Shifting Vegetation Belts

The distribution and patterns of vegetation are significantly influenced by changing climatic conditions, as increased temperatures might encourage plant species to migrate to more suitable environments. In higher altitudinal places where alpine treelines are rising due to warming, this phenomenon of "shifting vegetation belts" disrupts ecosystem equilibrium (Bhardwaj et al., 2021; Chauhan et al., 2022; Kumar et al., 2022a). Native plant species compete with exotic ones as they enter native habitats for resources, including light, water, and nutrients, resulting in biodiversity loss and affecting herbivore food sources, nesting habitats, and soil nutrient cycling (Sonwani, 2016). Moreover, climate change encourages tree species to migrate, a time-consuming process hampered by natural seed distribution constraints, making them vulnerable. Assisted migration is emerging as an important conservation strategy to support tree populations during adjustment. Climate change impacts forest structure, resulting in biodiversity loss and alterations to native habitats, and fragmented landscapes further hinder natural migration patterns (Xu & Prescott, 2024). This worsens the risk of population loss, extinction, and invasive species relationships. Different historical studies demonstrate that forests adapted to past temperature shifts through migration; nevertheless, the current rapid rate of climate change raises concerns about the ability of natural migration to keep pace (VanWeelden et al., 2021). Conservation measures such as assisted migration, wildlife corridors, and adaptive management are critical to maintaining forest resilience, biodiversity, and ecological integrity.

4.1.1.1.5 Alterations in Phenological Pattern

The flowering phenology of plants is directly influenced by climate change, which results in reduced reproductive growth. It is an important biological indicator of climate change, and alterations in phenology affect the functioning of forest ecosystems, thus plant fitness and biodiversity by influencing biological events, including pollination mechanisms and seed establishment (Lindnet et al., 2010). Additionally, changes in the phenology of plants highlight plant-pollinator mismatches affecting community stability and consequently impacting ecosystem structure and functioning (Muscarella et al., 2020).

4.1.1.1.6 Effect of Climate Change on Forest Processes

Climate change influences various forest processes by changing the photosynthesis rate and growth patterns, by increasing temperature and atmospheric CO_2

concentrations. Light, heat, CO_2, and water are basic requirements of plants for photosynthesis (Gebeyehu & Hirpo, 2019). Generally, plants have a mechanism to tolerate certain levels of increased temperatures. As the temperature increased, the soil decomposition rate also increased, which further increased nutrient mineralization and nutrient uptake. Nutrient availability is the main factor for forest ecosystem functioning, and climate change affects nutrient availability, resulting in a less productive ecosystem (Chauhan et al., 2022). Nitrogen is essential for the growth and development of trees. Generally, the organic matter in the form of leaves releases nitrogen to the soil and is broken down by various microorganisms; however, when temperature rises due to climate change, CO_2 concentration increases in the atmosphere, and more microbes become nitrogen-limited. Rising temperatures mainly affect the rate of photosynthesis, macronutrient concentration in soils, and nutrient uptake capacity.

4.1.1.1.7 Reduced Carbon Storage

Climate change significantly affects forest ecosystem services, such as air purification, water supply, carbon storage, recreational services, and habitat provision for wildlife (Paoletti et al., 2007). The main function of the forest ecosystem is to sequester carbon and store it in the roots, tree biomass, and forest floor (Sharma et al., 2023). However, climate change can interfere with carbon sequestration in several ways. Changing precipitation patterns, including altered rainfall intensity in certain regions, can result in soil erosion, thereby releasing sequestered carbon into the atmosphere. Additionally, increasing temperatures have resulted in wildfires and forest degradation, resulting in various insect infestations and disease outbreaks that can exacerbate carbon emissions (Ontl et al., 2020).

4.1.1.2 Positive Impacts

4.1.1.2.1 Enhanced Biomass Production

Climate change has also benefited forests. For example, more biomass can be produced in plants with an increase in temperature, where frosty or cold weather limits the growing seasons, which further results in a denser forest ecosystem that potentially increases carbon sequestration (Gomez-Guerrero & Doane, 2018; Lindner et al., 2010). Moreover, increased temperatures also vitalize insect survival and growth.

4.1.1.2.2 Extended Growing Season

Vegetation growth is directly affected by climate change. Temperature is one of the main factors responsible for vegetation growth for any species. A temperature rise is responsible for the early arrival of spring and the late arrival of winter. This lengthening of the season enables trees to grow for longer periods and produce higher biomass (Bonan, 2008; Hogan et al., 2024). The influence of climate change on different growing seasons may vary depending on vegetation type. In a forest ecosystem, which is mainly dominated by trees, increased temperature results in increased photosynthesis.

4.1.1.2.3 New Habitat for Species

Some species adapt to climate change by changing their behavior, body function, and physical characteristics to new environments as they find more favorable conditions (Shovon et al., 2024). In warmer situations due to climate change, some species can migrate to new areas where they can find more suitable habitats (Shanin et al., 2024). This movement helps in the survival of those species that face challenges in adapting to warmer conditions in their natural environment.

4.1.1.2.4 Habitat Expansion for Thermophilic Species

Thermophilic species, which are suited to dry conditions, can expand their habitats in warmer areas, resulting in enhanced biodiversity and species richness (Bosch-Belmar et al., 2024).

4.1.2 Silviculture treatments/practices in Climate Change Scenario

Forests are primarily viewed as sources of timber but are now being acknowledged for their ecological significance (Figure 4.3). This silvicultural advancement is mainly a response to changing societal needs and perceptions (Gustafson et al., 2020). Stable habitat conditions have allowed for predictable forest management practices in the past, but current rapid climate scenarios have challenged this approach. To mitigate the effects of climate change, silvicultural strategies should be holistic, focusing on identifying risks, assessing vulnerability, and implementing targeted and adaptable actions with ongoing monitoring to mitigate climate impacts (Szmyt & Dering, 2024). These adaptive principles will help to maintain forest continuity and sustainability amid global environmental challenges. An active and planned adaptation strategy will always prove to be the most effective and logical approach for promoting forest resilience and adaptation to new and uncertain future periods.

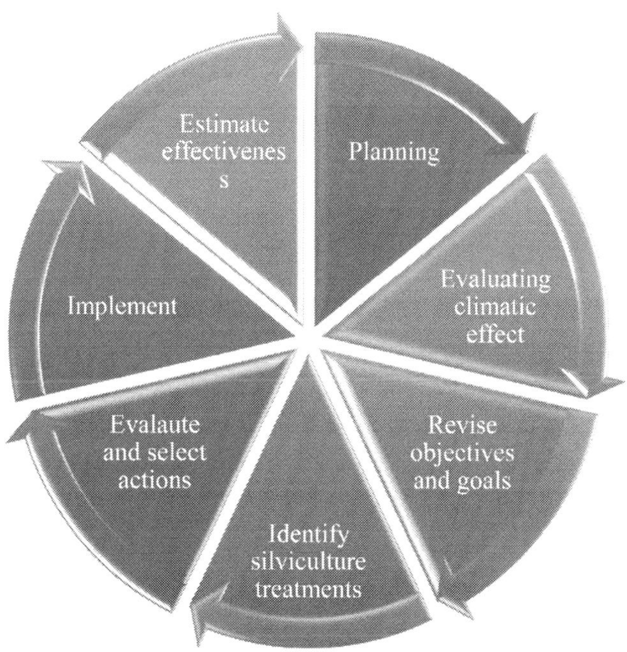

FIGURE 4.3 Climate changes adaptive silviculture cycle.

4.1.2.1 Incorporating Diverse Species Compositions

In forest ecosystems, multiple species integration helps enhance their adaptability to climate change. Different species exhibit varying sensitivities to climatic, topographic, and environmental conditions (Cosofret & Bouriaud, 2019; Hlasny et al., 2014), which offers ecological insurance by reducing risks linked to climate change by showing superior tolerance to species-specific stressors such as pest infestations compared to monocultures. Species selection must account for functional features such as drought tolerance and post-disturbance regenerative capacity, which enhance adaptive capacity (Forrester & Bauhus, 2016). Research has indicated that mixed forests, especially those comprising coniferous and broadleaf species, have enhanced resilience and productivity, which may mitigate economic losses (Haberstroh & Werner, 2022). Various silvicultural treatments are advised to achieve this principle using various methods (selection and shelterwood), fostering natural regeneration, facilitating the assisted migration of non-native species, and artificial planting of resilient species (Fischer et al., 2019). Rigorous thinning can enhance the growth of shade-tolerant understory vegetation in fertile regions. Efficient management of herbivore populations is crucial for effective regeneration of various tree types.

4.1.2.2 Structurally Complex Forests

High species diversity enhances the structural complexity of forest ecosystems, and diverse growth patterns enhance resilience to climate-related stressors. Techniques such as selection cutting, irregular shelterwood cutting, and intensive thinning create canopy gaps, supporting growth across generations (Carina & Keskitalo, 2011; Dusenge et al., 2019). Selective thinning and variable density thinning further maintain healthy individuals across canopy layers, thereby improving adaptability (Cosofret & Bouriaud, 2019). In monocultures, underplanting adds structural diversity; however, in fire-prone regions, multilayered forests require careful management to mitigate fire risk.

4.1.2.3 Enhancing Genetic Diversity

The adaptive capacity of forests is closely associated with their genetic diversity. Forestry practices that promote natural as well as artificial regeneration, retain older or ecologically valuable trees during thinning, and extend rotation cycles can enhance the genetic diversity within stands (Ennos et al., 2018). Given the climate change challenges, special attention should be given to assisted migration, focusing on not only tree species but also specific provenances (Gomory et al., 2020; Yousefpour et al., 2017). Adjusting selection criteria in thinning practices—currently focused on timber traits—can further improve genetic variability, strengthening forests' adaptability to changing conditions.

4.1.2.4 Enhancing Stand Stability

To improve stand stability and mitigate susceptibility to climate-related stressors, suggested silvicultural approaches emphasize diversifying forest structure and diminishing tree competition. (Corlett & Lafrankie, 1998). Regeneration cuts that promote mixed-species regeneration enhance forest sustainability. Resilience varies among animals based on their tolerance to stress. (Becknell & Powers, 2014). Thinning practices, such as selective thinning and crown thinning, can enhance the ability of the stand to fight various biotic and abiotic stressors. The more taper the tree, the greater the wind resistance (Jactel et al., 2017). Furthermore, effective stand management methods, reduced competition, and optimal spacing also favor the growth, productivity, and stability of stands, making them more resilient to environmental stresses.

4.1.2.5 Limiting Growing Stock

Maintaining an optimal growing stock is an effective approach for reducing economic losses related to climate change. Forest ecosystems with very high tree densities are continuously at wildfire risk due to elevated fuel loads, while wind damage risk escalates with stand age and height (Brang et al., 2014). In drought-prone regions, a limited growing stock can help mitigate productivity losses (Puettmann, 2011). Shortened rotation cycles, intensive thinning, reduced initial tree density, and diverse species compositions are key adaptive silvicultural practices that can effectively reduce climate-related damage and losses. Although different adaptive management and silvicultural techniques have been established in response to this changing climate, concerns about forest sustainability still exist. To combat these repercussions, increased awareness among foresters, researchers, and the public is necessary to enhance forest resilience and sustainability. This evolving approach in forest management prioritizes functional complexity, making it essential to achieve management goals in unpredictable climates.

4.1.2.6 Future Scenario

In future scenarios, if the same trend of climate change continues to persist, forest communities may undergo substantial changes due to increased temperature, irregular patterns of precipitation (heavy flood and drought), and forest disturbances, such as insect pest outbreaks and forest fires, which will further disrupt the natural dynamics of forests as well as hinder regeneration and seedling development, and may also disrupt existing forest compositions and trigger significant species migration. In the pursuit of favorable conditions, species migrate from lower altitudes to higher altitudes, which may result in the shifting of forest communities that could affect biodiversity and essential ecosystem services, such as carbon sequestration, water regulation, nutrient cycling, soil stabilization, and destabilization of the interactions among forest ecosystems. This future scenario highlights the need to adopt proactive conservation strategies and adaptive management approaches that anticipate these changes and work to support forest resilience in the face of unprecedented environmental change. In this envisaged scenario, resilient species may gradually dominate forest landscapes, resulting in new community structures that diverge significantly from existing ecosystems. Species that require specific habitats and have limited adaptability may face significant population declines and eventual local extinctions. Conservation initiatives should be prioritized to encourage species migration. Targeted reforestation and gene flow enhancement should be performed to improve forest resilience. Moreover, accurate modeling and ongoing monitoring are crucial for forecasting species migration patterns and identifying vital intervention zones.

This future necessitates proactive, scientifically informed methods to safeguard forest communities, equilibrate ecological processes, and enhance species resilience during severe climatic changes.

4.2 CONCLUSION

Forest ecosystems are dynamic, and changes in climatic conditions cause disturbance to them; however, changes in biodiversity over the past centuries have been significantly accelerated by anthropogenic land use and potentially by climate change produced by humans. Forest community structure changes with higher temperatures, altered precipitation patterns, and increased anthropogenic activities were major factors affecting the distribution of species throughout Florida. This is important as climate change may provide feedback through the biodiversity of the system, disrupting ecological interactions and therefore impacting major ecosystem services such as carbon storage, water regulation, and soil stability. This highlights the importance of implementing conservation strategies that are adaptable and capable of supporting resilient species populations along with broader ecosystem health.

REFERENCES

Agarwal, S., Sairorkham, B., Sakitram, P., & Lambin, E. F. (2022). Effectiveness of community forests for forest conservation in Nan province, Thailand. *Journal of Land Use Science*, *17*, 307–323.

Ao, A., Changkija, S., & Tripathi, S. K. (2024). Patterns of forest community diversity, regeneration potential and carbon storages along an altitudinal gradient in Eastern Himalaya, India. *Environmental and Sustainability Indicators*, *22*, 100399.

Battles, J. J., Robards, T., Das, A., Waring, K., & Gilless, J. K. (2008). Climate change impacts on forest growth and tree mortality. *Climatic Change*, *87*, s193–s213.

Becknell, J. M., & Powers, J. S. (2014). Stand age and soils as drivers of plant functional traits and aboveground biomass in a secondary tropical dry forest. *Canadian Journal of Forest Research*, *44*, 604–613.

Bhardwaj, D. R., Tahiry, H., Sharma, P., Pala, N. A., Kumar, D., & Kumar, A. (2021). Influence of aspect and elevational gradient on vegetation pattern, tree characteristics, and ecosystem carbon density in Northwestern Himalayas. *Land*, *10*(11), 1109.

Bonan, G. B. (2008). Forests and climate change: Forcings, feedbacks, and the climate benefits of forests. *Science*, *320*, 1444–1449.

Bosch-Belmar, M., Tantillo, M. F., & Sara, G. (2024). Impacts of increasing temperature due to global warming on key habitat-forming species in the Mediterranean Sea: Unveiling negative biotic interactions. *Global Ecology and Conservation*, *50*, 02844.

Brang, P., Spathelf, P., Larsen, J. B., Bauhus, J., Ina, A. B., Chauvin, C., Drössler, L., García-Güemes, C., Heiri, C., & Kerr, G. et al (2014). Suitability of close-to-nature silviculture for adapting temperate European forests to climate change. *Forestry: An International Journal of Forest Research*, *87*, 492–503.

Brockerhoff, E. G., Barbaro, L., Castagneyrol, B., Forrester, D. I., Gardiner, B., Gonzalez, J. R., Lyver, P. O. B., Meurisse, N., Oxbrough, A., & Taki, H. (2017). Forest biodiversity, ecosystem functioning, and the provision of ecosystem services. *Biodiversity and Conservation*, *26*, 3005–3035.

Carina, H., & Keskitalo, E. (2011). How can forest management adapt to climate change? Possibilities in different forestry systems. *Forests*, *2*, 415–430.

Chauhan, M. S., Sharma, A., Trivedi, A., Kumar, K., Ferguson, D. K., & Rathore, P. S.. 2022. Late Quaternary vegetation shifts and climate change in the subalpine belt of the Parvati Valley, Himachal Pradesh, India. *Quaternary International, 629*, 53–64. https://doi.org/:10.1016/j.quaint.2020.12.029

Chen S. J., Chen, C. Jiang, R. T. Yao, J. Xue, Y. Bai, H. Wang, C. Jiang, S. Wang, Y. Zhong. 2022. Trends in Research on Forest Ecosystem Services in the Most Recent 20 Years: A Bibliometric Analysis. *Forests* 13:1087.

Corlett, R. T., & Lafrankie, J. J. V. (1998). Potential impacts of climate change on tropical Asian forests through an influence on phenology. *Climatic Change, 39*, 439–453.

Cosofret, C., & Bouriaud, L. (2019). Which silvicultural measures are recommended to adapt forests to climate change? A literature review. *Bulletin of the Transilvania University of Brasov, Series II: Forestry, Wood Industry, Agricultural Food Engineering, 12*, 13–34.

Dusenge, M. E., Duarte, A. G., & Way, D. A. (2019). Plant carbon metabolism and climate change: Elevated CO_2 and temperature impacts on photosynthesis, photorespiration and respiration. *New Phytologist, 221*, 32–49.

Ennos, R., Cottrell, J., Hall, J., & O'Brien, D. (2018). Is the introduction of novel exotic forest tree species a rational response to rapid environmental change? – A British perspective. *Forest Ecology and Management, 432*, 718–728.

Fischer, C., Leimer, S., Roscher, C., Ravenek, J., de Kroon, H., Kreutziger, Y., Baade, J., Beßler, H., Eisenhauer, N., Weigelt, A., et al. (2019). Plant species richness and functional groups have different effects on soil water content in a decade-long grassland experiment. *Journal of Ecology, 107*, 127–141.

Forner, A., Valladares, F., Bonal, D., Granier, A., Grossiord, C., & Aranda, I. (2018). Extreme droughts affecting Mediterranean tree species' growth and water-use efficiency: The importance of timing. *Tree Physiology, 38*, 1127–113.

Forrester, D. I., & Bauhus, J. (2016). A review of processes behind diversity—Productivity relationships in forests. *Current Forestry Reports, 2*, 45–61.

Gebeyehu, M. N., & Hirpo, F. H. (2019). Review on effect of climate change on forest ecosystem. *International Journal of Environmental Sciences & Natural Resources, 17*, 126–129.

Gomez-Guerrero, A., & Doane, T. (2018). The response of forest ecosystems to climate change. In *Developments in soil science* (vol. 35, pp. 185–206). Elsevier.

Gomory, D., Krajmerová, D., Hrivnák, M., & Longauer, R. (2020). Assisted migration vs. close-to-nature forestry: What are the prospects for tree populations under climate change? *Central European Forestry Journal, 66*, 63–70.

Gopalakrishnan, R., Jayaraman, M., Bala, G., & Ravindranath, N. H. (2011). Climate change and Indian forests. *Current Science, 101*, 348–355.

Grimm, N. B., Chapin, F. S. III, Bierwagen, B., Gonzalez, P., Groffman, P. M., Luo, Y., Melton, F., Nadelhoffer, K., Pairis, A., Raymond, P. A., & Schimel, J. (2013). The impacts of climate change on ecosystem structure and function. *Frontiers in Ecology and the Environment, 11*, 474–482.

Guo, K., Wang, B., & Niu, X. (2023). A review of research on forest ecosystem quality assessment and prediction methods. *Forests, 14*, 317.

Gustafson, E. J., Kern, C. C., Miranda, B. R., Sturtevant, B. R., Bronson, D. R., & Kabrick, J. M. (2020). Climate adaptive silviculture strategies: How do they impact growth, yield, diversity and value in forested landscapes?. *Forest Ecology and Management, 470*, 118208.

Haberstroh, S., & Werner, C. (2022). The role of species interactions for forest resilience to drought. *Plant Biology, 24*, 1098–1107.

Hellmann, J. J., Byers, J. E., Bierwagen, B. G., & Dukes, J. S. (2008). Five potential consequences of climate change for invasive species. *Conservation Biology, 22*, 534–543.

Hlasny, T., Mátyás, C., Seidl, R., Kulla, L., Merganičová, K., Trombik, J., Dobor, L., Barcza, Z., & Konôpka, B. (2014). Climate change increases the drought risk in Central European forests: What are the options for adaptation? *Central European Forestry Journal, 60*, 5–18.

Hogan, J. A., Domke, G. M., Zhu, K., Johnson, D. J., & Lichstein, J. W. (2024). Climate change determines the sign of productivity trends in US forests. *Proceedings of the National Academy of Sciences*, *121*, 2311132121.

Hui, D., Deng, Q., Tian, H., & Luo, Y. (2017). Climate change and carbon sequestration in forest ecosystems. *Handbook of Climate Change Mitigation and Adaptation*, *555*, 594.

Hussain, S., Hussain, E., Saxena, P., Sharma, A., Thathola, P., & Sonwani, S. (2024). Navigating the impact of climate change in India: A perspective on climate action (SDG13) and sustainable cities and communities (SDG11). *Frontiers in Sustainable Cities*, *5*, 1308684.

Jactel, H., Bauhus, J., Boberg, J., Bonal, D., Castagneyrol, B., Gardiner, B., Gonzalez-Olabarria, J. R., Koricheva, J., Meurisse, N., & Brockerhoff, E. G. (2017). Tree diversity drives forest stand resistance to natural disturbances. *Current Forestry Reports*, *3*, 223–243.

Jain, M., Saxena, P., Sharma, S., & Sonwani, S. (2021). Investigation of forest fire activity changes over the central India domain using satellite observations during 2001–2020. *GeoHealth*, *5*, e2021GH000528.

Kam, J., Knutson, T. R., Zeng, F., & Wittenberg, A. T. (2016). Multimodel assessment of anthropogenic influence on record global and regional warmth during 2015. *Bulletin of the American Meteorological Society*, *97*, 4–8.

Khaine, I., & Woo, S. Y. (2015). An overview of interrelationship between climate change and forests. *Forest Science and Technology*, *11*, 11–18.

Kumar, D., Bhardwaj, D. R., Sharma, P., Bharti, N., Sankhyan, N., Al-Ansari, N., & Linh. N.T.T (2022a). Population dynamics of *Juniperus macropoda* bossier forest ecosystem in relation to soil physico-chemical characteristics in the cold desert of North-Western Himalaya. *Forests*, *13*(10), 1624.

Kumar, D., Bhardwaj, D. R., Thakur, C. L., Sharma, P., & Ayele, G. T. (2022b). "Vegetation shift of *Juniperus macropoda* Boisser Forest in response to climate change in North-Western Himalayas, India. *Forests*, *13*(12), 2088.

Lasco, R. D., Pulhin, F., Sanchez, P., Villamor, G. B., & Villegas, K. A. L. (2008). Climate change and forest ecosystems in the Philippines: Vulnerability, adaptation and mitigation. *Journal of Environmental Science and Management*, *11*, 1–14.

Lindner, M., Maroschek, M., Netherer, S., Kremer, A., Barbati, A., Garcia-Gonzalo, J., Seidl, R., Delzon, S., Corona, P., Kolström, M., & Lexer, M. J. (2010). Climate change impacts, adaptive capacity, and vulnerability of European forest ecosystems. *Forest Ecology and Management*, *259*, 698–709.

Linnakoski, R., Kasanen, R., Dounavi, A., & Forbes, K. M. (2019). Forest health under climate change: Effects on tree, and pest and pathogen dynamics. *Frontiers in Plant Science*, *10*, 1157.

Maji, S., & Sonwani, S. (2022). *Nature of sand and dust storm in South Asian region: Extremities and environmental impacts in extremes in atmospheric processes and phenomenon: Assessment, impacts and mitigation* (pp. 113–139). Springer Nature Singapore.

Marshet, N. G., & Fekadu, F. H. (2019). Review on effect of climate change on forest ecosystem. *International Journal of Environmental Sciences & Natural Resources*, *17*, 555968.

MD Annual Report. (2021). India Meteorological Department, New Delhi [Information Science & Knowledge Resource Development Division (IS&KRDD), (Formerly Publication Section)] oES/IMD/Annual Report -2022/(01)2023/02.

Mukherjee, A., Bhanja, S. N., & Wada, Y. (2018). Groundwater depletion causing reduction of baseflow triggering Ganges river summer drying. *Scientific Reports*, *8*, 12049.

Muscarella, R., Kolyaie, S., Morton, D. C., Zimmerman, J. K., & Uriarte, M. (2020). Effects of topography on tropical forest structure depend on climate context. *Journal of Ecology*, *108*, 145–159.

Nandi, S., & Swain, S. (2022). Analysis of heatwave characteristics under climate change over three highly populated cities of South India: A CMIP6-based assessment. *Environmental Science and Pollution Research*, *30*, 1–13.

Nanditha, J. S., van der Wiel, K., Bhatia, U., Stone, D., Selton, F., & Mishra, V. (2020). A seven-fold rise in the probability of exceeding the observed hottest summer in India in a 2°C warmer world. *Environmental Research Letters*, *15*, 044028.

Neilson, R. P., Pitelka, L. F., Solomon, A. M., Nathan, R. A. N., Midgley, G. F., Fragoso, J. M., Lischke, H., & Thompson, K. E. N. (2005). Forecasting regional to global plant migration in response to climate change. *Bioscience*, *55*, 749–759.

Nurdiati, S., Bukhari, F., Julianto, M. T., Sopaheluwakan, A., Aprilia, M., & Fajar, I. (2022). The impact of El Niño Southern Oscillation and Indian Ocean Dipole on the burned area in Indonesia. *Terrestrial, Atmospheric and Oceanic Sciences*, *33*, 16.

Ontl, T. A., Janowiak, M. K., Swanston, C. W., Daley, J., Handler, S., Cornett, M., Hagenbuch, S., Handrick, C., McCarthy, L., & Patch, N. (2020). Forest management for carbon sequestration and climate adaptation. *Journal of Forestry*, *118*, 86–101.

Pai, D. S., & Nair, S. (2022). Impact of El-Niño-Southern Oscillation (ENSO) on extreme temperature events over India. *Mausam*, *73*, 597–606.

Pandey, A. C., Kaushik, K., & Parida, B. R. (2022). Google Earth Engine for large-scale flood mapping using SAR data and impact assessment on agriculture and population of the Ganga-Brahmaputra basin. *Sustainability*, *14*, 4210.

Paoletti, E., Bytnerowicz, A., Andersen, C., Augustaitis, A., Ferretti, M., Grulke, N., Gunthardt-Goerg, M. S., Innes, J., Johnson, D., Karnosky, D., & Luangjame, J. (2007). Impacts of air pollution and climate change on forest ecosystems—Emerging research needs. *The Scientific World Journal*, *7*, 1–8.

Patel, L., Conlon, K. C., Sorensen, C., McEachin, S., Nadeau, K., & Kakkad, K. (2022). Climate change and extreme heat events: How health systems should prepare. *NEJM Catalyst Innovations in Care Delivery*, *3*, CAT-21.

Pathak, H. (2023). Impact, adaptation, and mitigation of climate change in Indian agriculture. *Environmental Monitoring and Assessment*, *195*, 52.

Penuelas, J., Sardans, J., & Estiarte, M. (2013). Evidence of current impact of climate change on life: A walk from genes to the biosphere. *Global Change Biology*, *19*, 2303–2338.

Puettmann, K. J. (2011). Silvicultural challenges and options in the context of global change: "Simple" fixes and opportunities for new management approaches. *Journal of Forestry*, *31*, 321–331.

Rahman, M. H., & Alam, K. (2016). Forest dependent Indigenous communities' perception and adaptation to climate change through local knowledge in the protected area—A Bangladesh case study. *Climate*, *4*(1), 12.

Rao, R. J., Vishwanath, V., Venkateshwaran, R., Selwyn, M. A., & Ramakrishnappa, V. (2007). Community management of forests in India: Its impact on environment and livelihood. *Environment and Sustainable Development*, *1*, 133.

Roy, P., Pal, S. C., Chakrabortty, R., Chowdhuri, I., Saha, A., & Shit, M. (2022). Climate change and groundwater overdraft impacts on agricultural drought in India: Vulnerability assessment, food security measures and policy recommendation. *Science of Total Environment*, *849*, 157850.

Rustad, L., Campbell, J., Dukes, J. S., Huntington, T., & Lambert, K. F. (2012). Changing climate, changing forests: The impacts of climate change on forests of The Northeastern United States and Eastern Canada. *Forests*, *7*, 56.

Salunke, A. Y. (2022). *Unravelling the Genotoxic Potential of Agrochemicals on Fish Cell Line* (Doctoral dissertation). Vadodara: Maharaja Sayajirao University of Baroda (India).

Sannigrahi, S., Pilla, F., Basu, B., Basu, A. S., Sarkar, K., & Chakraborti, S. (2020). Examining the effects of forest fire on terrestrial carbon emission and ecosystem production in India using remote sensing approaches. *Science of Total Environment*, *725*, 138331.

Scheffer, M., & Carpenter, S. R. (2003). Catastrophic regime shifts in ecosystems: Linking theory to observation. *Trends in Ecology and Evolution*, *18*, 648–656.

Shanin, V., Chumachenko, S., Frolov, P., Priputina, I., Tebenkova, D., & Kolycheva, A. (2024). Predicting the effect of climate change and management on net carbon sequestration in the forest ecosystems of the European part of Russia with the complex of models. *Ecological Modelling*, *496*, 110835.

Sharma, V., Bhardwaj, D. R., Kumar, D., Verma, S., & Bishist, R. (2023). Defoliator attack (*Diorhabda lusca maulik*) on Kharik (*Celtis australis* L.): Effect on nutritive value, phenology, and biomass productivity. *Agroforestry Systems*, *98*(3), 1–16.

Shovon, T. A., Auge, H., Haase, J., & Nock, C. A. (2024). Positive effects of tree species diversity on productivity switch to negative after severe drought mortality in a temperate forest experiment. *Global Change Biology*, *30*(3), e17252.

Singh, D., Horton, D. E., Tsiang, M., Haugen, M., Ashfaq, M., & Mei, R. (2014). Severe precipitation in Northern India in June 2013: Causes, historical context, and changes in probability. *Bulletin of the American Meteorological Society*, *95*, S58–S61.

Sonwani, S. (2016). Source apportionment of polycyclic aromatic hydrocarbons in urban atmosphere of South Delhi, India. In 2nd International Conference on Atmospheric Dust-DUST2016, Pro Science Conference Proceedings of Scientific Events (Bari) (pp. 111–116).

Subramanian, A., Nagarajan, A. M., Vinod, S., Chakraborty, S., Sivagami, K., & Theodore, T. (2023). Long-term impacts of climate change on coastal and transitional eco-systems in India: An overview of its current status, future projections, solutions, and policies. *RSC Advances*, *13*, 12204–12228.

Szmyt, J., & Dering, M. (2024). Adaptive silviculture and climate change—A forced marriage of the 21st century? *Sustainability*, *16*, 2703.

Thomas, C. D., Cameron, A., Green, R. E., Bakkenes, M., Beaumont, L. J., Collingham, Y. C., Erasmus, B. F., De Siqueira, M. F., Grainger, A., Hannah, L., & Hughes, L. (2004). Extinction risk from climate change. *Nature*, *427*, 145–148.

Thornton, P. E., Law, B. E., Gholz, H. L., Clark, K. L., Falge, E., Ellsworth, D. S., Goldstein, A. H., Monson, R. K., Hollinger, D., Falk, M., & Chen, J. (2002). Modeling and measuring the effects of disturbance history and climate on carbon and water budgets in evergreen needleleaf forests. *Agricultural and Forest Meteorology*, *113*, 185–222.

Toledo, M., Poorter, L., Peña-Claros, M., Alarcón, A., Balcazar, J., Leaño, C., Licona, C., Lianque, O., Vroomans, V., Zuidema, P., & Bongers, F. (2011). Climate is a stronger driver of tree and forest growth rates than soil and disturbance. *Journal of Ecology*, *99*, 254–264.

Van Oldenborgh, G. J., Otto, F. E. L., Haustein, K., & Achuta, R. K. (2016). The heavy precipitation event of December 2015 in Chennai, India. *Bulletin of the American Meteorological Society*, *97*, S87–S91.

VanWeelden, C., Towers, J. R., & Bosker, T. (2021). Impacts of climate change on cetacean distribution, habitat and migration. *Climate Change Ecology*, *1*, 100009.

Wang, L. (2024). Assessment of land use change and carbon emission: A Log Mean Divisa (LMDI) approach. *Heliyon*, *10*.

Xu, W., & Prescott, C. E. (2024). Can assisted migration mitigate climate-change impacts on forests?. *Forest Ecology and Management*, *556*, 121738.

Yousefpour, R., Temperli, C., Jacobsen, J. B., Thorsen, B. J., Meilby, H., Lexer, M. J., Lindner, M., Bugmann, H., Borges, J. G., & Palma, J. H. N. (2017). A framework for modeling adaptive forest management and decision making under climate change. *Ecology and Society*, *22*, 40.

Zhao, J., Ma, J., & Zhu, Y. (2019). Evaluating impacts of climate change on net ecosystem productivity (NEP) of global different forest types based on an individual tree-based model FORCCHN and remote sensing. *Global and Planetary Change*, *182*, 103010.

Zuidema, P. A., Baker, P. J., & Groenendijk, P. (2013). Tropical forests and global change: Filling knowledge gaps. *Trends in Plant Science*, *18*, 413–419.

5 Significance of Forest Biodiversity in Fostering Eco-resilience

Amitava Roy and Komal Rani

5.1 INTRODUCTION

Twenty-first century is welcoming the globe with the plethora of challenges in terms of climate change, higher pressure and exploitation of natural resource materials and newer diseases and threats. Therefore, there is an alarming requirement to address such challenges with urgent notice from ecologists for fostering climate change resilience in the environment (Calvin et al., 2023). Forests play a very crucial role in terms of tackling such challenges through carbon sequestration, providing sustainability of livelihood, habitat for wide range of species and plethora of other ecosystem services (Turner-Skoff & Cavender, 2019). Therefore, safeguarding of forests is an aspect which should be taken care of with utmost priority to ensure maintenance of environmental health against the threats of climate change. Conversely, forests are getting impacted through frequent turbulences such as wildfires, severe cases of droughts, changes in land use pattern, induction of new diseases and pests and changes in gene flow through mutations and other changes in the genetic makeup (Senf & Seidl, 2021). Due to these disturbances there is a severe question standing on forests' ability to cope up the fight against the threats of climate change and other factors. Additionally, the anthropogenic dynamics along with ever-increasing global population is further diminishing the space with greater and higher speed (Steffen et al., 2011).

One of the chief response strategies to overcome the challenges associated earlier is to persuade and foster ecological resilience in forests. There are several ways to enhance eco-resilience in forest systems such as engineering resilience; a process, through which environmental variables may return to the earlier equilibrium state before disturbance (Pimm, 1984). In terms of ecological resilience, the focus is majorly on quantification of the perseverance of the systems along with their capacity to soak up the change and maintain relationships between population state variables and thus escape the change to an unconventional state (Holling, 1973). Ecological resilience plays an important role for tackling the challenges of climate through shifting a particular variable in such a manner so that it may compensate for the change in the other variable and adapt the system. Therefore, it remains to be an important aspect to induce and foster ecological resilience, specifically in terms of forest ecosystem. Forest biodiversity plays a crucial role in the advancement of eco-resilience in forest ecosystems. Forests are one of the most important fighters in terms of biodiversity crisis through harbouring more than 80% of entire global biodiversity (FAO, WFP and IFAD, 2012).

DOI: 10.1201/9781003588580-5

5.1.1 Role of Forest Biodiversity in Enhancing Eco-Resilience

Forest biodiversity plays an important role in development and fostering of eco-resilience through augmentation of various processes such as maintenance of C cycle, development of adaptive capacity through the available genetic diversity, development of species co-existence and functional redundancy. These processes help the ecosystem to develop a compensatory mechanism against the changing climatic threats and thereby help the population to achieve a "steady state" to cope up with the adversities associated with those changes. The major processes through which forest biodiversity helps in attaining eco-resilience are discussed in the next section.

5.1.2 Forest Biodiversity and Maintenance of Carbon Cycle

Forest carbon dynamics are hugely impelled by climatic inputs which regulate the rate and intensity of primary production via photosynthesis along with maintaining balance between photosynthesis and respiratory mechanisms. With the greater richness in the forest biodiversity, there will be the presence of specialized species which have evolved with the aid of natural selection in the distinctive forest ecosystems. Evolution of these species in terms of adaptive new traits for the particular ecosystem resulted in the species to perform and function most favourably under specific climatic requirements and disturbances prevalent under different ecosystems (Law et al., 2002).

In terms of different forest systems, tropical forests have lower soil organic C due to least dead biomass as a result of higher rates of respiration and turnover with the rise in temperature. However, reverse is true for the boreal forest systems which can store significantly higher amounts of carbon in both living and dead biomass as a result of less ecosystem change due to insignificant change in the land use pattern (Keith et al., 2009). In case of tropical forests, alpha diversity is found to be much lower than the gamma diversity which reflects complex ecosystem behaviour in case of tropical forests. Therefore, it can be reflected that high biodiversity among ecosystems has a powerful operational value in case of tropical forests catering to its species richness. However, lesser diversity at the stand level may also contribute to its low C sequestration potential.

5.1.3 Biodiversity in Development of Adaptive Capacity of Population

Intra-specific biodiversity plays a much crucial role in building resilience against temperature fluctuations and incumbent disturbances through various biotic factors. Species with newer traits play a pivotal role in adaptation to the altered environments. Ecosystems prevalent in the temperate and tropical forests' resilience against different biotic factors such as pests and diseases were found to be facilitated through the existence of genetic diversity in such ecosystems (Ennos, 2015; Schaberg et al., 2008).

The presence of intra-specific genetic diversity also observed to be contributing to population fitness. Populations with higher individualistic diversity were exhibiting superior performance in terms of long-term survival of the populations in case of large meta-populations of butterflies in Europe (Vandewoestijne et al., 2008).

It was further deciphered while analysing the amphibian population that loss of genetic diversity was actually accompanying the loss of fecundity and per cent reproductive success and led to the extinction of several species (Allentoft & O'Brien, 2010). The existence of huge genotypic and phenotypic variation led to the insurance of the fitness of the populations under varieties of environmental extremities and thereby guarantees the resilience under the environmental change due to climate change and extinction risks as the fitness of a population decides the ability of a particular species to return to the equilibrium state after the disturbances due to various environmental and anthropogenic factors (Barton & Keightley, 2002).

5.1.4 Role of Landscape Diversity in Ensuring Eco-Resilience

Heterogeneity among ecosystem due to inter-specific diversity has been found to drive eco-resilience in case of several meta-populations of various species. A study report on mussels' meta-population demonstrated the development of resilience against the slow disturbances in the meta-populations of mussels due to the heterogeneity present in the sea shore landscape (Gouhier, 2005). Landscape diversity provides spatial heterogeneity which becomes a crucial factor in trimming down inter-specific competition and as a result ensures community stability and helps in co-existence of various competing species from the same functional groups and thus facilitates considerable levels of eco-resilience (Loreau et al., 2003; Miller & Chesson, 2009). Ecosystem diversity also helps in development of spatial resistance which helps in formation of favourable state for various ecosystems once those are disturbed by various natural and evolutionary forces also known as "Pulse disturbances" (West & Salm, 2003). Another factor which is probably of greater importance in terms of development of eco-resilience is formation of habitat refugia which allows the population diversity to persist and expand in the scenarios of climatic extremities due to various disturbing forces (Bengtsson et al., 2003). Such a system may also be recognized as historical "ecosystem memory" which provides recovery and reorganization of the disturbed ecosystems and helps in maintenance of population equilibrium and stability at the level of pre-disturbed periods (Johnstone et al., 2016). There have been multiple reports of forest ecosystems along with various other bionetworks to gain recovery subsequent to the natural and evolutionary disturbances as a result of incursion of new populations and species from the undisturbed environments to recuperate the stability of the altered ecosystem (Drever et al., 2006; Duncan, 1993; Foster et al., 1998). However, it is also to be considered that most of the stated studies were predominantly performed at a very minor scale spatially and as a result cannot be regarded as conclusive evidences (Spasojevic et al., 2016).

5.1.5 Role of Inter-Specific Diversity and Functional Redundancy in Development of Resilience in the Ecosystems

The presence of sufficient functional diversity helps in preserving important traits and thus provides resilience to the recurrent changes in various ecosystems. The two major mechanisms which operate for functional diversity are response diversity and functional redundancy. It was evidenced from an experiment performed in the Australian rangelands that grassy species which belonged to similar functional groups were varying in their resilience to the environmental extremities (Walker et al., 1999). Another study executed with the microcosms utilizing three shredder species revealed that a few detritivores in the stream living ecologies were much more tolerant to the acidification, whereas the decomposers in the same environment were found to be resilient against organic pollution (Jónsson et al., 2002). Another line of evidences for the impact of species diversity in eco-resilience come from the studies carried out on species co-existence. There is considerable variation observed in herbivores for size and grazing intensity in case of coral reef systems. The over-fishing for the large species in such systems was observed to be compensated by the influx of small grazers to provide balance of the overall population (Elmqvist et al., 2003). In another study spatial distribution of *Copernicia alba* was assessed in the natural stands and the results revealed that there are different scales of fruit dispersers operating in order to ensure resilience in the rainforests (Fragoso, 1997). In many cases declining populations of the keystone species in the different forest ecosystems resulted in the significant shifts in terms of ecological balance (Ellison et al., 2005). Such an example was found in Hawaiian Islands where the loss in the pollinator species resulted in the surge of extinction of local flora (Cox & Elmqvist, 2000).

5.1.6 Role of Forest Biodiversity in Habitat Provisioning

One of the most pivotal roles played by forest biodiversity is via habitat provisioning to a variety of flora and fauna (Figure 5.1). As an instance species with meagre dispersal such as lichens or rodents hugely suffer in terms of declining population in landscape fragmentations (Jónsson et al., 2017). The observable diversity in the forest specialist plants has a favourable influence for the proximity to the natural vegetation patches delivering colonization opportunities (Coote et al., 2013). Forest biodiversity encompasses the multitude of life forms which inhabit forest areas along with the associated allelic diversity. In total tropical, temperate and boreal forests together account for more than 80% of overall global population of plants, insects and animals (UN DESA, 2017). Increased plant diversity provides copious and quality food resources to variety of animals and thereby contributes to the overabundance of various animals in numerous sections (Côté et al., 2004).

5.1.7 Limitations and Challenges: Potential Risks for Forest Biodiversity

There are several limitations and challenges associated with fostering eco-resilience aided through forest biodiversity. Majority of such challenges are due to lack of policy formulation and monitoring along with the changing climatic extremities. These challenges are to be taken care of in consideration of protecting forest biodiversity and management of forest health.

5.1.8 Absence of Available Baseline Data for Monitoring

The most important challenge in the conservation of forest biodiversity is the absence of any extensive baseline data along with a proper monitoring system. Unavailability

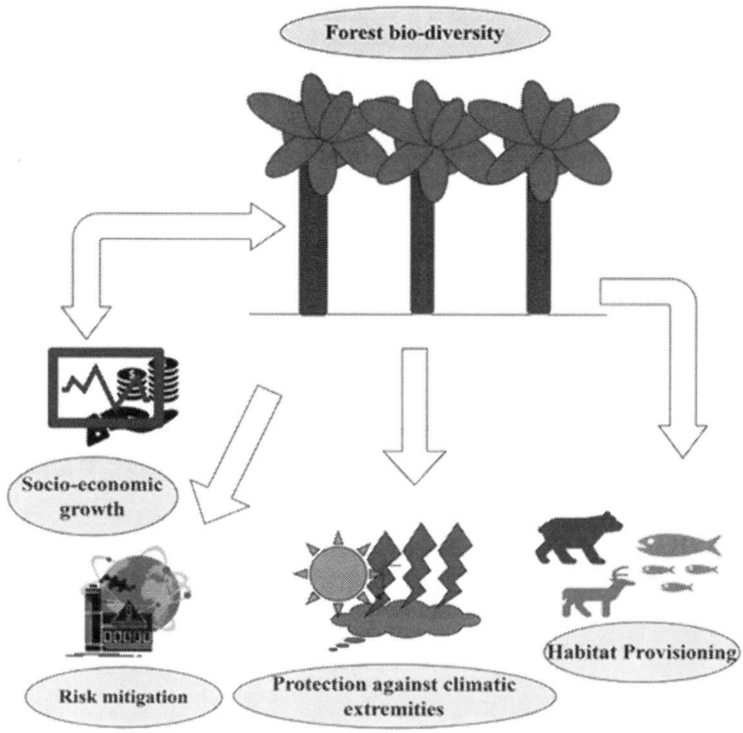

FIGURE 5.1 Role of forest biodiversity in fostering eco-resilience.

of precise data on species diversity, species distribution, population dynamics and other ecological factors has led to the improper monitoring of the species in forest ecosystems (Bueno & Peres, 2020). Without accurate information on various population parameters, it becomes very difficult to strategize effective conservation techniques. On the other hand, monitoring is a resource-intensive process which requires recurring monetary investment and human intervention (Moersberger et al., 2024). Such complexities slow down the process of human-induced conservation efforts and sometimes result in the development of faulty conservation strategy which may not yield any valuable results. Therefore, designing effective conservation and sustainable management strategies becomes much more challenging for the scientists and the conservationists which further hinder the protection of biodiversity.

5.1.9 Change in Land Use Pattern and Habitat Fragmentation

Change in land use pattern due to urbanization and extending agricultural practices (Figure 5.2) to the forest's pretence is a serious threat to the present forest biodiversity throughout the globe. Land use change has played major role in altering the global carbon cycle and thereby global climatic successions (Hussain et al., 2024). Interference in the natural hydrological cycle in order to supply freshwater

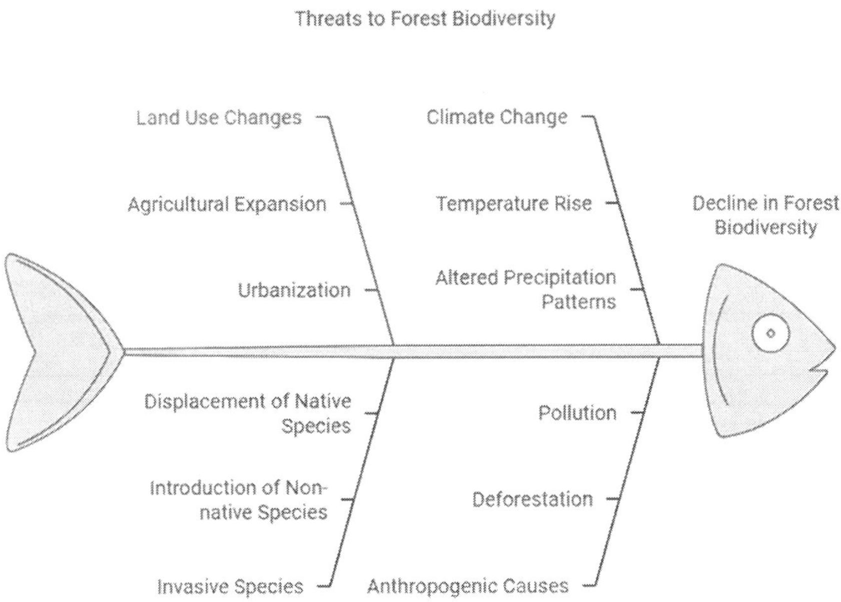

FIGURE 5.2 Major threats to forest biodiversity.

to agricultural systems, industries and other human requirements also led to consequences in the forest systems. In addition to this artificial nutrient supply to the plants in the form of fertilizers and other inorganic chemicals further resulted in deterioration in the quality of freshwater and coastal water ecosystems. Changes in the land use pattern also contribute to the decline in the forest biodiversity through alteration, fragmentation and loss of habitats, deteriorating soil quality and degradation in the utilizable freshwater and overexertion of the indigenous species (Foley et al., 2009). The majority of global dense forests have already been converted to the agricultural land holdings which also holds true for other substantially C dense ecosystems (Thompson et al., 2009). In case of large-scale alterations in the land use patterns, the species experienced loss of its capacity to disperse to the newer areas and therefore natural migration of different genotypes was unable to match the rapidity of changing climatic circumstances (Johnston & McIntyre, 2019).

5.1.10 Change in the Species Composition: Introduction of Invasive Species

There is an estimate that 80% of endangered species are threatened to experience extinction due to the pressure of non-native invasive species (Armstrong, 2005). In the last few decades, this threat of invasive species has increased gigantically due to globalization, rapid movement, ever-increasing growth of human population and changing climate. Trading and trafficking of plant materials throughout the globe has also increased by several folds in the past which is one of prime causes for unintentional introduction of the species and increase in the number of invasive weeds in many countries (Bryan & Metaxas, 2006). One of such unintentional introductions of invasive species is *Lantana camara* in India which is still challenging and posing a serious threat to the various ecosystems prevalent in the country over the last two centuries (Bhagwat et al., 2012). The lack of natural enemy of the invasive species such as predators and pathogens in the newer region also contributes to its growth and widespread distribution into the various ecosystems and ultimately increases severity of damage through the invasives by multiple folds. Finally, the changing climatic circumstances can become much more favourable for these introduced invasive species due to the higher physiological and genetic tolerance against the climate change consequences in comparison to the native species which further enhances the damage opportunities by these unintentional introductions (Karl et al., 2009).

5.1.11 Impact of Climate Change on Forest Biodiversity

Climate change may be defined as the long-term alteration in temperature and weather variables (Wang et al., 2004). These changes may either be natural or human-made but ultimately result in the shifts in climatic cycle. Climate change directly persuades the survival and fitness of disease pests and other pathogens and thus increases the susceptibility of forest ecosystems to such agents (Dale et al., 2001). On the other hand, it indirectly hampers the growth and abundance of natural enemies and predator insects and pathogens of such disease-causing agents which could regulate their population. Therefore, climate change ultimately causes further diminution in the quality and profusion of forest biodiversity which can potentially become a huge threat to address in the future (Dale et al., 2001). Environmental alterations also disturb species distribution, increase the vulnerability of forest ecosystems to the different biotic and abiotic factors and interrupt the sound functioning of the ecosystem. Therefore, climate change may pose a serious threat to the development of resilience in the forest ecosystems globally (Thompson et al., 2009).

5.1.12 Overcoming the Challenges

There are several approaches to overcome the above-mentioned challenges to keep the forest biodiversity healthy, thriving and prospering. These include appropriate assessment and monitoring forests, conservation of forest diversity and sustainable utilization of forest produce along with a strong policy perspective. Some of these approaches have been discussed in detail in the following sections.

5.1.13 Formulation of Baseline Data, Assessment and Monitoring

Alleles present in a single individual as against the large-scale forest landscapes and ultimately to overall global biodiversity, it becomes important to incorporate factors from each of these levels. Therefore, it is required to classify, monitor and report from all of these scales and to involve all the stakeholders, i.e., the local indigenous forest community along with the scientific fraternity to formulate a holistic approach for proper assessment and monitoring of forest biodiversity (Rykowski, 1999; Sala et al., 2000).

5.1.14 Conservation of Forest Biodiversity and Sustainable Utilization

Conservation of existing forest biodiversity along with sustainable utilization is hugely important for developing forest resilience in all types of forests from primary protected forests to agroforestry systems and all the other forest-based ecosystems. Convention on Biological Diversity (United Nations Environment Programme, 1992) was an important event in this regard which provided the guiding principles for conservation and sustainable utilization of all types of bio-resources and thereby to be administered in the forest resources too ranging from protected forests to the human-made plantations. The forest management system should involve both scientific and adaptive experience-based knowledge for effective conservation of forest resources and securing insurance for the available forest biodiversity.

5.1.15 Policy Formulation and Institutional Protection of Forest Biodiversity

To identify and execute the measures in order to reverse the biodiversity loss in the forests, both direct causes and underlying parameters of threats are to be considered. Socio-economic and political resolutions taken in forestry sector must safeguard the needs of forests in terms of available bio-resources and thereby execute a fair and equitable disbursement of costs and benefits among the stakeholders and resource users (Neugarten et al., 2024). Formulation of a legal, institutional and economic policy framework to tackle biodiversity loss is an essential prerequisite for conservation and protection of forest biodiversity. The emphasis should be provided on the development of economic ecology conducive for the protection of biological diversity and sustainable utilization of its various components which is also covered under Convention on Biological Diversity (1992). It is also to be ensured that there is sufficient engrossment of all the local communities and indigenous stakeholders at the levels of forest management to develop a holistic framework of conservation. There is requirement of increased political will along with massive institutional support and adequate human and financial resources for effective management and conservation of forest bio-resources. Finally, an appropriate legal policy is required to be formulated along with enforcement of suitable legislative framework in order to protect the forest lands from further exploitation in future and abuse of forest resources (Angelstam et al., 2011).

5.2 CONCLUSION

One of the chief response strategies against the future threats of climate change, alterations in land use pattern and changing socio-economic ecological purview is to develop and foster eco-resilience in the forest systems through enhancing capacity of the forest to support varieties of ecosystem services. Enrichment of forest biodiversity not only enhances forest's capacity to support various ecosystem services but also helps to tackle future challenges and threats of climate change and associated factors. However, lack of any legislative and operational framework hinders the monitoring and evaluation of forests in terms of these threats and therefore delays the progress of effective forest management. Absence of any base line data for monitoring further challenges the formulation of a sound policy framework.

Development of eco-resilience is certainly the key strategy to deal with such challenges; however, it requires huge institutional and legislative support to become competent at the operational level. The goal is to develop an effective management strategy which can transform according to the changing requirements for the unpredictable developments in the environment. Management practices should combine the conditioning of forest produce along with sustainable utilization and conservation of bio-resources (Lecina-Diaz et al., 2024). Finally, the policy framework should engage and involve local communities and indigenous forest dwellers to participate and incorporate the strategies in order to shape a holistic management system for the maintenance of forest ecosystems and make it capable of tackling future challenges of changing environment and climate.

5.3 FUTURE PROSPECTS AND WAY FORWARD

Challenging climate change and development of resilience against the climate is one of the hardest arenas to research upon. Unpredictability of the variables along with the unintended consequences may lead to a complete transformation of the rescue forces against it. Forests and forest-based ecosystems have been playing a pivotal role in shaping Earth's future from time immemorial. Therefore, it is of utmost importance to protect such resilient forces to secure the future of humanity and the other living organisms of the globe. With the advancement in technologies like artificial intelligence and machine learning along with the progress in our current knowledge of forestry sciences, it may be hoped that there will be huge opportunities to strategize and execute conservation efforts for effective administration of forests. At last, there is cooperation required from all the participating forces against climate change to connect together and save the globe from uncertain and unavoidable challenges of future which may become impossible to challenge otherwise.

REFERENCES

Allentoft, M. E., & O'Brien, J. (2010). Global amphibian declines, loss of genetic diversity and fitness: A review. *Diversity*, *2*(1), 47–71.

Angelstam, P., Andersson, K., Axelsson, R., Elbakidze, M., Jonsson, B. G., & Roberge, J. M. (2011). Protecting forest areas for biodiversity in Sweden 1991–2010: Policy implementation process and outcomes on the ground. *Silva Fennica*, *45*(5), 1111–1133.

Armstrong, D. P. (2005). Integrating the metapopulation and habitat paradigms for understanding broad-scale declines of species. *Conservation Biology*, *19*(5), 1402–1410.

Barton, N. H., & Keightley, P. D. (2002). Understanding quantitative genetic variation. *Nature Reviews Genetics*, *3*(1), 11–21.

Bengtsson, J., Angelstam, P., Elmqvist, T., Emanuelsson, U., Folke, C., Ihse, M., …, & Nyström, M. (2003). Reserves, resilience and dynamic landscapes. *AMBIO: A Journal of the Human Environment*, *32*(6), 389–396.

Bhagwat, S. A., Breman, E., Thekaekara, T., Thornton, T. F., & Willis, K. J. (2012). A battle lost? Report on two centuries of Invasion and management of *Lantana camara* L. in Australia, India and South Africa. *PLoS One*, *7*(3), e32407.

Bryan, T. L., & Metaxas, A. (2006). Distribution of deep-water corals along the North American continental margins: Relationships with environmental factors. *Deep Sea Research Part I: Oceanographic Research Papers*, *53*(12), 1865–1879.

Bueno, A. S., & Peres, C. A. (2020). The role of baseline suitability in assessing the impacts of land-use change on biodiversity. *Biological Conservation*, *243*, 108396.

Calvin, K., Dasgupta, D., Krinner, G., Mukherji, A., Thorne, P. W., Trisos, C., Romero, J., Aldunce, P., Barrett, K., Blanco, G., Cheung, W. W., Connors, S., Denton, F., Diongue-Niang, A., Dodman, D., Garschagen, M., Geden, O., Hayward, B., Jones, C.,. . . Ha, M. (2023). *IPCC, 2023: Climate Change 2023: Synthesis Report. Contribution of Working Groups I, II and III to the Sixth Assessment Report of the Intergovernmental Panel on Climate Change [Core Writing Team, H. Lee and J. Romero (eds.)]. IPCC, Geneva, Switzerland.* https://doi.org/10.59327/ipcc/ar6-9789291691647

Cootè, L., Dietzsch, A. C., Wilson, M. W., Graham, C. T., Fuller, L., Walsh, A. T., …, & O'Halloran, J. (2013). Testing indicators of biodiversity for plantation forests. *Ecological Indicators*, *32*, 107–115.

Côté, S. D., Rooney, T. P., Tremblay, J. P., Dussault, C., & Waller, D. M. (2004). Ecological impacts of deer overabundance. *Annual Review of Ecology, Evolution, and Systematics*, *35*(1), 113–147.

Cox, P. A., & Elmqvist, T. (2000). Pollinator extinction in the Pacific Islands. *Conservation Biology*, *14*(5), 1237–1239.

Dale, V. H., Joyce, L. A., McNulty, S., Neilson, R. P., Ayres, M. P., Flannigan, M. D., …, & Wotton, B. M. (2001). Climate change and forest disturbances: Climate change can affect forests by altering the frequency, intensity, duration, and timing of fire, drought, introduced species, insect and pathogen outbreaks, hurricanes, windstorms, ice storms, or landslides. *BioScience*, *51*(9), 723–734.

Drever, C. R., Messier, C., Bergeron, Y., & Doyon, F. (2006). Fire and canopy species composition in the Great Lakes-St. Lawrence Forest of Témiscamingue, Québec. *Forest Ecology and Management*, *231*(1–3), 27–37.

Duncan, R. P. (1993). Flood disturbance and the coexistence of species in a lowland Podocarp forest, South Westland, New Zealand. *Journal of Ecology*, *81*(3), 403. https://doi.org/10.2307/2261519

Ellison, A. M., Bank, M. S., Clinton, B. D., Colburn, E. A., Elliott, K., Ford, C. R., …, & Webster, J. R. (2005). Loss of foundation species: Consequences for the structure and dynamics of forested ecosystems. *Frontiers in Ecology and the Environment*, *3*(9), 479–486.

Elmqvist, T., Folke, C., Nyström, M., Peterson, G., Bengtsson, J., Walker, B., & Norberg, J. (2003). Response diversity, ecosystem change, and resilience. *Frontiers in Ecology and the Environment*, *1*(9), 488–494.

Ennos, R. A. (2015). Resilience of forests to pathogens: An evolutionary ecology perspective. *Forestry: An International Journal of Forest Research*, *88*(1), 41–52.

FAO, WFP and IFAD (2012). *The State of Food Insecurity in the World 2012. Economic growth is necessary but not sufficient to accelerate reduction of hunger and malnutrition.* FAO.

Foley, R. J., Chornock, R., Filippenko, A. V., Ganeshalingam, M., Kirshner, R. P., Li, W., …, & Wood-Vasey, W. M. (2009). SN 2008ha: An extremely low luminosity and exceptionally low energy supernova. *The Astronomical Journal*, *138*(2), 376.

Foster, D. R., Knight, D. H., & Franklin, J. F. (1998). Landscape patterns and legacies resulting from large, infrequent forest disturbances. *Ecosystems*, *1*, 497–510.

Fragoso, J. M. V., & Huffman, J. M. (2000). Seed-dispersal and seedling recruitment patterns by the last Neotropical megafaunal element in Amazonia, the tapir. *Journal of Tropical Ecology*, *16*(3), 369–385. https://doi.org/10.1017/s0266467400001462.

Gouhier, T. C. (2005). *The role of disturbance cycles in marine mussel metapopulations.* McGill University.

Holling, C. S. (1973). Resilience and stability of ecological systems. In L. Robin, S. Sörlin, & P. Warde (Eds.), *The future of nature: Documents of global change* (pp. 245–260). Yale University Press, 2013. https://doi.org/10.12987/9780300188479-023

Hussain, K., Mehmood, K., Anees, S. A., Ding, Z., Muhammad, S., Badshah, T., …, & Khan, W. R.. (2024). Assessing forest fragmentation due to land use changes from 1992 to 2023: A spatio-temporal analysis using remote sensing data. *Heliyon*, *10*(14).

Johnston, C. A., & McIntyre, N. E. (2019). Effects of cropland encroachment on prairie pothole wetlands: Numbers, density, size, shape, and structural connectivity. *Landscape Ecology*, *34*, 827–841.

Johnstone, J. F., Allen, C. D., Franklin, J. F., Frelich, L. E., Harvey, B. J., Higuera, P. E., …, & Turner, M. G. (2016). Changing disturbance regimes, ecological memory, and forest resilience. *Frontiers in Ecology and the Environment*, *14*(7), 369–378.

Jónsson, H., Sulem, P., Kehr, B., Kristmundsdottir, S., Zink, F., Hjartarson, E., …, & Stefansson, K. (2017). Whole genome characterization of sequence diversity of 15,220 Icelanders. *Scientific Data*, *4*(1), 1–9.

Jónsson, S., Zebker, H., Segall, P., & Amelung, F. (2002). Fault slip distribution of the 1999 Mw 7.1 Hector Mine, California, earthquake, estimated from satellite radar and GPS measurements. *Bulletin of the Seismological Society of America*, *92*(4), 1377–1389.

Karl, T. R., Melillo, J. M., & Peterson, T. C. (2009). *Global climate change impacts in the United States: A state of knowledge report from the US Global Change Research Program*. Cambridge University Press.

Keith, H., Mackey, B. G., & Lindenmayer, D. B. (2009). Re-evaluation of forest biomass carbon stocks and lessons from the world's most carbon-dense forests. *Proceedings of the National Academy of Sciences*, *106*(28), 11635–11640.

Law, B. E., Falge, E., Gu, L., Baldocchi, D. D., Bakwin, P., Berbigier, P., …, & Wofsy, S. (2002). Environmental controls over carbon dioxide and water vapor exchange of terrestrial vegetation. *Agricultural and Forest Meteorology*, *113*(1–4), 97–120.

Lecina-Diaz, J., Martínez-Vilalta, J., Lloret, F., & Seidl, R. (2024). Resilience and vulnerability: Distinct concepts to address global change in forests. *Trends in Ecology & Evolution, 39(8), 706–715. https://doi.org/10.1016/j.tree.2024.03.003*

Loreau, M., Mouquet, N., & Gonzalez, A. (2003). Biodiversity as spatial insurance in heterogeneous landscapes. *Proceedings of the National Academy of Sciences*, *100*(22), 12765–12770.

Miller, A. D., & Chesson, P. (2009). Coexistence in disturbance-prone communities: How a resistance-resilience trade-off generates coexistence via the storage effect. *The American Naturalist*, *173*(2), E30–E43.

Moersberger, H., Valdez, J., Martin, J. G. C., Junker, J., Georgieva, I., Bauer, S., Beja, P., Breeze, T. D., Fernandez, M., Fernández, N., Brotons, L., Jandt, U., Bruelheide, H., Kissling, W. D., Langer, C., Liquete, C., Lumbierres, M., Solheim, A. L., Maes, J., … Bonn, A. (2024). Biodiversity monitoring in Europe: User and policy needs. *Conservation Letters*, 17(5). https://doi.org/10.1111/conl.13038

Neugarten, R. A., Rasolofoson, R. A., Barrett, C. B., Vieilledent, G., & Rodewald, A. D. (2024). The effect of a political crisis on performance of community forests and protected areas in Madagascar. *Nature Communications*, *15*(1), 2963.

Pimm, S. L. (1984). The complexity and stability of ecosystems. *Nature*, *307*(5949), 321–326.

Rykowski, K. (1999). Assessment of the importance of forests in reducing global climatic change (counteracting and mitigating the impacts of greenhouse gases). *Geographia Polonica*, *72*, 27–44.

Sala, O. E., Stuart Chapin, F. I. I. I., Armesto, J. J., Berlow, E., Bloomfield, J., Dirzo, R., …, & Wall, D. H. (2000). Global biodiversity scenarios for the year 2100. *Science*, *287*(5459), 17.

Schaberg, P. G., DeHayes, D. H., Hawley, G. J., & Nijensohn, S. E. (2008). Anthropogenic alterations of genetic diversity within tree populations: Implications for forest ecosystem resilience. *Forest Ecology and Management*, *256*(5), 855–862.

Senf, C., & Seidl, R. (2021). Mapping the forest disturbance regimes of Europe. *Nature Sustainability*, *4*(1), 63–70.

Spasojevic, M. J., Turner, B. L., & Myers, J. A. (2016). When does intraspecific trait variation contribute to functional beta-diversity? *Journal of Ecology*, *104*(2), 487–496.

Steffen, W., Persson, Å, Deutsch, L., Zalasiewicz, J., Williams, M., Richardson, K., …, & Svedin, U. (2011). The Anthropocene: From global change to planetary stewardship. *Ambio*, *40*, 739–761.

Thompson, I., Mackey, B., McNulty, S., & Mosseler, A. (2009). Forest resilience, biodiversity, and climate change. In A synthesis of the biodiversity/resilience/stability relationship in forest ecosystems (Vol. 43, No. 1, pp. 1–67). Secretariat of the Convention on Biological Diversity, Montreal. *Technical Series.*

Turner-Skoff, J. B., & Cavender, N. (2019). The benefits of trees for livable and sustainable communities. *Plants, People, Planet*, *1*(4), 323–335.

UN DESA. (2017). *The Sustainable Development Goals Report 2017 – July 2017.* UN DESA. © UN DESA. https://unstats.un.org/sdgs/report/2017/

United Nations Environment Programme. (1992). *Convention on Biological Diversity, June 1992.* UNEP Document Repository.70-1774.

Vandewoestijne, S., Schtickzelle, N., & Baguette, M. (2008). Positive correlation between genetic diversity and fitness in a large, well-connected metapopulation. *BMC Biology*, *6*, 1–11.

Walker, B. H., Steffen, W. L., & Langridge, J. (1999). Interactive and integrated effects of global change on terrestrial ecosystems. *The terrestrial biosphere and global change: implications for natural and managed ecosystems*, 329–375.

Wang, X. L., Zwiers, F. W., & Swail, V. R. (2004). North Atlantic ocean wave climate change scenarios for the twenty-first century. *Journal of climate*, *17*(12), 2368–2383.

West, J. M., & Salm, R. V. (2003). Resistance and resilience to coral bleaching: Implications for coral reef conservation and management. *Conservation Biology*, *17*(4), 956–967.

6 Significance of Plant Functional Traits for Ecosystem Regulation and Climate Change Mitigation

Komal Rani, Gargi V. Navpute, Shambhavi Yadav, Abhishek Kumar, Yashaswi Singh, Ajay Thakur, and Amitava Roy

6.1 INTRODUCTION

Climate change is a predominantly anthropogenic activity that drastically impacts the environmental processes on the Earth. The last century witnessed extreme changes in environmental factors because of global warming, increasing CO_2 in the atmosphere, and changing climate conditions (Wang & Gu, 2021). Extreme weather events like heat waves, storms, floods, droughts, and wildfires have been increasing and are mainly caused by climatic fluctuations and atmospheric warming. Forests are crucial for controlling the climatic conditions of the Earth as they alter the dynamics of carbon, water, and nutrients. Understanding the roles of forests, especially in the context of plant functional traits (PFTs), sheds light on how individual species aid the ecological processes that moderate climate change. PFTs are critical indicators of how plants interplay with their environment, regulate ecosystem functions, and respond to changes (Figure 6.1). These are broadly classified into morphological, physiological, phenological, and reproductive traits. Morphological traits like the area, size, shape, thickness, and density of leaves, roots, and stems. Phenological traits like the emergence, fall, and duration of leaves and flowers greatly influence seasonal adaptations related to photosynthesis and temperature responses—reproductive traits like frequency and duration of flowering and fruit development from flowers. Dormancy periods reflect growth and rest patterns in response to environmental cues to ensure survival in seasonal climates (Singh et al., 2024).

PFTs are essential as they regulate interactions at different ecological scales. For example, leaf area influences the photosynthetic capacity and water-use efficiency (WUE) of plants, which are crucial when the water supply fluctuates. Similarly, traits like wood density can affect the capability of plants to store carbon and cope with mechanical stress from wind; denser wood is frequently associated with slower growth but higher long-term carbon retention (Poorter & Bongers, 2006). PFTs are

DOI: 10.1201/9781003588580-6

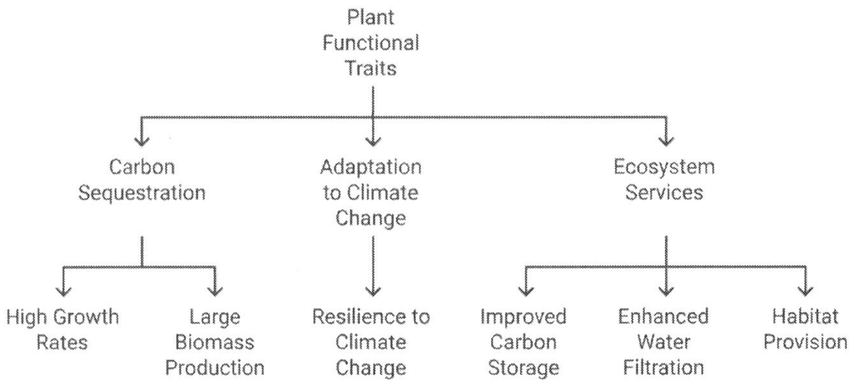

FIGURE 6.1 The role and functions of plant functional traits.

crucial in ecosystem regulation, particularly in light of climate change. The functional array of plant species determines how well ecosystems can adjust to these changes. This chapter emphasises the essentiality of these traits in ecosystem regulation, carbon sequestration, adaptation to, and mitigation of climate change. It explores the role of PFTs in anticipating how plant communities will react to shifting environmental conditions and describing the behaviour of individual plants. Understanding these functional traits (FTs) in detail will help us devise ecosystem-based mitigation measures that will be more successful in reducing the effects of climate change.

6.1.1 Plant Functional Traits and Ecosystem Regulation

FTs are critical in regulating ecosystems since they directly affect the nutrients, hydrological cycling, and climate modulation processes that work together towards stability and resilience. Influencing ecosystem primary productivity, nutrient cycling, and biomass allocation, these ecological and physiological traits play a very significant role in ecosystem functioning.

6.1.1.1 Role in Primary Productivity

Primary productivity can be defined as the rate at which plants and other photosynthetic organisms transform solar energy into chemical energy in the form of organic molecules (Barber, 2007). In environments where plants and algae are the leading producers, it forms the basis of the trophic architecture, producing and transferring the biomass and energy for higher trophic levels (Hu et al., 2024). Both the photosynthetic efficiency and the capacity of plants to absorb light, water, and nutrients determine the efficiency of primary productivity. Leaf area index (LAI), stomatal conductance, and chlorophyll concentration are important functional characteristics that impact photosynthetic rates and, in turn, biomass output. DuBois et al. (2021) reported that these characteristics influence interspecific variations in productivity, niche differentiation, and ecological stability in addition to the capacity of plants to perform photosynthesis.

The surface area of leaves per unit ground area, or LAI, is another crucial factor influencing the plant's capacity for photosynthetic activity (Li et al., 2018). Higher LAI values are linked with better light capture, which boosts photosynthesis and biomass development, assuming that other resources, such as water or nutrients, are not scarce (Paul et al., 2021). Asner et al. (2015) demonstrated the significance of LAI in promoting ecosystem production by revealing that tropical forests with high LAI had significantly greater primary productivity compared to habitats with lower leaf areas, such as temperate grasslands (Crous, 2008). However, increased leaf area may eventually lead to self-shading, reducing the overall efficiency of photosynthesis per leaf. This trade-off shows the need to optimise light-absorption efficiency to balance light-use efficiency (LUE).

Another important component in influencing a plant's ability to photosynthesise is its chlorophyll content. Plant species have different carbon fixation rates, and plants with higher chlorophyll concentrations often exhibit higher rates (Loreau & De Mazancourt, 2013). Evolutionary reactions to specific environmental circumstances may result in variations in the quantity of chlorophyll. Understory species in tropical forests are accustomed to low light conditions and generally have greater chlorophyll concentrations per unit leaf area than plants in open environments (Gómez-Sagasti et al., 2023). This enables them to gather as much light as possible in darkened areas. However, due to the quantity of light, species in highlight situations could have greater LUE but lower chlorophyll concentrations (Ustin & Jacquemoud, 2020).

Stomatal conductance balances the rate at which water vapour exits the leaf through stomata and carbon dioxide enters, influencing photosynthetic rates. WUE measures the direct effect of stomatal management on the equilibrium between water loss and carbon uptake (Hu et al., 2024). High stomatal conductance plants may assimilate more carbon dioxide for photosynthesis but drop more water through transpiration. Therefore, plants must balance carbon gain and water conservation, particularly in water-limited places (Shiferaw et al., 2020). Common characteristics of drought-adapted species allow them to maintain production in water-stressed conditions, including reduced stomatal conductance and improved WUE. A meta-analysis by Sun et al. (2020) elaborated that plants with higher WUE usually win out in dry and semi-arid environments where primary productivity is constrained by water availability. Conversely, mesic environments enable organisms to maintain greater stomatal conductance, maximising carbon intake and photosynthesis. Reduced stomatal conductance and enhanced WUE are traits of drought-adapted plants that enable them to continue producing underwater constraints.

Functional traits like LAI, chlorophyll content, and stomatal conductance, which aid in resource partitioning and niche differentiation in ecosystems, vary widely among plant species. Species can survive because niche differences reduce direct competition for resources like water, soil, nutrients, and light. To maximise light intake, species above the canopy of forest ecosystems, for example, may have large leaf areas and high stomatal conductance. In contrast, species in the understory may rely more on low light saturation points and greater chlorophyll concentrations (Loreau et al., 2001). It is empirically supported that plant species diversity and FTs variation positively affect ecosystem productivity. Tilman et al. (2001) demonstrated that plant communities with a broader range exhibit better primary productivity than monocultures. Species may use a more comprehensive range of resources in a varied environment, reducing competition and improving overall ecosystem performance. The ability of PFTs to sustain productivity under varied environmental circumstances makes ecosystems more stable (Hooper et al., 2005).

6.1.1.2 Influence on Biomass Production and Allocation between Shoots and Roots

PFTs control how efficiently plants absorb, distribute, and use essential resources like water and minerals. They affect the composition and efficiency of entire ecosystems and influence the performance of individual plants. The primary productivity and, consequently, the movement of matter and energy across ecosystems are determined by the interactions of resource acquisition tactics, biomass allocation patterns, and ecosystem processes. A major factor in biomass production is the capacity of plants to take up nutrients and water from their surroundings. PFTs affecting root depth and structure, nutrient absorption efficiency, and symbiotic relationships with soil microbes dominate how well plants absorb resources.

In regions where surface water is scarce, and seasonal fluctuations are common, plants with deeper root systems efficiently extract water and nutrients from deeper soil layers. Because of this, plants in dry and semi-arid regions develop deep roots that enable them to store biomass even under challenging circumstances (Jackson et al., 1996). Mediterranean environments increase the biomass production of *Quercus ilex* (Canadell et al., 1996). Similarly, root systems with many branches or those with few branches increase the surface area available for nutrient absorption, particularly for nitrogen and phosphorus, which are often scarce in many environments (Lynch, 2019). Plants with high nitrogen usage efficiency in nitrogen-poor soils can generate more biomass per unit of nitrogen absorbed, providing a competitive edge (Vierling, 1991). Mycorrhizal-associated plants benefit from the vast hyphal networks of fungi, which boost phosphorus uptake and effective root surface area (Kafle et al., 2019).

Another significant predictor of ecosystem structure and productivity is the distribution of biomass between roots (belowground) and shoots (aboveground) (Figure 6.2). Light is captured for photosynthesis by aboveground biomass, which includes leaves, stems, and reproductive organs. Greater allocation to shoot biomass, especially in the form of leaves, is often associated with higher light interception and photosynthetic capability in plants, which increases biomass output and carbon absorption (Khan et al., 2024). How much light a plant or a plant community intercepts depends largely on canopy structure, defined by LAI, height, and leaf arrangement (Loomis & Williams, 1969). Enough root biomass is required to promote water and nutrient

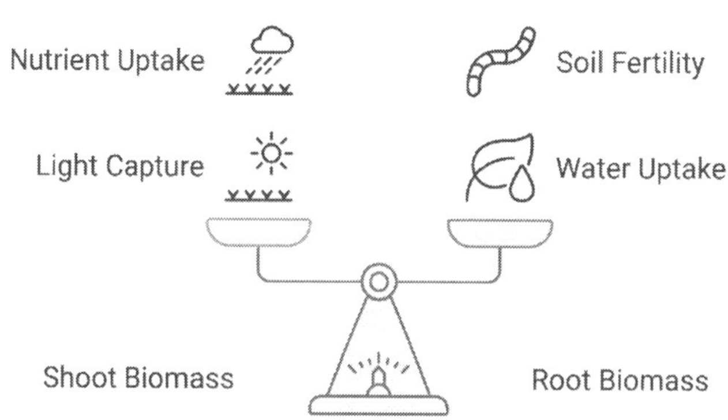

FIGURE 6.2 Balancing root and shoot biomass for ecosystem productivity.

absorption, although the investment in shoot biomass is frequently weighed against this. In nutrient-poor conditions, species devote a more significant percentage of biomass to roots to improve their ability to absorb nutrients and guarantee sustained productivity (Kagawa, 2022). Compared to those in more productive environments, grass species in nutrient-limited grasslands are excellent illustrations of this trade-off, often committing more biomass to roots than to shoots (Craine et al., 2005).

To acquire water and nutrition, plants require root biomass, and plants with well-developed root systems thrive efficiently when water or nutrients are scarce. According to Bardgett et al. (2014), the organic matter that roots contribute to the soil through root turnover and exudation affects the breakdown of organic matter and microbial activity. These systems are required to maintain soil fertility and long-term ecosystem production. However, this increased allocation to roots can come at the expense of shoot growth, potentially limiting the plant's ability to capture light and produce aboveground biomass. This trade-off between shoot and root allocation is a central component of plant ecological strategies and varies widely across species and ecosystems (Poorter et al., 2012). Because they devote more biomass to their roots than to their shoots, desert shrubs and certain perennial grasses in dry environments continue to be productive during times of water constraint. To optimise light absorption and photosynthetic rates, plants in humid regions, on the other hand, distribute more biomass to their shoots (McCarthy & Enquist, 2007). With an impact on both aboveground processes (canopy dynamics, light competition) and belowground processes (soil carbon storage, root-microbe interactions), this variation in biomass allocation patterns is a significant driver of ecosystem structure. By controlling the movement of matter and energy across ecosystems, the characteristics associated with biomass allocation and resource acquisition work together to control ecological processes. Plants with varying methods of acquiring resources and allocating biomass may live in ecosystems with high functional trait variety and diversification levels. This results in complementary resource utilisation and increased ecosystem production (Tilman et al., 2001). This process, known as niche complementarity, enables the effective utilisation of resources such as obtaining nutrients, water, and light from various soil levels or strata (Loreau et al., 2001).

Empirical studies in various ecosystems demonstrate that PFTs are essential to regulating biomass production (Table 6.1). Reich et al. (2001) observed that, in contrast, species with greater root allocation performed better in nutrient-poor conditions; those with higher nitrogen-use efficiency and more aboveground biomass allocation dominated high-nutrient settings. By enabling various plant species to obtain water and nutrients from various soil layers, lowering competition, and raising total biomass production, the coexistence of deep-rooted trees and shallow-rooted grasses in tropical savannas also improves ecosystem productivity (Sankaran et al., 2004). For ecosystem productivity and sustainability, it is crucial to comprehend the trade-offs between the allocation of shoot and root biomass (Lynch, 2019).

6.1.2 Role of PFTs in Nutrient Cycling

PFTs that control nutrient acquisition, storage, and the quality of litter returned to the soil significantly influence the effectiveness of nutrient cycling (Figure 6.3).

TABLE 6.1
Influence of Plant Functional Traits on Biomass Production and Resource Use Efficiency

Factor	Influence on Biomass Production	References
Light absorption efficiency	Larger-leafed plants, such as *Helianthus annuus*, absorb more light, which raises photosynthetic rates and increases biomass accumulation. These factors are particularly crucial for species suited to open environments.	Zhu et al. (2010a)
Photosynthetic efficiency	Higher chlorophyll content in species like *Zea mays* increases the rate of carbon fixation, thus boosting growth and biomass production, especially under nutrient-rich conditions.	Evans and Poorter (2001)
Water uptake efficiency	Plants with deep roots, like *Medicago sativa*, can extract water from deeper soil layers. This is crucial during dry spells and allows for consistent biomass production even when water is scarce.	Wu et al. (2021)
Water-use efficiency	In drought-stressed situations, plants with lower stomatal conductance, like *Triticum aestivum*, can maximise water use, improving growth and biomass accumulation.	Blum (2017)
Nutrient uptake efficiency	Species like *Pinus sylvestris* that form associations with arbuscular mycorrhizal fungi exhibit increased phosphorus uptake, significantly enhancing biomass under nutrient-poor conditions.	van Der Heijden et al. (2019)
Carbon allocation	A lower shoot-to-root ratio in *Glycine max* enhances nutrient and water uptake, which supports higher biomass production, especially under stress conditions like nutrient limitation.	Lopez et al. (2023)
Growth rate	High-specific leaf area plants, like *Arabidopsis thaliana*, exhibit quick leaf development and expansion, speeding up biomass accumulation. However, they may not be as resilient over the long term in nutrient-poor conditions.	Weraduwage et al. (2015)
Stress tolerance	Increased antioxidant activity (such as catalase and superoxide dismutase) in drought-tolerant species like *Olea europaea* lessens oxidative damage under stress, allowing biomass production to continue despite heat and drought.	Mittler (2002)

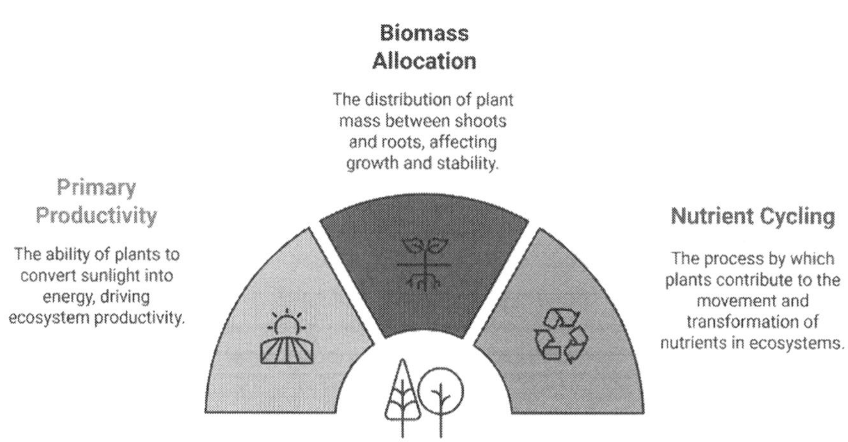

FIGURE 6.3 The role of plant functional traits in ecosystem regulation.

Microbial activity and plant characteristics regulate nutrient turnover and availability rates (Singh, 2024). Since nitrogen is a limiting resource for decomposer organisms, litter with a high nitrogen content often decomposes more quickly (Kanta et al., 2024). Fast-growing species typically create nitrogen-rich litter, which increases microbial activity and promotes faster breakdown and nutrient turnover. In contrast, leaf litter with high C:N ratios disintegrates more slowly (Zhou et al., 2018). Physical characteristics like thickness, surface area, and toughness influence leaf litter decomposition rates. By making litter less accessible to decomposers, leaf toughness acts as a physical barrier and is correlated with high lignin and cellulose content (Giweta, 2020). Rapid nitrogen release from rapidly decomposing litter supports high primary production and rapid nutrient turnover in environments like grasslands and tropical forests (Li et al., 2023). However, slow-decomposing litter, common in temperate and boreal forests, slows down the cycling of nutrients and encourages the buildup of organic matter in soils, which can help store carbon over time. This demonstrates how the functional characteristics of plants control ecosystem processes, which has significant ramifications for both local biodiversity and global carbon dynamics (Allen et al., 2020).

Alkaloids, phenolics, and terpenoids are secondary metabolites that affect ecosystem-level processes, including carbon and nutrient cycle and offer defence mechanisms against biotic stressors (Bardgett & Wardle, 2010). Tannins create complexes impervious to microbial breakdown by binding with proteins and other organic materials in the litter (Adamczyk et al., 2017). By delaying the pace at which nutrients are released from leaf litter back into the soil, this chemical restriction of decomposition effectively slows down the ecosystem's nutrient cycle. Ecosystems dominated by species with low tannin litter, such as grasses or legumes, show poorer nutrient availability and slower rates of decomposition than *Quercus acutissima* and *Quercus glauca* and conifers with high tannin concentrations in their leaf litter (Park et al., 2021). Because carbon is trapped in the resistant organic matter for extended periods, the slower decomposition rates linked to high-tannin species may result in higher soil carbon sequestration, which helps long-term carbon storage. This demonstrates the dual function of secondary metabolites, which control breakdown and nutrient cycling to influence ecosystem processes and act as a defence mechanism against herbivores (Shah et al., 2024). However, in some ecological circumstances, this characteristic can also be advantageous. Large-scale secondary metabolite production by plants may reduce nutrient losses from the ecosystem in nutrient-poor or stressful conditions by slowing the pace at which nutrients are released from decaying plant matter. Plants preserve a competitive edge in low-nutrient situations, where slower nutrient cycling might lessen competition from species that need high nutrient availability for development (Coq et al., 2022). Thus, secondary metabolites shape plant competitiveness and ecosystem functioning by acting as a regulatory mechanism in nutrient cycling.

Plants can impact the cycling of nutrients through their symbiotic associations with soil organisms, especially nitrogen-fixing bacteria and mycorrhizal fungi. Most herbaceous plants and certain trees are home to arbuscular mycorrhizal (AM) fungus, which improves phosphorus absorption by affecting nutrient flow dynamics (Richardson et al., 2009). However, compared to AM fungi, ectomycorrhizal (ECM) fungi, found on many temperate and boreal trees, slow down nutrient cycling by encouraging the breakdown of organic materials and releasing nutrients more slowly (Yan et al., 2024). The observed variance in nutrient dynamics between ecosystems dominated by various mycorrhizal species is partly explained by the differential in nutrient cycling rates between AM and ECM fungi (Tunlid et al., 2022). The functional characteristics of the prevalent plant species profoundly affect the pace of nitrogen cycling in ecosystems. Fast-growing species produce high-quality litter that breaks down quickly, which speeds up the nutrient cycle. These species usually predominate in early-successional habitats, where high production levels depend on quick nutrient turnover (Mayer et al., 2023).

6.1.3 Plant Functional Traits and Carbon Sequestration

A key mechanism for sequestering carbon is photosynthesis. Functional attributes like stomatal density, leaf thickness, and photosynthetic pathways (C3 and C4) affect how well a plant absorbs and uses CO_2. C4 plants have evolved to photosynthesise more efficiently under high temperatures and low CO_2 concentrations, making them highly effective at sequestering carbon in certain ecosystems. By optimising the ecological traits that maximise photosynthesis, we can enhance carbon sequestration at a global scale (Singh et al., 2022). The carbon cycle depends on the breakdown of plant matter because it influences the quantity and duration of time that carbon is retained in the atmosphere. Functional attributes of the leaves and roots, including lignin content, cellulose concentration, and carbon-to-nitrogen ratios, influence how quickly plant matter breaks down. Carbon is stored in soils for extended periods by plants with characteristics that generate slow-decomposing litter (such as high lignin or low nitrogen content), which helps with long-term carbon sequestration (Ghosh & Mishra, 2019). Long-term carbon storage depends on characteristics that make plant biomass stable and long-lived. Compared to annual plants, which may release stored carbon during decomposition, perennial plants are better suited for carbon sequestration because of their characteristics that enable longer lifespans and continued biomass production (Cadisch & Giller, 2001). In order to reduce the likelihood of carbon being released back into the atmosphere, deep-rooted organisms move carbon deeper into the soil profile. For reforestation and afforestation operations, these characteristics highlight the significance of choosing species with significant carbon storage capacity (Aponte et al., 2012).

6.1.4 Plant Functional Traits and Their Role in Climate Change Adaptation and Mitigation

Deep-rooted organisms contribute more to sequestering carbon as they reside deep inside the soil and thus move carbon deeper into the soil profile, where it is less likely to be released back into the atmosphere. These characteristics highlight the importance of choosing species with a high capacity to assimilate carbon for afforestation and reforestation initiatives (Nair et al., 2015). Understanding these characteristics makes selecting suitable species for carbon sequestration, ecosystem restoration, and climate change mitigation possible (Shahzad et al., 2018).

6.1.5 Traits that Enhance Resilience to Rising Temperatures and Drought Stress

Climate change is expected to increase the frequency and severity of extreme weather phenomena like heat waves and droughts. Functional characteristics that increase resilience are necessary for plants to respond to these challenges. Specific Leaf Area (SLA), which calculates the leaf area about its dry mass, is a crucial attribute in this context. Plants with lower SLA often have more robust, thicker leaves that withstand high temperatures and water stress better. Because of their more frugal water consumption, these species minimise transpiration and lower the chance of water loss during dry spells. Another critical trait is root system architecture (Figure 6.4). Species with large, deep root systems benefit drought-prone areas by efficiently reaching water in deeper soil layers (Comas et al., 2013). Because their roots may reach groundwater reserves during extended dry periods, species with deep taproots, like *Acacia* sp., are ideally suited to arid regions, guaranteeing their survival and ongoing carbon absorption (Bot & Benites, 2005). Reduced stomatal conductance and improved photosynthetic efficiency lead to high WUE, enabling plants to grow and produce even in water-limited environments (Zhu et al., 2010b). Thermotolerance is a crucial characteristic of plants that helps them be resilient in warmer climates. This characteristic is often connected to the physiological and biochemical defence mechanism and involves synthesising heat-shock proteins that shield cellular components of the plants from intense heat (Vierling, 1991). Furthermore, plants with traits that allow them to regulate their internal temperatures, such as reflective leaf surfaces or changes in leaf orientation to reduce heat absorption, are better adapted to survive in hotter environments (Ustin & Jacquemoud, 2020).

6.1.6 Selection of Species and Traits for Ecosystem Restoration and Carbon Sequestration

Ecosystem restoration efforts aim to reverse the degradation of ecosystems and enhance carbon storage, depending on the selection of plant species with functional

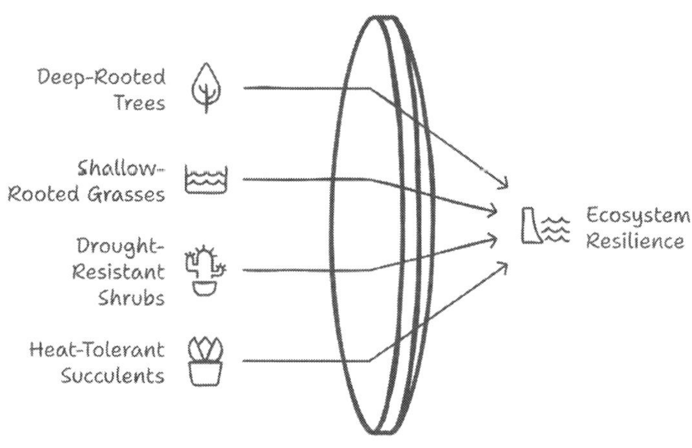

FIGURE 6.4 Ecosystem resilience to climate change.

attributes suited to the explicit environmental conditions and restoration objectives (Allen et al., 2020). The effectiveness of restoration efforts and their role in mitigating the effects of climate change depends heavily on characteristics like growth rate, stress tolerance, and nutrient cycle efficiency (Carlucci et al., 2020). High biomass-producing species are frequently selected for carbon sequestration and restoration because they quickly store carbon in their tissues. Fast-growing pioneer species like *Populus* and *Salix* are used to establish canopy cover quickly and initiate soil carbon sequestration processes. To ensure the long-term sustainability of afforestation endeavours, drought-tolerant species with deep root systems and high WUE are given priority in arid and semi-arid locations (Poorter & Bongers, 2006). Nitrogen-fixing species also play a decisive role in restoration projects, especially in degraded landscapes with poor soil fertility. These species enhance the soil nutrient levels through their symbiotic relationships with nitrogen-fixing bacteria, facilitate the establishment of other plant species, and promote ecosystem recovery (Nizamani et al., 2024). Leguminous trees like *Acacia* and *Prosopis* sp. are commonly used in dryland restoration to improve soil fertility and water retention (Maiti et al., 2021).

Plant species with higher wood density are prioritised for carbon sequestration because they accumulate more carbon per unit volume than species with lower wood density. High wood density is associated with greater structural integrity and longer-lasting carbon storage in living biomass and deadwood, making these species important for long-term carbon sequestration in forest ecosystems (Frank et al., 2015). Evergreen species, which maintain their leaves year-round, also contribute to continuous carbon uptake, especially in temperate and tropical regions where photosynthesis can occur throughout the year (Pawson et al., 2013). Tree species with rapid growth rates and high wood density are particularly effective in sequestering carbon because they accumulate biomass rapidly and assimilate carbon over long periods, like *Pinus radiata* and *Eucalyptus globulus* (Paul et al., 2021).

6.1.7 Future Directions in Terms of Research

FTs play a pivotal role in the recycling of nutrients through sequestration, increasing nutrient availability to the trees. Therefore, these traits provide many ecosystem services and are massively crucial for proficient ecosystem management and eco-restoration. Further research on understanding the physiological and genetic basis of such traits may prove very useful for identifying and characterising key genes and metabolites responsible for driving the ecosystem services provision. Genetic characterisation of these traits thus can effectively be utilised to strategise individual trait-based conservation efforts. The knowledge of the exact relationship between FTs and implicative ecosystem services will ensure the sustainability of such ecosystem services in future scenarios of climatic abnormalities.

With the advancements in functional genomics research, it is now possible to identify and characterise the genes associated with the traits. Analysing their expression patterns and regulation will be necessary for deciphering their response to a particular stress or changing environment. Further, it will be a guide for designing plant traits as a tool for combating and developing resilience against the threats of climate change. Such plants will perform much better under variable soil and climatic regimes and will aid in achieving sustainability over more extended periods. Analysis of metabolic pathways in response to stress adaptation and resistance will also be very useful for the identification and role of essential nutrients and enzymes in response to a particular stress trigger and selection of species and plant types which can effectively cope with such pressure and achieve stable performance in variable ecological extremities.

Remote sensing-based approaches can be implied to dissect the role of PFTs in ecosystem services. The developments of digital sensors and artificial intelligence-based software have accelerated imagery analysis in the last few decades. It has opened up a rapid, reproducible, and reliable methodology for detecting vegetation health and the effect of biotic and abiotic factors on various species groups. It is now possible to illustrate the differences between varieties of ecotypes with a species and further study the individual plant trait contribution to the state of vegetation health. With the help of airborne Light Detection and Ranging (LiDAR) analysis of three-dimensional canopy architecture, it is now possible to track and characterise differences even at the level of leaf attributes. However, further research must be conducted to track the entire vegetation stand and accurately provide precise results. Combining LiDAR with multi-position imaging spectroscopy will be very helpful for canopy-level studies. It can deliver newer discernment in analysing spatial patterns of the physiological functioning of such traits.

6.2 CONCLUSION

PFTs have become a powerful tool for understanding the mechanisms underlying plant community dynamics and ecosystem functioning. Specific leaf area, leaf thickness, stomatal conductance, and others are traits that respond to environmental factors such as temperature, light, soil composition, and water availability. These traits reflect a plant's adaptation to its environment and influence ecosystem processes such as nutrient cycling and productivity. This chapter discusses the role of PFTs in predicting species coexistence within a community and strengthening the community structure. Community-weighted means of functional traits effectively predict ecosystem functions such as plant primary productivity, decomposition rates, and photosynthesis to respiration ratio. Variability of these traits within a species is also important since it reflects the competence of plants to adjust to environmental change. Scaling up to the ecosystem level, it is possible to use traits-based research in plants in developing integrative functions to elucidate the response of ecosystems and biomes to climatic change. Hence, including PFTs in environmental planning and decision-making policy suggestions will need further understanding of the traits-based approaches in mitigating climate change.

REFERENCES

Adamczyk, B., Simon, J., Kitunen, V., Adamczyk, S., & Smolander, A. (2017). Tannins and their complex interaction with different organic nitrogen compounds and enzymes: Old paradigms versus recent advances. *ChemistryOpen*, *6*(5), 610–614.

Allen, J., Maunoury-Danger, F., Felten, V., Danger, M., Legout, A., & Guérold, F. (2020). Liming of acidified forests changes leaf litter traits but does not improve leaf litter decomposability in forest streams. *Forest Ecology and Management*, *475*, 118431.

Aponte, C., García, L. V., & Maranon, T. (2012). Tree species effect on litter decomposition and nutrient release in Mediterranean oak forests changes over time. *Ecosystems, 15*, 1204–1218.

Asner, G. P., Ustin, S. L., Townsend, P. A., Martin, R. E., & Chadwick, K. D. (2015). Forest biophysical and biochemical properties from Hyperspectral and LiDAR Remote Sensing. In *land resources monitoring, modeling, and mapping with remote sensing* (p. 20). 1st Ed., CRC Press.

Barber, J. (2007). Biological solar energy. *Philosophical Transactions of the Royal Society A: Mathematical, Physical and Engineering Sciences, 365*(1853), 1007–1023.

Bardgett, R. D., Mommer, L., & De Vries, F. T. (2014). Going underground: Root traits as drivers of ecosystem processes. *Trends in Ecology & Evolution, 29*(12), 692–699.

Bardgett, R. D., & Wardle, D. A. (2010). *Aboveground-belowground linkages: Biotic interactions, ecosystem processes, and global change.* Oxford University Press.

Blum, A. (2017). Osmotic adjustment is a prime drought stress adaptive engine in support of plant production. *Plant, Cell & Environment, 40*(1), 4–10.

Bot, A., & Benites, J. (2005). *The importance of soil organic matter: Key to drought-resistant soil and sustained food production (no. 80).* Food & Agriculture Org.

Cadisch, G., & Giller, K. (2001). Soil organic matter management: The roles of residue quality in C sequestration and N supply. In *Sustainable management of soil organic matter* (pp. 97–219). CABI.

Canadell, J., Jackson, R. B., Ehleringer, J. B., Mooney, H. A., Sala, O. E., & Schulze, E. D. (1996). Maximum rooting depth of vegetation types at the global scale. *Oecologia, 108*, 583–595.

Carlucci, M. B., Brancalion, P. H., Rodrigues, R. R., Loyola, R., & Cianciaruso, M. V. (2020). Functional traits and ecosystem services in ecological restoration. *Restoration Ecology, 28*(6), 1372–1383.

Comas, L. H., Becker, S. R., Cruz, V. M. V., Byrne, P. F., & Dierig, D. A. (2013). Root traits contributing to plant productivity under drought. *Frontiers in Plant Science, 4*, 442.

Coq, S., Cárdenas, R. E., Mousain, D., Selosse, M. A., Richard, F., & Hättenschwiler, S. (2022). Acquisition of nitrogen from tannin protein complexes in ectomycorrhizal pine seedlings. *Pedobiologia, 93*, 150817.

Craine, J. M., Lee, W. G., Bond, W. J., Williams, R. J., & Johnson, L. C. (2005). Environmental constraints on a global relationship among leaf and root traits of grasses. *Ecology, 86*(1), 12–19.

Crous, K. Y. (2008). *Nitrogen availability alters species photosynthetic responses to elevated atmospheric carbon dioxide* [Doctoral dissertation, University of Michigan].

DuBois, K., Williams, S. L., & Stachowicz, J. J. (2021). Experimental warming enhances effects of eelgrass genetic diversity via temperature-induced niche differentiation. *Estuaries and Coasts, 44*(2), 545–557.

Evans, J., & Poorter, H. J. P. C. (2001). Photosynthetic acclimation of plants to growth irradiance: The relative importance of specific leaf area and nitrogen partitioning in maximizing carbon gain. *Plant, Cell & Environment, 24*(8), 755–767.

Frank, D., Reichstein, M., Bahn, M., Thonicke, K., Frank, D., Mahecha, M.D., Smith, P., van der Velde, M., Vicca, S., Babst, F., Beer, C., Buchmann, N., Canadell, J.G., Ciais, P., Cramer, W., Ibrom, A., Miglietta, F., Poulter, B., Rammig, A. … Zscheischler, J. (2015). Effects of climate extremes on the terrestrial carbon cycle: Concepts, processes and potential future impacts. *Global Change Biology, 21*(8), 2861–2880.

Ghosh, B. N., & Mishra, P. K. (2019). Effects of productivity and soil carbon storage in mixed forests. In Ghosh, P. K., Mahanta, S. K., Mandal, D., Mandal, B., & Ramakrishnan, S. (Eds.), *Carbon management in tropical and sub-tropical terrestrial systems* (pp. 349–361). Springer Singapore.

Giweta, M. (2020). Role of litter production and its decomposition, and factors affecting the processes in A tropical forest ecosystem: A review. *Journal of Ecology and Environment, 44*(1), 11.

Gómez-Sagasti, M. T., López-Pozo, M., Artetxe, U., Becerril, J. M., Hernández, A., García-Plazaola, J. I., & Esteban, R. (2023). Carotenoids and their derivatives: A "Swiss Army knife-like" multifunctional tool for fine-tuning plant-environment interactions. *Environmental and Experimental Botany*, *207*, 105229.

Hooper, D.U., Chapin, F.S. III, Ewel, J.J., Hector, A., Inchausti, P., Lavorel, S., Lawton, J. H., Lodge, D. M., Loreau, M., Naeem, S., Schmid, B., Setälä, H., Symstad, A. J., Vandermeer, J. and Wardle, D. A. (2005). Effects of biodiversity on ecosystem functioning: A consensus of current knowledge. *Ecological Monographs*, *75*(1), 3–35. https://doi.org/10.1890/04-0922

Hu, N., Bourdeau, P. E., & Hollander, J. (2024). Responses of marine trophic levels to the combined effects of ocean acidification and warming. *Nature Communications*, *15*(1), 3400.

Jackson, R. B., Canadell, J., Ehleringer, J. R., Mooney, H. A., Sala, O. E., & Schulze, E. D. (1996). A global analysis of root distributions for terrestrial biomes. *Oecologia*, *108*, 389–411.

Kafle, A., Cope, K. R., Raths, R., Krishna Yakha, J., Subramanian, S., Bücking, H., & Garcia, K. (2019). Harnessing soil microbes to improve plant phosphate efficiency in cropping systems. *Agronomy*, *9*(3), 127.

Kagawa, A. (2022). Foliar water uptake as a source of hydrogen and oxygen in plant biomass. *Tree Physiology*, *42*(11), 2153–2173.

Kanta, C., Kumar, A., Chauhan, A., Singh, H., & Sharma, I. P. (2024). The interplay between plant functional traits and climate change. In *Plant functional traits for improving productivity* (pp. 41–58). Springer Nature Singapore.

Khan, N., Choi, S. H., Lee, C. H., Qu, M., & Jeon, J. S. (2024). Photosynthesis: Genetic strategies adopted to gain higher efficiency. *International Journal of Molecular Sciences*, *25*(16), 8933.

Li, Q., Lu, X., Wang, Y., Huang, X., Cox, P. M., & Luo, Y. (2018). Leaf area index identified as a major source of variability in modeled CO_2 fertilization. *Biogeosciences*, *15*(22), 6909–6925.

Li, W., Liu, Z., Zhao, J., Ma, L., Wu, J., Qi, J., & Wang, H. (2023). Leaf mechanical properties as potential predictors of leaf-litter decomposability. *Forestry Research*, *3*(1).

Loomis, R. S., & Williams, W. A. (1969). Productivity and the morphology of crop stands: Patterns with leaves. *Physiological aspects of crop yield* (pp. 27–47). ASA, CSSA, and SSSA Books, American Society of Agronomy. https://doi.org/10.2135/1969.physiologicalaspects.c3

Lopez, G., Ahmadi, S. H., Amelung, W., Athmann, M., Ewert, F., Gaiser, T., Gocke, M. I., Kautz, T., Postma, J., Rachmilevitch, S., Schaaf, G., Schnepf, A., Stoschus, A., Watt, M., Yu, P., Seidel, S. J. (2023). Nutrient deficiency effects on root architecture and root-to-shoot ratio in arable crops. *Frontiers in Plant Science*, *13*, 1067498. https://doi.org/10.3389/fpls.2022.1067498

Loreau, M., & De Mazancourt, C. (2013). Biodiversity and ecosystem stability: A synthesis of underlying mechanisms. *Ecology Letters*, *16*, 106–115.

Loreau, M., Naeem, S., Inchausti, P., Bengtsson, J., Grime, J. P., Hector, A., Hooper, D. U., Huston, M. A., Raffaelli, D., Schmid, B., Tilman, D. and Wardle, D. A. (2001). Biodiversity and ecosystem functioning: Current knowledge and future challenges. *Science*, 294(5543), 804–808. https://doi.org/10.1126/science.1064088

Lynch, J. P. (2019). Root phenotypes for improved nutrient capture: An underexploited opportunity for global agriculture. *New Phytologist*, *223*(2), 548–564.

Maiti, S. K., Bandyopadhyay, S., & Mukhopadhyay, S. (2021). Importance of selection of plant species for successful ecological restoration program in coal mine degraded land. In *Phytorestoration of abandoned mining and oil drilling sites* (pp. 325–357). Elsevier.

Mayer, M., Matthews, B., Sandén, H., Katzensteiner, K., Hagedorn, F., Gorfer, M., Berger, H., Berger, T. W., Godbold, D. L. and Rewald, B. (2023). Soil fertility determines whether ectomycorrhizal fungi accelerate or decelerate decomposition in a temperate forest. *The New Phytologist*, *239*(1), 325–339. https://doi.org/10.1111/nph.18811

McCarthy, M. C., & Enquist, B. J. (2007). Consistency between an allometric approach and optimal partitioning theory in global patterns of plant biomass allocation. *Functional Ecology*, *21*(4): 713–720.

Mittler, R. (2002). Oxidative stress, antioxidants and stress tolerance. *Trends in Plant Science*, *7*(9), 405–410.

Nair, R., Mehta, C. R., & Sharma, S. (2015). Carbon sequestration in soils – A review. *Agricultural Reviews*, *36*(2), 81–99.

Nizamani, M.M., Tarafder, E., Das, D., Hughes, A.C., Qureshi, S., Acharya, K., Wang, Y., Zhang, Q. & Zhang, Z.G. (2024). Microbial biodiversity and plant functional trait interactions in multifunctional ecosystems. *Applied Soil Ecology*, *201*, 105515.

Park, B. B., Han, S. H., Hernandez, J. O., An, J. Y., Youn, W. B., Choi, H. S., & Jung, S. (2021). Leaf litter decomposition of deciduous *Quercus acutissima* Carruth. and evergreen *Quercus glauca* Thunb. in an inter-site experiment in three contrasting temperate forest stands in South Korea. *Annals of Forest Science*, *78*, 1–11.

Paul, M., Rajib, A., Negahban-Azar, M., Shirmohammadi, A., & Srivastava, P. (2021). Improved agricultural water management in data-scarce semi-arid watersheds: Value of integrating remotely sensed leaf area index in hydrological modeling. *Science of the Total Environment*, *791*, 148177.

Pawson, S. M., Brin, A., Brockerhoff, E. G., Lamb, D., Payn, T. W., Paquette, A., & Parrotta, J. A. (2013). Plantation forests, climate change and biodiversity. *Biodiversity and Conservation*, *22*, 1203–1227.

Poorter, H., Niklas, K. J., Reich, P. B., Oleksyn, J., Poot, P., & Mommer, L. (2012). Biomass allocation to leaves, stems and roots: Meta-analyses of interspecific variation and environmental control. *New Phytologist*, *193*(1), 30–50.

Poorter, L., & Bongers, F. (2006). Leaf traits are good predictors of plant performance across 53 rain forest species. *Ecology*, *87*(7), 1733–1743.

Reich, P.B., Knops, J., Tilman, D., Craine, J., Ellsworth, D., Tjoelker, M., Lee, T., Wedin, D., Naeem, S., Bahauddin, D., Hendrey, G., Jose, S., Wrage, K., Goth, J. and Bengtson, W. (2001). Plant diversity enhances ecosystem responses to elevated CO_2 and nitrogen deposition. *Nature*, *410*(6830): 809–810. https://doi.org/10.1038/35071062

Richardson, A.E., Barea, J.M., McNeill, A.M. and Prigent-Combaret, C. (2009). Acquisition of phosphorus and nitrogen in the rhizosphere and plant growth promotion by microorganisms. *Plant and Soil*, *321*(1–2): 305–339. https://doi.org/10.1007/s11104-009-9895-2

Sankaran, M., Ratnam, J., & Hanan, N. P. (2004). Tree–grass coexistence in savannas revisited–insights from an examination of assumptions and mechanisms invoked in existing models. *Ecology Letters*, *7*(6), 480–490.

Shah, J.A., Liu, W., Ullah, S., Duan, H., Shen, F., Liao, Y., Huang, G. and Wu, J. (2024). Linkages among leaf nutrient concentration, resorption efficiency, litter decomposition and their stoichiometry to canopy nitrogen addition and understory removal in subtropical plantation. *Ecological Processes, 13*(1), 27. https://doi.org/10.1186/s13717-024-00454-4

Shahzad, T., Rashid, M.I., Maire, V., Barot, S., Perveen, N., Alvarez, G., Mougin, C., & Fontaine, S. (2018). Root penetration in deep soil layers stimulates mineralization of millennia-old organic carbon. *Soil Biology and Biochemistry*, *124*, 150–160.

Shiferaw, H., Gebremedhin, A., Solomon, N., Taye, G., & Zenebe, A. (2020). Comparative study on spatial variability of topsoil moisture content and carbon stocks as influenced by land use and soil and water conservation structures in the semi-arid Ethiopian highlands. *Physical Geography*, *41*(5), 389–406.

Singh, H. (2024). Exploring adaptive modulation in plant functional traits and its impact on the productivity of Acacia auriculiformis under CO_2-enriched environment. *Industrial Crops and Products*, *210*, 118186.

Singh, H., Kumar, N., & Kumar, A. (2024). Enhancing resource use efficiency in crops through plant functional traits. In *Plant functional traits for improving productivity* (pp. 97–117). Springer Nature Singapore.

Singh, P., Ghosh, A. K., Kumar, S., Kumar, M., & Sinha, P. K. (2022). Influence of input litter quality and quantity on carbon storage in post-mining forest soil after 14 years of reclamation. *Ecological Engineering, 178*, 106575.

Sun, Y., Wang, C., Chen, H. Y., & Ruan, H. (2020). Response of plants to water stress: A meta-analysis. *Frontiers in Plant Science, 11*, 978.

Tilman, D., Reich, P. B., Knops, J., Wedin, D., Mielke, T., & Lehman, C. (2001). Diversity and productivity in a long-term grassland experiment. *Science, 294*(5543), 843–845.

Tunlid, A., Floudas, D., Op De Beeck, M., Wang, T., & Persson, P. (2022). Decomposition of soil organic matter by ectomycorrhizal fungi: Mechanisms and consequences for organic nitrogen uptake and soil carbon stabilization. *Frontiers in Forests and Global Change, 5*, 934409.

Ustin, S. L., & Jacquemoud, S. (2020). How the optical properties of leaves modify the absorption and scattering of energy and enhance leaf functionality. In N. Cavender-Bares, J. A. Gamon, & P. A. Townsend (Eds.), *Remote sensing of plant biodiversity* (pp. 349–384). Springer.

van Der Heijden, G., Hinz, A., Didier, S., Nys, C., Dambrine, E., & Legout, A. (2019). Quantifying the uncertainty in modeled water drainage and nutrient leaching fluxes in forest ecosystems. *Ecosystems, 22*(3), 677–698.

Vierling, E. (1991). The roles of heat shock proteins in plants. *Annual Review of Plant Physiology and Plant Molecular Biology, 42*(1), 579–620.

Wang, Y. S., & Gu, J. D. (2021). Ecological responses, adaptation and mechanisms of mangrove wetland ecosystem to global climate change and anthropogenic activities. *International Biodeterioration & Biodegradation, 162*, 105248.

Weraduwage, S. M., Chen, J., Anozie, F. C., Morales, A., Weise, S. E., & Sharkey, T. D. (2015). The relationship between leaf area growth and biomass accumulation in *Arabidopsis thaliana*. *Frontiers in Plant Science, 6*, 167.

Wu, Z. Z., Zhang, Y. X., Yang, J. Y., Zhou, Y., & Wang, C. Q. (2021). Effect of vanadium on testa, seed germination, and subsequent seedling growth of alfalfa (*Medicago sativa* L.). *Journal of Plant Growth Regulation, 40*, 1566–1578.

Yan, G., Fan, C., Zheng, J., Liu, G., Yu, J., Guo, Z., Cao, W., Wang, L., Wang, W., Meng, Q., Zhang, J., Li, Y., Zheng, J., Cui, X., Wang, X., Xu, L., Sun, Y., Zhang, Z., Lü, X.-T.,et al. (2024). Forest carbon stocks increase with higher dominance of ectomycorrhizal trees in high latitude forests. *Nature Communications, 15*(1), 5959. https://doi.org/10.1038/s41467-024-41637-z

Zhou, G., Zhang, J., Qiu, X., Wei, F., & Xu, X. (2018). Decomposing litter and associated microbial activity responses to nitrogen deposition in two subtropical forests containing nitrogen-fixing or non-nitrogen-fixing tree species. *Scientific Reports, 8*(1), 12934.

Zhu, J. Y., Pan, X., & Zalesny, R. S. (2010b). Pretreatment of woody biomass for biofuel production: Energy efficiency, technologies, and recalcitrance. *Applied Microbiology and Biotechnology, 87*, 847–857.

Zhu, X. G., Long, S. P., & Ort, D. R. (2010a). Improving photosynthetic efficiency for greater yield. *Annual Review of Plant Biology, 61*(1), 235–261.

7 Carbon Mitigation Potential and Carbon Sequestration in Forest Ecosystem

Abhay Kumar, M. S. Malik, P. R. Oraon, Swati Shabnam, Sunita Kumari, Ram Gopal, Muneshwar Prasad, Dinesh Mahto, A. K. Chanchal, and Jaya Kumari

7.1 INTRODUCTION

Due to industrialization and urbanization, as humans continue to rely on fossil fuels, carbon dioxide (CO_2) will inevitably be generated and result in an increase of global warming gases (GWGs). However, their prospect is misted up because of the environmental and economic intimidation posed by probable climate shift, generally called the "greenhouse effect". Among all GWGs, the major contributor to greenhouse effect is CO_2. The increase in concentration of carbon dioxide by anthropogenic activities such as change in land use pattern, deforestation, industrialization, transportation and modern life style leads to global climate shift, which is one of the critical environmental challenges that the world is facing today. The atmospheric concentration of CO_2 is presently 422 ppm "NOAA (2024)". As one of the most potent GWGs, CO_2 has increased by approximately 43% since the Industrial Revolution and is expected to further increase to 60% in 2100 if the current trend continues (Kumar et al., 2016c).

Mitigation of CO_2 can be attained mainly by three possible ways: primarily by improving the energy efficiency of existing engine technology and proper fossil fuel utilization, secondly by CO_2 sequestration and lastly by facilitating the uses of unconventional fuels such as bio-hydrocarbon and biodiesel (Bharti et al., 2014a). Mitigation strategies that include capture and storage of CO_2 by biological means may reduce the impact of CO_2 emissions on environment. Global climate warming is one of the greatest threats to human health and animal and plant survival. Thus, atmospheric CO_2 mitigation has attracted much attention. The carbon capture and storage schemes have been developed (Boot-Handford et al., 2014; Von der Assen et al., 2013). Yet, biological CO_2 disposal is more beneficial for reducing side effects compared to chemical and engineering techniques, such as treatment techniques that intrinsically produce CO_2 and consume energy (Boot-Handford et al., 2014). One biological carbon capture scenario involves sequestering atmospheric CO_2 in terrestrial ecosystem carbon pools, including plant carbon pools and soil carbon storage with secure storage and long residence

 DOI: 10.1201/9781003588580-7

times, to mitigate global warming. Pan et al. (2011) suggested that a third of annual CO_2 emissions is captured and stored by terrestrial ecosystems. Both CO_2 and nitrogen deposition may interact with other climate factors such as warming and precipitation to affect tree physiology, carbon allocation and ecosystem carbon sequestration (Allen et al., 2010; Norby & Luo, 2004; Tian et al., 2012). Forest sequesters a huge amount of carbon in above-ground biomass, below-ground biomass and soil carbon, which plays an important role in regulating the carbon cycle. The forest sector is both a source and a sink for natural carbon, and various forest policies and programs have implications on the standing carbon stock in the forests. Forest sequesters a huge amount of carbon in above-ground biomass, below-ground biomass and soil carbon, which plays an important role in regulating the carbon cycle. The forest sector is both a source and a sink for natural carbon, and various forest policies and programs have implications on the standing carbon stock in the forests. While forest deforestation, degradation and destruction caused almost 17% of emissions of CO_2 and other greenhouse gases (GHGs). According to India's State of Forest Report (ISFR), data indicate that total forest carbon stocks increased during the period from 2011 to 2021, from 6663 Mt in 2011 to 7204 Mt in 2021. Total carbon stocks increased 541 Mt in ten years, which means an average annual increment of 54.1 Mt of carbon stock. The total carbon sequestration annually by the forest in India $54.1 \times 3.666 = 198.33$ Mt of CO_2. Globally, forests sequester almost 16 billion metric tonnes of CO_2 per year and currently hold 861 Gt of carbon in their stems, branches, leaves, roots and soils. From the present study it may be concluded that forests have very high potential for carbon sequestration on a long-term basis; it will be a milestone for achieving the goal of net zero carbon emissions, controlling or reducing climate change, global warming, GHGs, environment protection and other benefits.

7.1.1 Global Scenario of Climate Change

India is the world's fourth biggest emitter of carbon dioxide after China, the United States and EU27 and the total carbon emissions 2648 Mt of India in 2021 (Crippa et al., 2022). According to the report "State of Global Climate 2020" published by the World Meteorological Organization, in 2019, GHG concentrations reached new highs, with globally averaged mole fractions of carbon dioxide (CO_2) at 410.5 ppm, methane (CH_4) at 1877 ppb and nitrous oxide (N_2O) at 332 ppb, respectively, which are increases of 148%, 260% and 123%, respectively, from pre-industrial (before 1750) levels. The State of the World's Forests 2022 reconnoitres the potential of three forest pathways – halting deforestation and maintaining forests, restoring degraded lands and expanding agroforestry and sustainably using forests and building green value chains – for achieving green recovery and tackling environmental crises, including climate change and biodiversity loss.

Intergovernmental Panel on Climate Change (IPCC, 2013), which includes more than 1300 scientists from the United States and other countries, forecasts a temperature rise of 2.5–10°F over the next century. The extent of climate change effects on individual regions will vary over time and with the ability of different societal and environmental systems to mitigate or adapt to change. The IPCC predicts that increases in global mean temperature of less than 1.8–5.4°F (1–3°C) above 1990 levels will produce beneficial impacts in some regions and harmful ones in others.

Net annual costs will increase over time as global temperatures increase. However, in the last 100 years or so, fossil fuel burning, deforestation, changes in land use patterns and other factors have disturbed this natural balance, consequentially increasing atmospheric CO_2 concentration (Sheikh et al., 2014).

7.1.2 Impact of Climate Change

The Earth's climate is getting warmer, and the signs are everywhere. Rain patterns are changing, sea level is rising and snow and ice are melting sooner in the spring. As global temperatures continue to rise, we'll see more changes in our climate and our environment. These changes will affect people, animals and ecosystems in many ways.

Less rain can mean less water for some places, while too much rain can cause terrible flooding. More hot days can dry up crops and make people and animals sick. In some places, people will struggle to cope with a changing environment. In other places, people may be able to successfully prepare for these changes. The negative impacts of global climate change will be less severe overall if people reduce the amount of GHGs we're putting into the atmosphere and worse if we continue producing these gases at current or faster rates.

- Agriculture and food security
- Water stress and water insecurity
- Ecosystem and biodiversity
- Climate change and health
- Higher temperatures
- Changing rain and snow patterns
- More droughts
- Warmer oceans
- Rising sea level
- Wilder weather
- Increased ocean acidity
- Shrinking sea ice
- Melting glaciers
- Less snowpack
- Thawing permafrost

Global temperature rise: The planet's average surface temperature has risen about 2.0°F (1.1°C) since the late 19th century, a change driven largely by increased carbon dioxide and other human-made emissions into the atmosphere. Latest climate change scenario projections for Europe suggest that by 2100 temperatures will increase between about 2 and 5°C and by 1.4 and 5.8°C globally, depending on the quantity of future GHG emissions (Peters et al., 2013; UNFCCC, 2015).

Warming oceans: The oceans have absorbed much of this increased heat, with the top 700 m of ocean showing warming of 0.30°F since 1969.

Shrinking ice sheets: The Greenland and Antarctic ice sheets have decreased in mass. Data from NASA's Gravity Recovery and Climate Experiment show Greenland lost 150–250 km^3 (36–60 mi^3) of ice per year between 2002 and 2006, while Antarctica lost about 152 km^3 (36 mi^3) of ice between 2002 and 2005.

Glacial retreat: Glaciers are retreating almost everywhere around the world – including in the Alps, Himalayas, Andes, Rockies, Alaska and Africa.

Decreased snow cover: Satellite observations reveal that the amount of spring snow cover in the Northern Hemisphere has decreased over the past five decades and that the snow is melting earlier.

Sea level rise: Global sea level rose about 8 in. in the last century. The rate in the last two decades, however, is nearly double that of the last century.

Declining Arctic sea ice: Both the extent and thickness of Arctic sea ice have declined rapidly over the last several decades.

Extreme events: The number of record high temperature events in the United States has been increasing, while the number of record low temperature events has been decreasing since 1950. The United States has also witnessed increasing numbers of intense rainfall events.

Ocean acidification: Since the beginning of the Industrial Revolution, the acidity of surface ocean waters has increased by about 30%. This increase is the result of humans emitting more carbon dioxide into the atmosphere and hence more being absorbed into the oceans. The amount of carbon dioxide absorbed by the upper layer of the oceans is increasing by about 2 billion tonnes per year.

In December 2015, the Paris Agreement was approved by nearly 200 countries at the United Nations Framework Convention on Climate Change (UNFCCC) 21st Conference of the Parties (COP 21). This agreement aims to limit global mean temperature increase to well below 2°C above pre-industrial levels and to pursue efforts to limit it to 1.5°C (UNFCCC, 2015). At the same time, the IPCC has accepted an invitation to prepare a special report on the 1.5°C target in 2018.

7.1.3 Carbon Source and Sink

Carbon dioxide is a colourless gas formed during the combustion of any material containing carbon and an important GHG. Processes that release CO_2 to the atmosphere are called carbon "sources", while processes that absorb it are called carbon "sinks". Forests, soil, oceans, the atmosphere and fossil fuels are important stores of carbon. Carbon is constantly moving among these different stores, which act as either "sinks" or "sources". A sink absorbs more carbon than it gives off, while a source emits more than it absorbs. The amount of carbon in the atmosphere at any one time depends on the balance that exists between the sinks and sources. This system of sinks and sources operates all over the planet and is known as the carbon cycle.

Source is any process, activity or mechanism which releases a GHG, an aerosol or a precursor of a GHG or aerosol into the atmosphere. "Source" means any process or activity which releases a GHG, an aerosol or a precursor of a GHG into the atmosphere. Processes that release CO_2 to the atmosphere are called carbon sources. List the sources on the board –

i. *We breathe out CO_2*: We are carbon sources along with all air-breathing organisms
ii. Volcanoes
iii. Fires
iv. Oceans and fresh water bodies (CO_2 dissolves easily in water)

v. *Agriculture*: Grazing animals (cows, sheep, pigs) produce methane
vi. Burning fossil fuels, gas energy plants, coal, oil, gas, tar sands
vii. *Fossil fuel production*: Coals, oil gas, tar sands – including the burning of natural gas
viii. Decomposing garbage in landfills/dumps (methane)
ix. Rice farms (produce methane)
x. *Degrading farm soils*: Release of carbon from soils when ploughed – causes it to oxidize creating CO_2
xi. Wetland destruction
xii. Deforestation
xiii. Urban sprawl
xiv. Decomposition of dead animals and plants with oxygen (aerobic) produces CO_2

"Sink" means any process, activity or mechanism which removes a GHG, an aerosol or a precursor of a GHG from the atmosphere. The term "sinks" is also used to refer to forests, croplands and grazing lands and their ability to sequester carbon. Therefore, a carbon sink occurs when carbon sequestration is greater than carbon releases over some time period.

Carbon sink: Any process or mechanism which removes a GHG, an aerosol or a precursor of a GHG from the atmosphere. Natural sinks for atmospheric CO_2 include photosynthesis, forests, oceans and freshwater bodies, fossil fuels and carbonate rocks. The forests of the world are big carbons sink; however, deforestation is reducing the size of this sink, allowing more carbon dioxide to remain in the atmosphere. Photosynthesis accounts for about half of the carbon extracted from the atmosphere. Land plants take most of their carbon dioxide from the air around them, while aquatic plants in lakes, seas and oceans use carbon dioxide dissolved in water. Phytoplankton are some of these important plants, as they produce up to 50% of the atmospheric oxygen through photosynthesis. Other important sinks are the world's oceans. Processes that absorb CO_2 are called carbon sinks. List the sinks on the board –

i. Point to a green plant – photosynthesis in plants. All green plants, peat bogs (these are green plants in storage), growing forests
ii. Diatoms, algae etc.
iii. Fossilized organic remains become fossil fuels like coal, oil and gas stored safely underground
iv. Oceans can absorb CO_2 directly from the atmosphere. CO_2 will stay in storage for a long time provided oceans stay cool and undisturbed. The carbon hydrate is in the slurry at the bottom in the form of sediments
v. Shells, coral etc.
vi. Carbonate rocks
vii. Permafrost
viii. Wetlands trorganic matter builds up at the bottom of ponds, providing long-term storage of carbon
ix. Soil
x. *Limestone/Dolomite sedimentary rocks*: Shale, mudstone and coal. The Rocky Mountains are a good example of this.

7.1.4 Level of Carbon in Environment and Its Consequences

When first started measuring atmospheric CO_2 consistently in 1958, at the pristine Mauna Loa mountaintop observatory in Hawaii, the CO_2 level stood at 316 ppm, just a little higher than the pre-industrial level of 280 ppm. In October 2024, at the time of the usual annual maximum of CO_2, the air briefly tipped over the 422.36 ppm mark for the first time in several million years (Figure 7.1).

7.1.5 Methodologies for Carbon Capture

Reducing GHG emissions is the key to solving global climate change. A major way these gases get into the atmosphere is when people burn coal, oil and natural gas for energy. Many of these technologies are already available today, while others are still being developed and tested. Clean energy technologies like wind and solar power produce energy without burning fossil fuels. Other technologies reduce GHG emissions through energy efficiency or by capturing these gases before they can enter the atmosphere. Explore clean energy and other technologies that reduce the amount of GHGs being put into the atmosphere:

i. Solar energy
ii. Wind energy
iii. Water energy
iv. Nuclear energy
v. Geothermal energy
vi. Biomass energy
vii. Methane capture and use
viii. Carbon capture and underground storage
ix. Green vehicles
x. Energy-efficient buildings

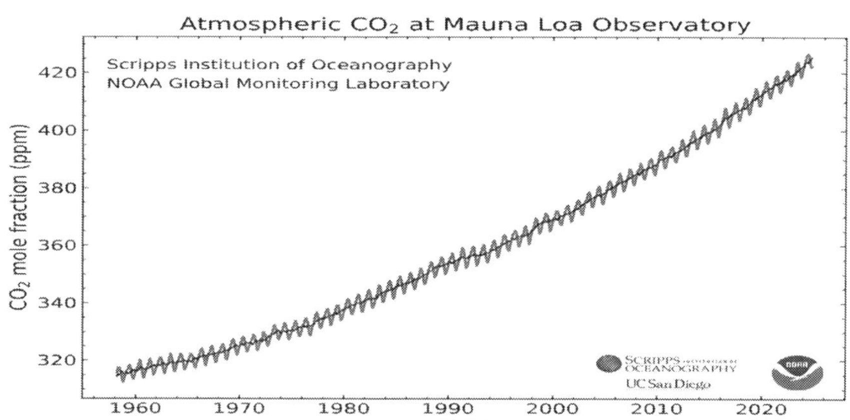

FIGURE 7.1 Carbon dioxide concentration (ppm).

According to Tribal Energy and Environmental Information Clearing House (2018): Ways to Reduce Carbon Dioxide Emissions, there are three primary methods for reducing the amount of carbon dioxide in the atmosphere: employing energy efficiency and conservation practices, using carbon-free or reduced-carbon energy resources and capturing and storing carbon either from fossil fuels or from the atmosphere.

i. Efficiency and conservation
ii. Carbon-free and reduced-carbon energy sources
iii. Carbon capture and sequestration
iv. Cap and trade and market-based controls
v. Voluntary carbon dioxide cap and trade approaches
vi. Clean development mechanism (CDM)
vii. *Agriculture*: Methane avoidance (manure management), biogas projects, agricultural residues for biomass energy
viii. *Forestry*: Reforestation, afforestation

Future technologies for carbon capture

i. *Metal-organic frameworks (MOFs)*: In recent years, a class of highly absorbent, nanoporous materials called MOFs has emerged as a promising material for carbon capture in power plants.
ii. *Nanosponges*: Cornell University materials scientists have invented low-toxicity, highly effective carbon-trapping "sponges" that could improve carbon capture economics. A research team led by Emmanuel Giannelis has invented a powder that performs as well as or better than industry benchmarks for carbon capture.
iii. *Hybrid membranes*: A new, highly permeable carbon capture membrane developed by scientists at the Lawrence Berkeley National Laboratory (Berkeley Lab) could lead to more efficient ways of separating CO_2 from power plant exhaust. The researchers focused on a hybrid membrane that is part polymer and part MOF.
iv. *Crystals*: Swedish scientists have created crystals that capture CO_2 much more efficiently than previously known materials, even in the presence of water. One way to mitigate climate change could be to capture CO_2 from the air. So far, this has been difficult since the presence of water prevents the adsorption of CO_2.
v. *Turning carbon to rock*: An international team of scientists reports they may have found a potentially permanent way to remove CO_2 emissions from the atmosphere – turn it into rock. The study, published in Science, has shown for the first time that CO_2 can be permanently and rapidly locked away from the atmosphere by injecting it into volcanic bedrock. The CO_2 reacts with the surrounding rock, forming environmentally benign minerals.
vi. *Turning carbon into fuel*: They're making fuel from thin air at the University of Southern California's Loker Hydrocarbon Research Institute. For the first time, researchers there have directly converted CO_2 from the air into methanol at relatively low temperatures.
vii. *Turning carbon into fibres*: A team of chemists says they have developed a technology to economically convert atmospheric CO_2 directly into highly valued carbon nano fibres for industrial and consumer products.

7.1.6 Carbon Sequestration Prospective of Different Land Use Systems

Carbon is released to the atmosphere as carbon dioxide through decomposition and microbial activity. The potential increase in storage through such methods is limited by soil type, which determines the carbon-holding capacity, and climate, which determines the rate of decomposition. Soil microbial activity increases with soil moisture and temperature, and increasing average temperatures due to climate change may be expected to increase the turnover rate of labile carbon in soils.

Plant sequestration: Plants in the presence of sunlight convert CO_2 from the atmosphere to carbohydrates for their growth and maintenance by the process of photosynthesis. Natural terrestrial biological sinks for CO_2 already sequester about one-third of CO_2 emissions from fossil fuel combustion. Biological storage could be enhanced through agricultural and forestry practices and reforestation, but the capacity is limited, and the longevity of storage depends on the final fate of the timber or plant material. However, carbon sequestration from reforestation and plantation programs could provide a significant shorter-term contribution to climate change mitigation.

Geo-sequestration: Geo-sequestration is the injection and storage of GHGs underground, out of contact with the atmosphere. The most suitable sites are deep geological formations, such as depleted oil and natural gas fields or deep natural reservoirs filled with saline water (saline aquifers). Geo-sequestration is part of the three-component scheme of carbon capture and storage, which involves capture of CO_2 either before or after combustion of the fuel, transport of the captured CO_2 to the site of storage and injection and storage of the CO_2. The concept of carbon sequestration may also be applied to other long-term storage options (ocean sequestration and mineral sequestration below). The storage option geo-sequestration is thought to be the most promising due to higher confidence in the longevity of storage, the large capacity of potential storage sites and a generally greater understanding of the mechanisms of storage. The largest geo-sequestration demonstration project, the Weyburn project in Canada, uses CO_2 injection to inject about 1.5 Mt of CO_2 per year. In Australia, the Otway Project in Victoria commenced in April 2008 and is injecting about 4500 tonnes of CO_2 per month into a depleted gas reservoir about 2 km underground.

Ocean sequestration: The ocean represents the largest carbon store on earth. The ocean has been a significant sink for anthropogenic CO_2 emissions of similar magnitude to the land sink, but, as with the land sink, the ocean sink will decrease in strength. Increasing CO_2 concentration in the upper layer of the oceans is also causing ocean acidification with potentially severe consequences for marine organisms and ecosystems. CO_2 dissolves in seawater by combining with carbonate ions, but the number of these ions is limited, and as their concentration decreases, this will limit the rate at which CO_2 is taken up by the ocean. A possible slow-down in ocean circulation may also reduce the ocean sink capacity. In addition to the dissolution process, phytoplankton in the surface layers perform photosynthesis and incorporate

CO_2 into biological material, but, as with terrestrial photosynthesis, there comes a saturation point where other factors restrict further photosynthesis.

Mineral sequestration: Mineral sequestration, also known as mineral carbonation involves reaction of CO_2 with metal oxides that are present in common, naturally occurring silicate rocks. The process mimics natural weathering phenomena and results in natural carbonate products that are stable on a geological time scale. There are sufficient reserves of magnesium and calcium silicate deposits to fix the CO_2 that could be produced from all fossil fuel resources.

7.1.7 Biomass Production and Carbon Sequestration through Forestry

The State of the World's Forests (2022) explores the potential of three forest pathways – halting deforestation and maintaining forests; restoring degraded lands and expanding agroforestry; and sustainably using forests and building green value chains – for achieving green recovery and tackling environmental crises, including climate change and biodiversity loss. There are three possible strategies for the management of forest carbon. The first is to increase the amount or rate of carbon accumulation by creating or enhancing carbon sinks (carbon sequestration). The second is to prevent or reduce the rate of release of carbon already fixed in existing carbon sinks (carbon conservation). The third strategy is to reduce the demand for fossil fuels by increasing the use of wood, either for durable wood products (substitution of energy-intensive materials such as steel and concrete) or for biofuel (carbon substitution). These strategies are not mutually exclusive. A number of carbon sequestration and carbon conservation initiatives have already been developed, including Activities Implemented Jointly (AIJ) under the UNFCCC and Land Use Change and Forestry (LUCF).

1. *Carbon sequestration*: The carbon sequestration potential of afforestation/reforestation is specific to the species, site and management involved, and it is therefore very variable. Typical sequestration rates for afforestation/reforestation, in tonnes of carbon per hectare per year, are 0.8–2.4 t in boreal forests, 0.7–7.5 t in temperate regions and 3.2–10 t in the tropics (Brown et al., 1996). The sequestration potential for agroforestry practices is even more variable, depending on the planting density and production objectives of the system.
2. *Carbon conservation*: The potential for carbon conservation through the maintenance of forest cover is dependent on the assumed baseline for non-project deforestation. In principle, 1.2–2.2 Gt of carbon could be conserved annually if deforestation were stopped completely Dixon et al., (1993). However, while carbon revenues could improve the economics of forest land, projects will also have to address the underlying causes of deforestation and unsustainable forest use to achieve effective carbon conservation. Brown et al. (1996) estimate that a reduction in deforestation in tropical regions could feasibly conserve 10–20 Gt of carbon by 2050 (0.2–0.4 Gt $year^{-1}$).
3. *Carbon substitution*: Biofuels currently provide 14% of the global primary energy supply. In developing countries, biofuels account for one-third of the total energy supply. If current biofuel use were to be replaced by fossil fuel-derived energy, an additional 1.1 Gt of carbon per year would be

released into the atmosphere (IPCC, 2000). The substitution of fossil fuels by sustainable biofuels will therefore result in a reduction of CO_2 emissions that is directly proportional to the volume of fossil fuel replaced.

4. *LUCF project-based activities*: LUCF projects currently under implementation, covering about 3.5 Mha (IPCC, 2000). Overall, 83% of this area is managed for the conservation of carbon in existing forests, either through forest protection (zero harvesting) or forest management (sustained production). Long-term carbon conservation by these projects varies from 40 to 108 t ha^{-1} from forest management and from 4 to 252 t ha^{-1} from forest protection. The estimated total lifetime sequestration effect of these projects is 5.7 Mt of carbon from forest management and 40–108 Mt from forest protection.
5. *Reducing emissions from deforestation and forest degradation (REDD+)*: It is the REDD+ and forest conservation, sustainable management of forests and enhancement of forest carbon stocks.

7.1.8 Potential of Climate Change Mitigation through Forest Sector

Climate change mitigation actions from the land-based sector, comprising agriculture, forestry and other land use (AFOLU), are of utmost relevance to this challenge. Forests have a key role in climate change mitigation, as they act as both sinks and sources of carbon. According to the Fifth Assessment Report of the IPCC, the AFOLU sector is responsible for 24% of global GHG emissions, while vegetation and soil account for approximately 30% of the sequestration of carbon (IPCC, 2014b). Although the contribution of the AFOLU sector to global emissions has remained constant, the mitigation potential in land-based activities is extremely important. Within the AFOLU sector, forestry and other land use change activities (of which forestry accounts for the greatest part) were responsible annually for net emissions amounting to 4.3–5.5 Gt CO_2 equivalents (CO_2e) in the period 2000–11, roughly 40%–50% of overall annual emissions from AFOLU and 10%–12% of global emissions.

Agriculture can contribute to climate change mitigation through reduced emissions from agricultural production and through enhanced carbon sequestration in soils and plants. There are many feasible and promising approaches to climate change mitigation in the AFOLU sector, and the technical potential is considerable. However, different options carry different costs in different regions. Two complementary goals to be pursued for mitigation are reducing the intensity of GHG emissions per unit of output and creating carbon-rich landscapes.

Potential and cost of mitigation in AFOLU:

- Hypothetical price of carbon that would induce farmers and foresters to apply practices for sequestering carbon and reducing emissions. Based on the combined mitigation potential of forestry and agriculture, estimated in the IPCC's Fourth Assessment Report, the IPCC suggests an economic potential in 2030 of between 3 and 7.2 Gt of CO_2e a year at carbon prices of US$20, US$50 and US$100 per tonnes, respectively.
- Most countries or constituencies considering climate mitigation policy instruments have considered the role of baseline-and-credit schemes. In a

baseline-and-credit scheme, an incentive is provided to an action that is deemed to reduce emissions from a reference scenario. The carbon credits represent emissions reductions and are mostly sold to buyers in compliance or voluntary markets to compensate for emissions, which would be more expensive to mitigate than the price of the credit. Crediting schemes have been considered and developed both at an international level and increasingly at a domestic level.

In the AFOLU sectors, different types of activities can help to reduce/avoid emissions and increase the removal of GHGs:

i. Forestry activities such as afforestation and reforestation, sustainable forest management, agroforestry and avoided deforestation/REDD+.
ii. Agricultural activities such as cropland and grazing land management, livestock management, peat land management and manure management.
iii. Energy activities such as increasing the energy efficiency at household level, sustainable biofuel production and the employment of integrated food and energy systems.
iv. Biodiversity-enhancing activities such as watershed and soil management, biodiversity conservation.
v. Agroforestry increases carbon sequestration and resilience to climate extremes through improved water retention and enriched soil fertility.
vi. *Restoration of degraded lands*: Restore degraded watersheds and reduce soil erosion.
vii. In general, enhance productivity and increase income and food security through a diversified production system.

Of 122 countries, INDCs submitted to UNFCCC mentioned forests (including mangroves). Forests were mentioned in INDCs for all regions but most frequently in Africa, where forests were mentioned in mitigation by 92% of countries and in adaptation by 75% of countries. REDD+, contributing to both adaptation and mitigation, was mentioned specifically by 28 countries; more than 69 countries list actions that would fall under REDD, such as avoiding emissions from deforestation and land degradation through sustainable forest management or the enhancement of carbon sinks. According to IPCC (2014b), A/R offers the best mitigation option because of its short timescale and ease of implementation. Reducing deforestation, forest management and forest restoration also offer good mitigation potential, especially because of the possibility for immediate action. By 2030, forestry mitigation options could contribute to reductions of 0.2–13.8 Gt CO_2e per year at carbon prices up to US$100 per tonne CO_2e and to reductions of 0.01–1.45 Gt CO_2e per year at prices below US$20 per tonne CO_2e (Smith et al., 2014).

Key features of mitigation through forest sector:

i. Progress in science and lessons learnt from implementation have paved the way to improvements in emission accounting under the Kyoto Protocol, favouring forest activities. Accounting for forest management has become mandatory, and carbon storage in harvested wood products can now be considered.

ii. The Paris Agreement recognizes the importance of forests for both mitigation and adaptation, singling out REDD+ activities in particular, and affirms the importance of non-carbon benefits, which could pave the way for reflection of the multiple benefits from forests in future carbon prices.
iii. While potentials and costs differ greatly by activity, region, system boundaries and time horizon, estimates indicate that the total economic mitigation potential of afforestation, reducing deforestation and forest management could range from 1.9 to 5.5 Gt CO_2e per year in 2040 at a carbon value of less than US$20 per tonne CO_2e. By region, these combined options have the largest economic potential in the Caribbean, Central and South America, followed by Africa, South Asia and North America.
iv. The wood product value chain (from the forest to the end of the product's life) contributes to the removal of CO_2 from the atmosphere, because CO_2 removed by trees is stored as carbon in products in use and in well-managed landfills.
v. The use of wood-based building materials avoids emissions of 483 Mt CO_2e annually via substitution effects. In addition, by displacing fossil fuels, the burning of used products at the end of the life cycle avoids the emission of more than 25 Mt CO_2e annually, which could be increased to 135 Mt by diverting material from landfills.
vi. Forest biomass-derived energy could reduce global emissions by between 400 Mt and 4.4 billion tonnes CO_2e per year.

7.1.9 Carbon Sequestration, Carbon Credit and Trading through Forestry

With the increase in international efforts to prevent climate change, participation of business and industrial sectors has become more relevant in carbon credit projects. In a drive to reduce their carbon footprints, industry is purchasing from projects that offset emissions by either sequestering or reducing GHGs. The regulations and corporate social responsibility (CSR) programs mean that companies are increasingly watching to carbon credit markets as part of the solution to achieving climate change mitigation and adaptation and their sustainability goals. The farmers are one of the main persons behind these markets, due to their huge potential for carbon sequestering in the way they use land. The convergence of agriculture and industry opens the door to new ways of beginning climate change as well as benefits farmers financially. A carbon credit is a tradable certificate or permit representing the right to emit one tonne of carbon or carbon dioxide equivalent (t CO_2e). One carbon credit is 1 t of carbon dioxide, or in some markets, carbon dioxide equivalent gases, after which it can be traded. An organization which produces 1 t less of carbon dioxide equivalent than the standard level of carbon emission allowed for its outfit or activity earns a carbon credit. Countries which are signatories to the Kyoto Protocol under the UNFCCC have laid down gas emission norms for their companies to be met by 2012. In such cases, a company has two ways to reduce emissions.

i. It can reduce the GHGs by adopting new technology or improving upon the existing technology to attain the new norms for emission of gases.

ii. It can tie up with developing nations and help them set up new technology that is eco-friendly, thereby helping the developing country or its companies earn credits. This credit becomes a permit for the company to emit GHGs in its own country. However, only a portion of carbon credits of the company in developing country can be transferred to the company in developed country.

Developing countries like India and China are likely to emerge as the biggest sellers, and Europe is going to be the biggest buyer of carbon credits. Presently, almost all global carbon credit trading was estimated at $5 billion, with India's contribution at around $1 billion. China is currently the largest seller of carbon credits, controlling about 70% of the market share. MCX has become the first exchange in Asia to trade carbon credits.

Two types of carbon market exist: the regulatory compliance and the voluntary markets. The compliance market is used by companies and governments that by law have to account for their GHG emissions. It is regulated by mandatory national, regional or international carbon reduction regimes. On the voluntary market, the trade of carbon credits is on a voluntary basis. The size of the two markets differs considerably. In 2008, on the regulated market, $119 billion was traded, and on the voluntary market, $704 million (Hamilton et al., 2009). The three Kyoto Protocol mechanisms are very important for the regulatory market: CDM, Joint Implementation (JI) and the EU Trading System (ETS). Some countries have not legally accepted the Kyoto Protocol but have other legally binding state and regional GHG reduction schemes. Developing countries can only participate in the CDM. In general, for small-scale AFOLU projects in developing countries, the voluntary market is more interesting than the regulatory market because the CDM market has quite complex procedures and methodologies for project registration, and the majority of agriculture and forestry and REDD+ projects are excluded. However, a brief introduction is given to the CDM, because some possibilities for small-scale projects (e.g. renewable energy) exist.

Carbon trading is a mechanism through which countries that have to achieve their emission targets are achieved. Carbon trading means trading of units of carbon dioxide reduced in the environment. Countries which reduce carbon emissions earn Carbon Emission Certificates, i.e. CER, which are traded and are called carbon trading. One carbon credit is equal to 1 t of carbon dioxide reduced in the environment. There are various exchanges in which carbon emissions are traded, like Chicago Climate Exchange, NASDAQ OMX, POWER NEXT and the European Energy Exchange; in India, MCX and NCDEX are two exchanges which deal with the trading of carbon emissions in the form of Carbon Emission Reduction Certificates, which countries earn through the completion of CDM projects, or in the form of ERUs.

7.1.10 Carbon Balance, Research Development and Future Perspectives in Forestry

Recent initiatives of the Government of India have taken a number of measures to promote sustainable development and address the threat of climate change at the national and sub-national levels.

The first step was revisiting the National Missions under the National Action Plan on Climate Change (NAPCC). Government is proposing to set up new missions on

Wind Energy, Health, Waste to Energy, Coastal Areas and redesigning the National Water Mission & National Mission on Sustainable Agriculture. Other initiatives include the following:

Mitigation Strategies

- *Green Generation for Clean & Energy Secure India*: More than five times increase in Renewable Capacity from 35 (up to March 2015) to 175 GW by 2022.
- National Solar Mission scaled up five-fold from 20 to 100 GW by 2022. Kochi Airport is the world's first airport to fully run on solar power.
- Solar powered toll plazas envisaged for all toll collection booths across the country.
- National Smart Grid Mission was launched for an efficient transmission and distribution network.
- Green Energy Corridor projects being rolled out to ensure evacuation from renewable energy plants.
- Nationwide Campaign for Energy Conservation launched with the target to save 10% of current energy consumption by the year 2018–19.
- Launched Smart Cities Mission to develop new generation cities by building a clean and sustainable environment.
- National Heritage City Development and Augmentation Yojana (HRIDAY) launched to bring together urban planning, economic growth and heritage conservation in an inclusive manner.
- Atal Mission for Rejuvenation and Urban Transformation (AMRUT) is a new urban renewal mission for 500 cities across India.
- Launched one-of-its-kind "Swachh Bharat Mission" (Clean India Mission) to make the country clean and litter-free by 2019.
- Zero Effect, Zero Defect (ZED) with the Make in India campaign to enhance energy and resource efficiency, pollution control, use of renewable energy, waste management etc.
- Formulated Green Highways (Plantation & Maintenance) Policy to develop 140,000 km long "tree-line" along both sides of national highways.
- Faster Adoption and Manufacturing of Hybrid and Electric Vehicles (FAME) to promote faster adoption and manufacturing of hybrid and electric vehicles.
- Country's first passenger vehicle fuel-efficiency standards finalized.
- Policies to increase production of energy-efficient 3-phase locomotives and switchover to 100% of these locos from 2016 to 2017 onwards.
- Policy directive issued to use 5% bio-diesel in traction fuel in diesel locomotives.
- National Air Quality Index launched with One Number, One Colour and One Description to give the status of air pollution in a particular city.

Adaptation Strategies

- Launched Soil Health Card Scheme. Additionally, 100 mobile soil-testing laboratories setup across the country.
- Paramparagat Krishi Vikas Yojana was launched to promote organic farming practices.

- The Pradhan Mantri Krishi Sinchayee Yojana launched to promote efficient irrigation practices.
- Neeranchal is a new program to give additional impetus to watershed development in the country.
- Launched National Mission for Clean Ganga (Namami Gange), which seeks to rejuvenate the river.
- National Bureau of Water Use Efficiency (NBWUE) proposed for promotion, regulation and control of efficient use of water.
- "Give It Up" Campaign launched to encourage citizens to give up subsidy on cooking gas to meet the needs of the truly needy citizens, thereby promoting a shift away from the inefficient use of biomass in rural areas.

Climate Finance Policies

- Setting up of INR 3500 million (US$55.6 million) National Adaptation Fund.
- Reduction in subsidies on fossil fuels, including diesel, kerosene and domestic LPG.
- Coal cess quadrupled from INR 50 to INR 200 per tonne to help finance clean energy projects and Ganga rejuvenation.
- Introduction of Tax Free Infrastructure Bonds for funding of renewable energy projects.

Illustrative list of some of the Technologies (Mitigation perspective)

Clean Coal Technologies (CCT)

- Pulverized Combustion Ultra Super Critical (PC USC)
- Pressurized Circulating Fluidized Bed Combustion, Super Critical, Combine Cycle (PCFBC SC CC)
- Integrated Gasifier Combined Cycle (IGCC)
- Solid Oxide Fuel Cell (SOFC), Integrated Gasifier Fuel Cell (IGFC) etc.
- Underground Coal Gasification (UCG)

Nuclear Power

- Pressurized water reactor, integral pressurized water reactor, Advanced Heavy Water Reactor (AHWR)
- Fast breeder reactor (FBR)
- Accelerated-driven systems in advanced nuclear fuel cycles

Renewable Energy:

- Yeast/Enzyme-based conversion to high-quality hydrocarbon fuels
- Conversion of pre-treated biomass to fuels and chemicals
- Gasification technologies like fluidized bed and plasma induced for power generation
- *Wind energy technologies*: Development of smaller and efficient turbines, wind turbines for low wind regime, designs of offshore wind power plants etc.

Solar PV technologies:

- Based on p-type silicon wafers and n-type silicon wafers
- Hetero junction with Thin Interfacial (HIT) Module, Back Contact Back Junction (BCBJ) Modules
- Crystalline silicon photovoltaic cells of >24% cell efficiency
- High efficiency Concentrating PV (CPV)
- Non-silicon-based solar PV technologies

7.1.11 Effects on People and the Environment

Global climate change will affect people and the environment in many ways. Some of these impacts, like stronger hurricanes and severe heat waves, could be life-threatening. Others, like spreading weeds, will be less serious. And some effects, like longer growing seasons for crops, might even be good! However, as the Earth keeps getting warmer, the negative effects are expected to outweigh the positive ones. The more we learn about how climate change will affect people and the environment, the more we can see why people need to take action to reduce the GHG emissions that are causing climate change. We can also take steps to prepare for the changes we know are coming.

Learn more about how climate change will affect people and the environment in the following ways:

i. Health
ii. Agriculture
iii. Energy
iv. Water supplies
v. Plants, animals and ecosystems
vi. Forest ecosystem
vii. Coastal areas
viii. Recreation

7.2 CONCLUSION

Carbon sequestration is one of the important aspects in the abatement of GHGs from the atmosphere and minimizing the pace of global warming. Forest sequesters a huge amount of carbon in above ground biomass, below ground biomass and soil carbon, which plays an important role in regulating the carbon cycle. The forest has high potential for carbon sequestration and lower carbon emission on a long-term basis. It is increasing the tree coverage, reducing carbon emissions, enhancing carbon sequestration, supporting carbon credits and carbon trading, supporting local peoples and strengthening the climate-resilient agriculture, thereby supporting global SDGs and climate change mitigation and adaptation. It will also support national and international goals and international commitments related to climate initiatives regarding green carbon, carbon credits and net zero carbon emissions goal.

REFERENCES

Allen, C. D., Macalady, A. K., Chenchouni, H. et al (2010). A global overview of drought and heat induced tree mortality reveals emerging climate change risks for forests. *Forest Ecology and Management*, *259*, 660–684.

Bharti, R. K., Srivastava, S., & Thakur, I. S. (2014a). Production and characterization of biodiesel from carbon dioxide concentrating chemolithotrophic bacteria, *Serratia* sp. ISTD04. *Bioresource Technology*, *153*, 189–197.

Boot-Handford, M. E., Abanades, J. C., Anthony, E. J., Blunt, M. J., Brandani, S., MacDowell, N., Fernandez, J. R., Ferrari, M. C., Gross, R., & Hallett, J. (2014). Carbon capture and storage update. *Energy & Environmental Science*, *7*(1), 130–189.

Brown, S., Sathaye, J., Cannell, M. G. R., & Kauppi, P. (1996). Management of forests for mitigation of greenhouse gas emissions. In: R. T. Watson, M. C. Zinyowera, R. H. Moss, & D. J. Dokken (Eds.), *Climate change 1995. Impacts, adaptations and mitigation of climate change, scientific Hi-technical analyses. Contribution of working group II to the second assessment report of the IPCC* (pp. 773–797). Cambridge University Press.

Crippa, M., Guizzardi, D., Banja, M., Solazzo, E., Muntean, M., Schaaf, E., Pagani, F., Monforti-Ferrario, F., Olivier, J. G. J., Quadrelli, R., Risquez Martin, A., Taghavi-Moharamli, P., Grassi, G., Rossi, S., Oom, D., Branco, A., San-Miguel, J., & Vignati, E. (2022). *CO2 emissions of all world countries-JRC/IEA/PBL 2022 report*. Publications Office of the European Union, 2022, http://dx.doi.org/10.2760/07904, JRC130363.

Dixon, R. K., Brown, S., Houghton, R. A., Solomon, A. M., & Trexler, M. C. (1993). Carbon pools and fluxes of global forest ecosystems. *Science*, *263*, 185–190.

Hamilton, K., Sjardin, M., Shapiro, A., & Marcello, T. (2009). *Fortifying the foundation: State of the voluntary carbon markets 2009*. Ecosystem Marketplace & New Carbon Finance. www.ecosystemmarketplace.com/documents/cms_documents/StateOfTheVoluntaryCarbonMarkets_2009.pdf.

IPCC. (2000). *Land use, land-use change, and forestry. A special report of the IPCC* (pp. 375). Cambridge University Press.

IPCC et al. (2013). *Climate change 2013: The physical science basis*. Cambridge University Press.

IPCC. (2014b). Agriculture, Forestry and Other Land Use (AFOLU). Chapter 11. Assessment Report 5: Technical Summary. Working Group III – Mitigation of Climate Change (p 181).

Kumar, M., Gupta, A., & Thakur, I. S. (2016c). Carbon dioxide sequestration by chemolithotrophic oleaginous bacteria for production and optimization of polyhydroxyalkanoate. *Bioresource Technology*, *213*, 249–256.

NOAA. (2024). *Global Monitoring Laboratory, Boulder, Colorado, USA*. https://gml.noaa.gov/ccgg/trends/

Norby, R. J., & Luo, Y. (2004). Evaluating ecosystem responses to rising atmospheric CO_2 and global warming in a multi-factor world. *New Phytologist*, *162*, 281–294.

Pan, Y. D., Birdsey, R. A., Fang, J. Y., et al (2011). A large and persistent carbon sink in the world's forests. *Science*, *333*, 988–993.

Sheikh, A. Q., Skinder, B. H., Pandit, A. K., & Ganai, B. A. (2014). Terrestrial carbon sequestration as a climate change mitigation activity. *Journal of Pollution Effects*, *2*, 110. http://dx.doi.org/10.4172/2375-4397.1000110

Smith, P., Bustamante, M., Ahammad, H., Clark, H., Dong, H., Elsiddig, E. A., Haberl, H., Harper, R., House, J., Jafari, M., Masera, O., Mbow, C., Ravindranath, N. H., Rice, C. W., Robledobad, C., Romanovskaya, A., Sperling, F., & Tubiello, F. (2014). Agriculture, forestry and other land use (AFOLU). In O. Edenhofer, R. Pichs-Madruga, Y. Sokona, E. Farahani, S. Kadner, K. Seyboth, A. Adler, I. Baum, S. Brunner, P. Eickemeier, B. Kriemann, J. Savolainen, S. Schlomer, C. Von Stechow, T. Zwickel, & J. C. Minx

(Eds.), *Climate change 2014: Mitigation of climate change. Contribution of working group III to the fifth assessment report of the intergovernmental panel on climate change* (pp. 811–922). Cambridge University Press.

TEEIC. (2018). https://matteroftrust.org/ways-to-reduce-carbon-dioxide-emissions/. https://teeic.indianaffairs.gov/er/carbon/carboninfo/reduce/index.htm

Tian, H. Q., Chen, G., & Zhang, C. et al. (2012). Century-scale response of ecosystem carbon storage and flux to multifactorial global change in the Southern United States. *Ecosystems*, *15*, 674–694.

UNFCCC. (2015). Adoption of the Paris Agreement. Report No. FCCC/CP/2015/L.9/Rev.1, http://unfccc.int/resource/docs/2015/cop21/eng/l09r01.pdf (UNFCCC, 2015).

Von der Assen, N., Jung, J., & Bardow, A. (2013). Life-cycle assessment of carbon dioxide capture and utilization: Avoiding the pitfalls. *Energy & Environmental Science*, *9*, 2721–2734.

8 Forestry and Agroforestry Interventions for Climate Change Mitigation

Garima, Yourmilla Kumari, Kashish Walia, Aruna Mehta, Meera Devi, and Jagriti Thakur

8.1 INTRODUCTION

Forestry and agroforestry are land management systems that offer agricultural enhancement and also the enhancement of the environment by providing biodiversity and tree resources for food and nutrition security on a global scale. Forestry provides more attention in the management of the forest and the conservation of forest resources for timber and also for non-timber goods and services, including many ecosystem services such as sequestering carbon and water supply. Forests account for around 31% of the earth's land area, or about 4.06 billion ha (FAO 2022), making them essential for both people and ecosystems. Kumar et al. (2023) said 12.5 million people are employed in the forestry sector across the world, out of which 6.23 million jobs are found in India. This shows considerable concern for employment creation and sustainable development.

Introduction of trees and shrubs into farms and rural terrain is essential for enhancing production, profitability, biodiversity and ecosystem resilience (Rizvi et al. 2020). Agroforestry consists of the deliberate integration of trees with crops and/or livestock so that a variety of outputs can be obtained from one management unit (Nair et al. 2021). Over 1.2 billion people practice agroforestry around the globe, and 1 billion ha of land are agroforestried; it is only around 25.32 million ha in India (Ghale et al. 2022). Compared to the enormous forest sector, agroforestry can be implemented by the use of tree-shaded farming methods such as alley cropping and silvopastoralism. In practice, such methods improve soil quality and biodiversity and enhance agriculture and forestry profitability, among other benefits. It also helps in mitigating climate change and in the preservation of carbon, which means it is useful in solving environmental issues and enhancing the economy of the rural area.

Forestry and agroforestry are important to society not only for carbon sequestration but also for ecosystem balance and sustainable land use practices. Pan et al. (2011) estimated that every year forests have a net absorption of 7.6 billion Mt of CO_2, which helps keep the greenhouse effect within limits. There is more reason to suggest that planting and growing trees helps keep carbon locked up, while ecosystem services such as water and nutrient cycles are maintained together with the world's 80% terrestrial biodiversity (FAO 2020). It is estimated that agroforestry practices are able to reduce atmospheric CO_2 depletion from 0.29 to 15.21 Mt ha^{-1} $year^{-1}$ (Nair

DOI: 10.1201/9781003588580-8

et al. 2010) and reduce soil erosion while improving organic matter and increasing soil carbon and greenhouse gas sinks (Fahad et al. 2022).

The upcoming chapter will explore the crucial role of forestry and agroforestry in combating climate change. The chapter will discuss strategies, such as reforestation, afforestation, sustainable forest management (SFM) and integrating trees into agriculture, emphasizing their benefits for biodiversity, soil health and rural livelihoods. Readers will gain insights into the scientific, economic and policy frameworks that support these interventions for global climate change mitigation.

8.1.1 Climate Change and Its Impact on Forest Ecosystems

Climate change is the most important global problem that has a negative effect on all ecosystems, while forests are essential in combating climate, conserving biodiversity and in the carbon cycle. They enhance the society's efforts toward sustainable development by bringing goods, services and resources. But climate changes alter the health and structure of forests, so fencing, tort and fire increase tree death and loss of diversity. The structure of increasing temperature, moistening centers and switching precipitation level also causes disturbances like pest outbreaks disrupting forest ecosystems as primary producers' food and protein value. These alterations impact vital resourcing services of forests and reduce the efficiency of forests as carbon sinks while changing the chemical composition and biological pattern of the ecosystem, which also threatens the sustainability of ecosystems. Expansion and preservation of forested areas is important in mitigating the impacts of climate change. Through the use of adaptive management strategies, it is possible to assist in assuring that the forests do not become a carbon source and worsen climate change (IUCN 2021).

8.1.2 Key Impacts

8.1.2.1 Forest Disturbances and Carbon Sequestration

Climate change has made disruptions in forest ecosystems worse, leading to more frequent and intense wildfires that result in habitat destruction and increased tree death rates. These wildfires undermine forests' roles as carbon sinks by releasing large amounts of carbon dioxide. Meanwhile, invasive species benefit from these disturbed areas. They outcompete native plants because they grow quickly and can thrive in various conditions, a situation that is worsened by higher levels of atmospheric CO_2 (Hellmann et al. 2008). By changing the makeup and structure of ecosystems, invasive species lower native biodiversity and disrupt ecosystem functions.

8.1.2.2 Alteration of Forest Health and Phenology

Climate change is causing significant changes in the phonological behavior of forests. Warmer temperatures are making trees bud and flower earlier than before. These changes interfere with the timing between trees and their associated species, like pollinators, which can potentially hinder forest regeneration (Gebeyehu and Hirpo 2019). Also, hotter weather makes it harder for trees to get enough water, lowering soil moisture and preventing tree seedling establishment and growth, while making trees more susceptible to diseases (Moore and Allard 2008). Insect and disease outbreaks are also intensified, as warmer weather speeds up insect life cycles, especially for bark beetles, leading to more tree damage. Fungi grow better in warm, wet places, making tree health and forest strength worse.

8.1.2.3 Nutrient Cycling and Soil Health

Nutrient cycling in forests is affected by climate change via altering decomposition rates of organic matter. Nutrient availability increases initially due to warmer temperatures and changing precipitation patterns; however, in the long run, they can deplete essential nutrients, reducing forest productivity and resilience. Nitrogen availability, critical for plant growth, is especially affected, leading to imbalances that deteriorate soil health and certainly decrease forest productivity in long term (Langley 2014). Disturbances in nutrient cycling menace the forest ecosystem's structure and function, reducing its capacity to adapt to changing circumstances.

8.1.2.4 Ecosystem Services and Carbon Balance

Forest ecosystems provide vital services, such as carbon sequestration, water regulation and biodiversity support, which are all threatened by climate change. Intensified fire frequency and outbreaks of pests reduce the ability of forests to store carbon, transforming them from carbon sinks to potential carbon sources, and this increases levels of carbon in the atmosphere, amplifying climate change (Rice 2010). In addition, disturbances such as wildfires degrade water quality by promoting soil erosion and sedimentation that affect water resources. Forest health and resilience are vital to sustaining biodiversity; however, climate change can drive loss of these species and change community composition dynamics, hampering ecosystems' ability to recover and sustain biodiversity across the temporal scales (Lasco et al. 2008).

8.1.3 Forestry Interventions for Climate Change Mitigation

Forests are essential for absorbing carbon dioxide (CO_2) from the atmosphere and storing it in their biomass and soils. Forestry interventions intentionally facilitate this process to tackle climate mitigation and adaptation.

8.1.3.1 Reforestation and Afforestation Strategies

Reforestation and afforestation are key carbon-sequestering strategies. Reforestation refers to planting trees where forests have been logged, and afforestation is planting trees in places where there were never forests. Both strategies store carbon on a significant scale, with tropical reforestation predicted to sequester 2–4 tCO_2 ha^{-1} $year^{-1}$ (Poorter et al. 2016). Reforestation and afforestation on a global scale are critical to combating climate change and biodiversity loss. Restoring ecosystems, expanding forest cover and enhancing carbon sequestration—these efforts mitigate greenhouse gas emissions and stabilize weather patterns (Table 8.1).

8.1.3.2 Sustainable Forest Management Practices

Forests are utilized for multiple purposes, and hence SFM practices are required to balance the needs of forest products with the conservation of forest ecosystems for

TABLE 8.1
Reforestation and Afforestation Projects in India and Abroad

Sr. No.	Project Name	Location	Main Objectives	Opportunities	Challenges	References
1	Bonn Challenge	Global	Restoration of 350 million ha of deforested and degraded land by 2030	Climate mitigation, biodiversity restoration, international collaboration	Funding and implementation gaps in developing countries	IUCN (2017)
2	Green India Mission	India	Enhance forest cover, ecosystem services and biodiversity	Empowering local communities, improving forest-based livelihoods	Slow implementation, lack of local capacity	MoEFCC (2014)
3	Cauvery Calling	India (Karnataka & Tamil Nadu)	Plant 2.4 billion trees to restore the Cauvery River basin	Revitalizing water bodies, enhancing soil health and community involvement	Land tenure issues, long-term sustainability	Isha Foundation (2021)
4	Great Green Wall	Sahel, Africa	Restore 100 million ha of degraded land, combat desertification	Economic development, poverty reduction, ecosystem resilience	Political instability, extreme weather conditions	UNCCD (2018)
6	Miyawaki Urban Forests	India (various cities)	Create dense, native urban forests using the Miyawaki method	Urban cooling, air quality improvement, community engagement	Space constraints, maintenance requirements	Miyawaki (2016)
8	China's Great Green Wall	China	Control desertification through afforestation and reforestation	Large-scale landscape restoration, economic benefits for rural areas	Water scarcity, soil degradation, species selection issues	NASA (2019)
9	Trillion Trees Initiative	Global	Conserve, restore and grow 1 trillion trees by 2030 to fight climate change	Global collaboration, corporate involvement, wide geographical scope	Deforestation rates, corporate greenwashing	Trillion Trees (2020)
10	Joint Forest Management (JFM)	India	Involve indigenous communities in forest protection and management	Local knowledge, co-management, livelihood improvements	Conflicts over resource use, weak institutional support	MoEFCC (2012)

future generations. Some of the methods of SFM are selective logging, agroforestry in forests and conservation of forest biodiversity. Many European countries practicing SFM have integrated carbon sequestration with timber harvesting in such a way that forests keep growing and offsetting large quantities of CO_2 emissions (Lindner et al. 2014).

8.1.3.2.1 Principles of Sustainable Forest Management

To ensure forest ecosystems remain resilient and productive, SFM directs to stabilize ecological stability, social equity and economic viability (UNFF 2022). Protection of biodiversity, soil and water resources through techniques like selective logging and natural regeneration refers to **ecological stability**. **Social equity means that all the rights of local and** indigenous communities are protected and promotes their involvement in decision-making (FAO 2019). **Economic viability** focuses on forestry practices that reinforce subsistence without deteriorating resources, via activities such as ecotourism and value addition in forest products (Krausmann et al. 2017).

8.1.3.2.2 REDD+

REDD+ (Reducing emissions from deforestation and forest degradation) is an initiative of the UN that remunerates developing countries for forest carbon stocks maintenance. The success of this program depends on effective governance and community involvement (Brockhaus et al. 2021).

8.1.3.2.3 Certification Systems

Forest Stewardship Council (FSC) and the Programme for the Endorsement of Forest Certification (PEFC)-like systems are the market-based, optional tools made to encourage SFM by certifying forests that can meet the certain sustainability criteria. Such systems can help different consumers to make conscious decisions, ensuring that the products they acquire are coming from rationally managed forests. 'FSC' mainly focuses on protecting biodiversity and indigenous rights of people (FSC 2020), whereas PEFC's main emphasis is on local adaptation and certification of entire landscapes (PEFC 2019).

8.1.3.2.4 Indigenous Knowledge

It plays a crucial role in SFM. Practices such as rotational agriculture and selective harvesting significantly reduce deforestation by assisting forest resilience and securing land tenure for tribal peoples (Nepstad et al. 2014). The Kayapo people in the Amazon, who have protected large rainforest areas through community management, are a key example of this (Schwartzman et al. 2019).

8.1.3.3 Forest Conservation

Protection of forests from deforestation and degradation is essential for climate change mitigation. Conservation strategies, such as reduction of illicit felling, establishment of protected areas and encouraging community-based forest management (CBFM) practices, have been playing a major role in decreasing deforestation rates globally. Approximately 1.8 Gt of CO_2 emissions per year can be prevented by forest conservation strategies (Griscom et al. 2017).

8.1.3.3.1 Forest Preservation in Protected Areas

Protected areas are essential for conservation, safeguarding water resources, conserving biodiversity and strengthening ecosystem resilience. Demarcation of conservation zones, monitoring, regulation enforcement and community involvement are some of the effective strategies. Around the world, protected areas cover approximately 15% of land but shelter over 80% of biodiversity (UNEP-WCMC 2020). Different tools like remote sensing enable monitoring of forests to track illegal activities (Hansen et al. 2013). In the Brazilian Amazon, despite pressures from illegal logging and agriculture, protected areas have lower deforestation rates.

8.1.3.3.2 Policy Frameworks

Legal frameworks exist in both national and international law to support forest protection and sustainable management. So, for example, Brazil's Forest Code requires maintaining at least 80% of forest cover on the private land in the Brazilian Amazon, while Indonesia's Forest Moratorium prohibits new concessions—that is, new approvals for clearing primary forests (Streck et al. 2019). International agreements such as the Convention on Biological Diversity (CBD) and REDD+ offer financial incentives for forest protection and carbon stock enhancement.

8.1.3.3.3 Community-Based Forest Management (CBFM)

It involves indigenous communities in forest management, recognizing them as effective stewards. Uniting devolution of management rights with community governance, as exemplified by Nepal's Community Forestry Program, decreases deforestation (Ojha et al. 2009). Such livelihood diversification through sustainable harvesting, ecotourism and agroforestry reduces pressures on forests. Mexico's Ejido system, in which communities' control large tracts of forest, shows low levels of deforestation.

8.1.3.3.4 Combatting Illegal Logging and Forest Degradation

Illegal logging, contributing to 15–30% of the global timber trade, jeopardizes forest preservation. Technological solutions, such as satellite monitoring and platforms such as Global Forest Watch, enable detection (Hansen et al. 2013). Laws like the U.S. Lacey Act and the EU Timber Regulation forbid trade in illegal timber, and initiatives like the EU's FLEGT Action Plan ensure only legally harvested timber enters the market.

8.1.3.4 Enhancing Carbon Storage in Forests

Maintaining forest ecosystems (e.g., through selective thinning, fire retardation and promoting tree species with greater carbon sequestration potential) can optimize carbon storage. Mangrove forests, for example, sequester five times the amount of carbon per hectare than terrestrial forests as a result of their unique soil carbon stores (Donato et al. 2011).

8.1.3.4.1 Forests as Carbon Sinks: Mechanisms and Dynamics

Forests function as carbon sinks, sequestering carbon from the atmosphere during photosynthesis. In the process, trees absorb CO_2 and transform it into organic compounds, sequestering carbon in their biomass (trunks, branches, leaves and roots).

TABLE 8.2
Comparison of Carbon Sequestration Rates in Different Forest Types

Forest Type	Carbon Sequestration Rate (tCO_2 ha^{-1} $year^{-1}$)	Primary Carbon Pool
Tropical rainforests	6–12	Aboveground biomass
Temperate forests	3–5	Aboveground biomass, soil carbon
Boreal forests	2–4	Soil carbon
Managed plantations	10–15	Aboveground biomass

The carbon stored in forests is classified into three main pools: aboveground biomass, belowground biomass (roots) and soil carbon. It is estimated that forests globally store approximately 861 Gt of carbon, with tropical forests accounting for the largest share (Pan et al. 2011). Approximately 44% of this carbon is stored in forest biomass, while the rest resides in soils and litter.

8.1.3.4.2 Carbon Stock in the Boreal, Temperate and Tropical Forests

Carbon stock varies significantly among different biomes. Tropical forests sequester more carbon with their high gross primary production (GPP) of 3351 g m^2 $year^{-1}$, compared to temperate forests. However, due to their long growing seasons and hot climates, they also experience high rates of respiration (Rh) and autotrophic respiration (Ra) (Table 8.2). In contrast, boreal forests, with cold climates and short growing seasons, have low carbon sequestration capacities and slower decomposition and growth rates (Malhi et al. 1999). On the other hand, temperate forests for both coniferous and broadleaf forests show the highest net ecosystem production (NEP) values, exceeding 200 g m^2 $year^{-1}$. This high NEP is probably affected by intensive forest management practices in temperate biomes, which aim to enhance woody biomass production (Luyssaert et al. 2007). Despite these variations, the differences between carbon sequestration potential among different biomes are relatively small. For example, when calculating production efficiency Net Primary Productivity (NPP/GPP) for evergreen forests (Goulden et al. 2011), the efficiency values are 34% for tropical forests, 38% for boreal forests and 46% for temperate forests. These contrasting carbon dynamics may influence the efficacy of forest management strategies (FMSs) to promote carbon sequestration and carbon storage in each biome.

8.1.3.4.3 Silvicultural Practices

Silvicultural practices, such as reforestation, afforestation, selective logging and thinning, play a critical role in managing forest carbon stocks. By maintaining forest structure, selective logging minimizes the impact on carbon sequestration, while thinning reduces competition, promoting tree growth and carbon absorption. Reforestation and afforestation store carbon; a mixture of species enhances both diversity and resilience (Putz et al. 2008). Carbon sequestration can be further increased by managing tree density and extending rotation lengths, as shown in

Canada's boreal forests, where these practices have significantly improved carbon storage in aboveground biomass (Kurz et al. 2008).

8.1.3.4.4 Long-Term Carbon Sequestration Potential of Forests

Forest type, age and management are some factors that affect the long-term carbon sequestration potential of forests. For example, in the Amazon, old-growth forests act as carbon sinks that store large amounts of carbon in their biomass and soils, with sequestration rates of 0.5–1.5 Gt annually (Lewis et al. 2011). Whereas young forests absorb carbon more quickly, they are more vulnerable to disturbances that can release stored carbon. Further, wood products from sustainable forestry also sequester carbon for decades, especially in building materials.

8.1.4 Agroforestry Interventions for Climate Change Mitigation

Agroforestry is a sustainable land use system that supports increased total production by growing food crops (annuals) with tree crops (perennials) and/or livestock on the same piece of land. Agroforestry systems aim to optimize the environmental and economic benefits of both tree and crop production while promoting ecosystem services, e.g., carbon sequestration and biodiversity conservation.

8.1.4.1 Trees in Agricultural Landscapes

Integration of trees and shrubs into agricultural farms and rural landscapes is important to enhance production, economic benefits, biodiversity and ecosystem perseverance (Rizvi et al. 2020). An agroforestry system is an extensive approach to land management that integrates trees and shrubs with crops and/or livestock on the same unit of land. Systems such as alley cropping, silvopasture and multilayered systems offer numerous benefits. By integrating trees, crops and/or livestock, agroforestry maximizes land use efficiency and increases carbon storage. Studies have shown that agroforestry systems have the potential of sequestering 2–9 tCO_2 ha^{-1} annually (Nair 2012).

8.1.4.1.1 Agroforestry Systems and Benefits

By incorporating trees, crops and livestock on the same piece of land, agroforestry systems aim to maximize the use of land while providing multiple economic, environmental and social benefits. These benefits include amplified agricultural output, improvements in soil health, enhanced biological diversity and increased carbon sequestration. Agroforestry systems are carefully designed to promote the interdependence of trees, crops and livestock, ensuring the long-term viability and adaptability of environments. Agroforestry systems are diverse in nature and can be classified based on their structure, function and management practices. The prevalent systems include the following:

1. *Alley Cropping/Hedge Row Intercropping*: Crops grown in the form of alley in between the trees planted in a row.
2. *Silvopasture*: Trees and grazing animals are integrated.
3. *Forest Farming*: Commercial are cultivated under forest canopies.
4. *Homegardens*: Diverse plant species are grown around homes.

8.1.4.1.2 Strategies for Tree Integration in Smallholder Farms

Integration of trees into farms requires special strategies considering local conditions, the needs of farmers and ecological principles. It includes the following:

- *Participatory Planning*: Involve farmers in decision-making to ensure alignment with their goals.
- *Species Selection*: Choose trees that provide multiple benefits.
- *Spatial Arrangement*: Optimize tree placement to maximize light interception and minimize competition.
- *Training and Extension Services*: Provide farmers with knowledge and support.
- *Financial Incentives*: Offer subsidies and market support to encourage adoption.

8.1.4.1.3 Optimizing Land Use in Agroforestry Practices

Optimizing land use in agroforestry involves balancing the spatial and temporal distribution of trees and crops to maximize productivity and ecosystem services. It includes the following:

- *Land Suitability Assessment*: Evaluate soil types, climate and topography.
- *Crop-Tree Compatibility*: Select compatible crops and trees.
- *Intensive Planting Techniques*: Maximize productivity through intercropping and high-density planting.
- *Rotation and Succession Planning*: Maintain soil health and reduce pest pressure.
- *Precision Agroforestry*: Use modern technologies for monitoring and decision-making.

8.1.5 Case Studies: Successful Forestry and Agroforestry Interventions

8.1.5.1 Costa Rica's Reforestation Program

Costa Rica has emerged as a global leader in reforestation and forest restoration. After extensive deforestation in the mid-20th century, the country implemented a programme called payment for ecosystem services (PES) programs, promoting landowners to protect and restore forests. As a result, forest cover of Costa Rica increased from 26% in 1983 to over 52% in 2020, sequestering millions of tons of carbon every year (Porras et al. 2013).

8.1.5.2 Kenya's Green Belt Movement

The Green Belt Movement in Kenya, started by Wangari Maathai, has been one successful agroforestry initiative. A total of 50 million trees were planted under this across Kenya, helping to restore barren/degraded lands, enhance soil fertility and provide fuelwood for indigenous communities. The project has also improved the local livelihoods and sequestered notable amounts of carbon, making it an agroforestry-based climate change mitigation model (Maathai 2006).

8.1.5.3 Carbon Sequestration in India's Deccan Plateau

Agroforestry interventions have shown significant advantages in improving soil health and sequestering carbon on India's Deccan Plateau. A study conducted for two decades revealed a 40% increase in soil carbon content and carbon sequestration of approximately 3–5 t ha^{-1} annually (Sharma et al. 2020). These productive outcomes have not only contributed to environmental stability but have also improved agricultural productivity, ensured food supply and offered a stable income source for local people.

8.1.6 Challenges and Barriers to Implementation

Despite notable evidence that supports the positive impact of forestry and agroforestry on climate change mitigation, e.g., soil erosion control, carbon sequestration and improved biodiversity, the application of these interventions is limited due to different challenges faced by farmers and stakeholders. These challenges can be grouped into three categories: (i) economic and policy barriers, (ii) technical challenges in scaling up interventions and (iii) social and cultural constraints (Tranchina et al. 2024). Addressing these challenges needs deep scientific scrutiny and targeted solutions to these problems.

8.1.6.1 Economic and Policy Barriers

8.1.6.1.1 Economic Barriers

There are multiple economic barriers that hinder the adoption of agroforestry, including high capital investments in initial stage for land conversion that restrict farm cash flows. Uncertainty lies in forestry systems about financial returns compared to traditional agriculture due to the longer production cycles (Garrity et al. 2010) that lead to the notion of irreversibility in land use changes. Moreover, for farmers dependent on stable outputs, agroforestry may threaten household food security, making it less appealing (Abdul-Salam et al. 2022). Small-scale farmers usually lack access to financial incentives, further obstructing adoption, whereas the irregularity of carbon markets in developing regions restricts the financial benefits of forestry interventions.

8.1.6.1.2 Weak Policy Support

For the successful adoption of forestry and agroforestry, there should be some supportive policies that provide financial incentives, remove regulatory barriers and give compensation provisions to farmers for delayed returns. Effective adoption requires a balance of economic incentives, regulatory frameworks and institutional support to underscore the economic and environmental benefits while reducing the risks for smallholders (FAO 2013). However, a lack of such supportive policies in many regions limits the agroforestry's widespread adoption. In India, despite its potential for carbon sequestration, inadequate financial incentives limit the implementation of agroforestry on a large scale (Chavan et al. 2015). Additionally, uncertain land tenure discourages farmers from investing in long-term agroforestry systems, as seen in Kenya (Place and Otsuka 2001).

8.1.6.2 Technical Challenges

Successful forestry and agroforestry practices require deep knowledge of tree species, crop integration, interaction of different components and long-term maintenance. This gap in technical knowledge, along with insufficient training and support for farmers, hinders implementation (Chaturvedi et al. 2015). In the Himalayan region, apple-based agroforestry faces challenges in tree-crop interaction and soil fertility management. Monitoring carbon sequestration is complex and costly, as seen in Indonesia's REDD+ program, where remote forests and high data collection costs hinder accurate assessments (Tacconi et al. 2013).

Additionally, selecting suitable tree species that meet local environmental and agricultural needs can be difficult. Farmers in sub-Saharan Africa struggle to balance species that provide agricultural benefits with those that sequester high carbon, as fast-growing species like eucalyptus may deplete water resources, while slower-growing options yield lower economic returns (Ajayi et al. 2009). Unpredictable weather incidences, such as droughts, further threaten the planting success of trees, and restoration of degraded soils via agroforestry can be expensive and complex, which limits its scalability.

8.1.6.3 Social and Cultural Constraints

8.1.6.3.1 Lack of Awareness and Education

Many farmers and communities are not aware of the long-term economic, environmental and climate benefits of agroforestry, leading to its low adoption rates.

8.1.6.3.2 Cultural Resistance

In many communities, traditional farming practices are deeply rooted, making the shift to agroforestry risky and incompatible with established methods. In Guatemala, farmers prefer monocultures and resist agroforestry, considering tree planting as conflicting with their traditional cropping systems (Milder et al. 2010).

8.1.6.3.3 Gender Inequality

Women play a pivotal role in forestry and agroforestry; however, efforts of women are often undervalued, resulting in poor working conditions and unequal pay (FAO 2013). Many women have limited access to land, training and credit, which prohibits their participation in agroforestry initiatives despite their role in agriculture. In Kenya, women's elimination from land ownership limits their involvement in such programs. As a result, their restricted access to resources and decision-making power prevents them from benefiting from agroforestry interventions (Rocheleau and Edmunds 1997).

8.1.6.3.4 Marginalized Communities

Despite possessing valuable traditional knowledge of sustainable land use practices, tribal and marginalized groups, who depend on forest resources for their livelihoods, are generally excluded from decision-making processes. In the Amazon, indigenous communities are frequently restricted in forestry intervention programs like REDD+, resulting in tensions and unproductive outcomes due to their lack of involvement (Larson and Dahal 2012).

8.1.7 Policies and Incentives for Promoting Forestry and Agroforestry

Policies and incentives are important for the success of forestry and agroforestry by providing frameworks as well as support to implement sustainable land management practices. One of the more effective is the reform of land tenure, which guarantees the rights of farmers to land and trees on it and gives them the impetus to invest in long-term forestry and agroforestry practices. Financial incentives like subsidies, tax breaks and low-interest loans help pay for planting and upkeep of trees. Moreover, programs for the PES provide landowners payment in exchange for ecological benefits such as carbon storage and biodiversity conservation. Policies that embed forestry and agroforestry within national climate frameworks, including within Nationally Determined Contributions (NDCs) under the Paris Agreement, enhance these practices by establishing them within an overarching context on climate and sustainability. Technical training is offered through capacity-building by governments and NGOs to ensure communities are equipped with adequate knowledge to successfully implement agroforestry. Combined, these measures help conserve forests, enhance ecosystems and increase rural well-being.

8.1.7.1 Global and National Policies

Agroforestry is recognized as a pathway to sustainable land management through global initiatives such as the Sustainable Development Goals (SDGs), the Green Climate Fund (GCF) and REDD+. The UN Decade for Ecosystem Restoration (2021–2030) explicitly supports agroforestry to restore degraded landscapes and improve soil health and carbon sequestration. Moreover, nations, such as Kenya, India and Nepal, have included agroforestry in their national programs, providing financial, technical and market resources to farmers who implement it (Awazi et al. 2024).

8.1.7.2 Sectoral Contributions

CIFOR-ICRAF, IUCN and WWF are among organizations that work to promote sustainable agroforestry models by providing research, training and funding on the ground. They are investing in farmer education about agroforestry benefits and policy advocacy to include agroforestry in national land use policies (Plieninger et al. 2020).

8.1.7.3 Financial Incentives

The potential of various financial incentives to stimulate agroforestry adoption among farmers was identified by Tavernier et al. (2024) as below:

- *Government-Based Incentives*: Subsidies for planting and maintaining agroforestry systems provide significant financial support.
- *Products Payment*: Market-based benefits are niche market opportunities and higher prices for agricultural products such as specialty fruit or wood products. But, due to low consumer awareness and high marketing costs, they have limited outreach, particularly for smallholders who do not have access to direct sales channels.
- *Payment for Ecosystem Services (PES)*: Agroforestry's environmental contributions, particularly carbon sequestration, offer potential revenue streams through carbon credits.
- *Community-Based Incentives*: Association with local communities, including Community Supported Agriculture and crowdfunding, promotes social and economic resilience. These collaborations can promote public awareness of agroforestry; however, they need substantial social capital and may deter farmers who prefer autonomy.

8.1.7.4 Role of Government

Community Forest Rights (CFR) are given to forest-dwelling communities under India's Forest Rights Act (FRA), aiming to promote SFM. The government supports this with funding, technical assistance and regulatory frameworks; however, bureaucratic complexities and conflicting community engagement can limit its effectiveness (Gupta et al. 2020).

8.1.7.5 Role of NGOs

Government efforts are supported by NGOs via provided technical expertise, mobilizing communities and supporting sustainable land management. They exemplify benefits of agroforestry through local projects, tie up stakeholders with international resources and share best practices on ecological and economic gains. NGOs also make communities aware of their forest rights, assist in leading government resources and build decision-making capacity and resilient forest governance (Gupta and Koontz 2019).

8.1.7.6 Synergistic Benefits

When NGOs and government agencies come together, they create synergies that improve forestry efforts of a community. NGOs help to shorten the gaps by guiding communities through bureaucratic processes, whereas the government's technical and financial resources expand NGOs' ground impact. This collaboration fosters collective action, enhances community participation and strengthens structures of local governance.

8.1.7.7 Role of Private Sector

Agroforestry is supported by the private sector via offering investment, technical expertise and market access with companies in food, cosmetics and pharmaceuticals, leading to sustainable supply chains and fair pricing. Further, they train farmers for sustainable practices, improving productivity and access to upscale markets, thus complementing the government and NGO efforts to strengthen agroforestry's viability (Gupta et al. 2020).

8.1.8 Innovations and Emerging Trends

One of the greatest features of forestry and agroforestry systems is their ability to cope with climate change. These systems can potentially provide effective and sustainable options for carbon sequestration, biodiversity protection and environmental change adaptation. Ultimately, the advanced technologies are on the rise in the areas of interest.

8.1.8.1 Technological Advances in Forestry and Agroforestry

One of the most crucial changes taking place in the forestry and agroforestry industries is the development of technology, which enhances their management, productivity and ecosystem service. Among the major progress are the following subheads

8.1.8.1.1 Remote Sensing and GIS

Satellite remote sensing and geographic information systems (GISs) support forestry and agroforestry by providing essential data for decision-making and conservation through checking forest cover, land use changes and health status. This technology makes it possible to map out in detail the species, age, density and health factors of trees in order to manage them properly. Agroforestry suitability mapping in Samastipur, India, classifies 48.22% of the area as 'very high suitable', 22.83% as 'highly suitable', 23.32% as 'moderate' and 5.63% as 'low suitable' (Ahmad et al. 2018).

Further, remote sensing and satellite imaging can evaluate crop health and the problem areas so quickly that resources can be managed accordingly. GIS promotes forest planning by simulating scenarios for adequate decision-making across social, ecological and economic factors. Fire management and deforestation monitoring are improved by GIS and remote sensing through determining risks of wildfire with a view to allowing rapid response and tracking illegal logging and land cover changes, ultimately supporting conservation and mitigating ecosystem impacts.

8.1.8.1.2 Tree Breeding and Genomics

The major objective of tree breeding programs is to develop genetic strains suitable for reforestation and afforestation while maximizing the economic, social and environmental revenues from the forest resources. Non-timber forest products are supplemented to the economic value of forests in a large array of programs around the globe. It has been possible to produce trees containing improved genetics for some attributes through forest tree breeding because of the repetition of selection, mating and testing cycles (Grattapaglia et al. 2018). Through genetic selection and breeding, better tree species have been developed due to the recent developments in biotechnology. Such trees have been shown to possess improved growth rates, e.g., eucalyptus has high resistance to pests (Avisar et al. 2024) and adapts well to various climatic conditions (Shikakura et al. 2022), thereby facilitating their use in the opportunities of carbon capture and climate change mitigation.

8.1.8.1.3 Agroforestry Design Tools

Digital platforms to help farmers make informed decisions about agroforestry practices combine data from multiple sources, including soil health, weather patterns and market prices. An agroforestry design tool offers guidance for designing multilayered agroforestry systems created for food or timber production and native habitat restoration, etc. Likewise, the RegenWorks software focuses on the process of designing agroforestry systems customized for a particular land area and business model.

8.1.8.2 Precision Forestry and Agroforestry Practices

Advanced tools and AI-driven analysis are helpful in precision forestry and agroforestry to support site-specific management, reducing illegal felling, improving wood quality, minimizing waste and enhancing profitability. These techniques favor productivity and sustainability across different stages like planning, operations, monitoring, processing and marketing.

- *Drones and UAVs*: Unmanned aerial vehicles (UAVs) are increasingly used for monitoring forest health, identifying pests or disease outbreaks and optimizing planting strategies. They help in precise data collection over vast areas, reducing the time and cost of manual surveys.
- *Precision Irrigation*: In water-scarce areas, precision irrigation using sensors and automated systems ensures water efficiency and enhances the

resilience of agroforestry systems. This technology is particularly relevant for integrating trees in drylands or semi-arid regions.

- *Site-Specific Nutrient Management*: Agroforestry systems, particularly in crop-tree combinations like your focus on apple and cabbage systems, benefit from the precision application of nutrients, which improves tree health and crop yield while reducing environmental impacts.
- *LiDAR*: A remote sensing technique using near-infrared laser and GPS enables precise land and forest mapping at both large scales (via aircraft or drones) and smaller ground-based scales (Kinhal 2023). It effectively captures forest structure details, such as canopy layers, tree shapes, and leaf density, while also detecting understory vegetation like shrubs. LiDAR is instrumental in forest inventories, offering accurate data on tree height, basal area and volume, and supports post-harvest assessments.

8.1.8.3 Integrating Agroforestry with Climate-Smart Agriculture

Agroforestry is key to climate-smart agriculture (CSA) by boosting resilience, productivity and carbon sequestration. Climate-smart agroforestry (CSAF) integrates trees into farms to support food security and climate adaptation, especially in developing regions, by enhancing soil health and productivity while reducing carbon emissions. Practices like alley cropping and silvopasture in agroforestry increase adaptability to climate stresses and sequester carbon. Further, selection of climate-resilient species promotes sustainability; however, integration of CSA with agroforestry improves ecosystem services, such as soil health, water-holding capacity and biodiversity, promoting agroecological adaptation for smallholder farmers.

8.1.8.3.1 Climate-Smart Agroforestry Models

CSAF models are created to improve the agricultural land use system's resilience to climate change and participate in mitigation efforts. These models are based on principles of sustainability, productivity and adaptation. Key components are included in Table 8.3.

TABLE 8.3
Components of Climate-Smart Agroforestry Models

Component	Description	Climate Mitigation and Adaptation
Diversification	Variety of tree species and crops	Reduces risk and enhances ecosystem services
Water management	Practices to improve water retention and reduce runoff	Ensures productivity during droughts, reduces flood risks
Energy efficiency	Use of trees for bioenergy	Lowers reliance on fossil fuels, reduces emissions
Enhanced carbon sequestration	Strategic tree placement and management	Increases long-term carbon storage
Adaptive management	Monitoring and adjusting practices based on climate data	Maintains effectiveness under changing conditions

In Kenya, by adopting CSAF models, farmers have boosted resilience to climate change. Integrating maize farming with drought-resistant tree species like *Faidherbia albida* has led to increased maize yields, improved soil fertility and enhanced water retention (Nair 2004). It combines modern agroforestry techniques with traditional knowledge for climate mitigation and adaptation.

8.1.8.4 Monitoring and Reporting of Carbon Sequestration

Accuracy in monitoring and reporting is important for analyzing the climate mitigation benefits of forestry and agroforestry. Carbon stock estimates are critical for agroforestry projects that involve carbon models. MRV (monitoring, reporting and verification) units combine ground measurements, carbon models and satellite images for quantification of carbon stocks. There are several tools that enable people to estimate carbon footprints, e.g., COMET-Farm and Cool Farm Tool. These tools help farmers to estimate the carbon sequestration potential of their systems, promoting agroforestry's integration into carbon markets and climate change practices (Hillier et al. 2011). Additionally, digital platforms for smallholder farmers simplify carbon accounting and access to global carbon credit markets, strengthening farmers to participate in and benefit from climate action.

8.1.9 Future Directions and Research Needs

8.1.9.1 Gaps in Knowledge and Research Priorities

Despite the advancements in forestry and agroforestry interventions for climate mitigation, their adoption still faces key gaps. More research is required to understand the carbon sequestration potential of diverse agroforestry systems, how biodiversity is helpful in enhancing climate resilience and the social extent like land tenure and equal access to carbon markets. Further, studies should focus on optimizing tree-crop-livestock interactions, developing carbon accounting frameworks and examining agroforestry's socio-economic impacts on rural livelihoods and food security in future.

8.1.9.2 Developing Climate-Resilient Forestry and Agroforestry Systems

For climate mitigation and long-term sustainability, strengthening of climate-resilient forestry and agroforestry systems is essential. Climate-resilient practices include the introduction of climate-resilient tree species with high carbon sequestration potential, drought-resistant agroforestry systems for arid regions, such as silvopasture with livestock integration and the adoption of adaptive management practices like flexible planting schedules and diverse species mixes.

8.1.9.3 Collaborative Approaches to Scaling Interventions

A collaborative approach is required for scaling up forestry and agroforestry for climate mitigation. Public-private partnerships (PPPs) are necessary to provide funding, expertise and policy support in large-scale projects, whereas farmer-led innovations ensure that agroforestry interventions are practical and sustainable. International cooperation, like the GCF and REDD+, can mobilize resources and incentives for CSAF.

8.2 CONCLUSION

For combating climate change, forestry and agroforestry interventions are potent strategies that provide environmental and socioeconomic benefits. Practices such as reforestation, afforestation and SFM enhance carbon storage, improve biodiversity and support livelihoods. However, their adoption is hindered by economic, technical and social barriers. To address these challenges, targeted policies, financial incentives and innovative practices are needed, fostering collaboration among governments, NGOs and the private sector. Future research should focus on filling knowledge gaps, developing climate-resilient systems and scaling up interventions to realize the capability of forestry and agroforestry in mitigating climate change.

REFERENCES

Abdul-Salam, Y., P. Ovando, and D. Roberts. 2022. "Understanding the Economic Barriers to the Adoption of Agroforestry: A Real Options Analysis." Journal of Environmental Management 302: 113955. https://doi.org/10.1016/j.jenvman.2021.113955.

Ahmad, F., M. M. Uddin, and L. Goparaju. 2018. "Assessment of Remote Sensing and GIS Application in Identification of Land Suitability for Agroforestry: A Case Study of Samastipur, Bihar, India." Contemporary Trends in Geosciences 7, no. 2: 214–28.

Ajayi, O. O., F. K. Akinnifesi, G. W. Sileshi, and W. Kanjipite. 2009. "Labour Inputs and Financial Profitability of Conventional and Agroforestry-Based Soil Fertility Management Practices in Zambia." Agrekon 48, no. 2: 109–25.

Avisar, D., A. Manoeli, A. A. Dos Santos, et al. 2024. "Genetically Engineered Eucalyptus Expressing Pesticidal Proteins from *Bacillus thuringiensis* for Insect Resistance: A Risk Assessment Evaluation Perspective." Frontiers in Bioengineering and Biotechnology 12: 1322985. https://doi.org/10.3389/fbioe.2024.1322985.

Awazi, N. P., T. F. Ambebe, N. A. Njamnjubo, and T. G. H. Zanguim. 2024. "Promoting Agroforestry Systems for Livelihoods and Ecosystem Management: The Role of Global, National, and Sectorial Initiatives." International Journal of Environmental Sciences & Natural Resources 34, no. 1: IJESNR.MS.ID.556379.

Brockhaus, M., M. D. Gregorio, and R. Carmenta. 2021. "REDD+: Limitations and Potential for Forest Governance in the Post-Paris Agreement World." Forest Policy and Economics 122: 102331.

Chaturvedi, O. P., A. K. Handa, and D. P. Nandal. 2015. "Agroforestry: Systems, Challenges, and Solutions." In Agroforestry in India, edited by K. R. Solanki, 23–41. New Delhi: Indian Council of Agricultural Research.

Chavan, S. B., A. Keerthika, S. K. Dhyani, et al. 2015. "National Agroforestry Policy in India: A Low Hanging Fruit." Current Science 108, no. 10: 1826–34.

Donato, D. C., J. B. Kauffman, D. Murdiyarso, S. Kurnianto, M. Stidham, and M. Kanninen. 2011. "Mangroves among the Most Carbon-Rich Forests in the Tropics." Nature Geoscience 4, no. 5: 293–97.

Fahad, S., S. B. Chavan, A. R. Chichaghare, et al. 2022. "Agroforestry Systems for Soil Health Improvement and Maintenance." Sustainability 14, no. 22: 14877. https://doi.org/10.3390/su142214877.

Food and Agriculture Organization of the United Nations (FAO). 2013. Advancing Agroforestry on the Policy Agenda: A Guide for Decision-Makers, by G. Buttoud, in collaboration with O. Ajayi, G. Detlefsen, F. Place, and E. Torquebiau. Agroforestry Working Paper No. 1. Rome: FAO.

Food and Agriculture Organization of the United Nations (FAO). 2019. Sustainable Forest Management: A Sound Forest-Based Economy. Rome: FAO. https://www.fao.org/sustainable-forest-management.

Food and Agriculture Organization of the United Nations (FAO). 2020. State of the World's Forests 2020: Forests, Biodiversity and People. Rome: FAO. https://doi.org/10.4060/ca8642en.

Food and Agriculture Organization of the United Nations (FAO). 2022. The State of the World's Forests 2022: Forest Pathways for Green Recovery and Building Inclusive, Resilient, and Sustainable Economies. Rome: FAO. https://openknowledge.fao.org/handle/20.500.14283/cb9360en.

Garrity, D. P., F. K. Akinnifesi, O. C. Ajayi, G. Sileshi, J. Mowo, A. Kalinganire, and M. Larwanou. 2010. "Evergreen Agriculture: A Robust Approach to Sustainable Food Security in Africa." Food Security 2, no. 3: 197–214.

Gebeyehu, M. N., and F. H. Hirpo. 2019. "Review on Effect of Climate Change on Forest Ecosystem." International Journal of Environmental Sciences & Natural Resources 17, no. 4: 126–29.

Ghale, B., E. Mitra, H. S. Sodhi, A. K. Verma, and S. Kumar. 2022. "Carbon Sequestration Potential of Agroforestry Systems and Its Potential in Climate Change Mitigation." Water, Air, & Soil Pollution 233, no. 228. https://doi.org/10.1007/s11270-022-05689-4.

Goulden, M. L., A. M. S. McMillan, A. V. Winston, et al. 2011. "Patterns of NPP, GPP, Respiration, and NEP during Boreal Forest Succession." Global Change Biology 17, no. 2: 855–71.

Grattapaglia, D., O. B. Silva-Junior, R. T. Resende, et al. 2018. "Quantitative Genetics and Genomics Converge to Accelerate Forest Tree Breeding." Frontiers in Plant Science 9: 1693. https://doi.org/10.3389/fpls.2018.01693.

Griscom, B. W., J. Adams, P. W. Ellis, R. A. Houghton, et al. 2017. "Natural Climate Solutions." Proceedings of the National Academy of Sciences 114, no. 44: 11645–50.

Gupta, D., and T. M. Koontz. 2019. "Working Together? Synergies in Government and NGO Roles for Community Forestry in the Indian Himalayas." World Development 114: 326–40. https://doi.org/10.1016/j.worlddev.2018.09.016.

Gupta, D., S. Lele, and G. Sahu. 2020. "Promoting a Responsive State: The Role of NGOs in Decentralized Forest Governance in India." Forest Policy and Economics 111: 102066. https://doi.org/10.1016/j.forpol.2019.102066.

Hansen, M. C., P. V. Potapov, R. Moore, M. Hancher, S. A. Turubanova, A. Tyukavina, and J. R. Townshend. 2013. "High-Resolution Global Maps of 21st-Century Forest Cover Change." Science 342, no. 6160: 850–53.

Hellmann, J. J., J. E. Byers, B. G. Bierwagen, and J. S. Dukes. 2008. "Five Potential Consequences of Climate Change for Invasive Species." Conservation Biology 22, no. 3: 534–43.

Hillier, J., C. Walter, D. Malin, T. Garcia-Suarez, L. Mila-i-Canals, and P. Smith. 2011. "A Farm-Focused Calculator for Emissions from Crop and Livestock Production." Environmental Modelling & Software 26, no. 9: 1070–78.

International Union for Conservation of Nature (IUCN). 2021. Forests and Climate Change: Issues Brief. Gland, Switzerland: IUCN, February 2021. https://www.iucn.org/issues-briefs.

Kinhal, V. 2023. Precision Forestry in 2023: A Revolution in Intensive Management.

Krausmann, F., K. H. Erb, S. Gingrich, C. Lauk, and H. Haberl. 2017. "Global Patterns of Socioeconomic Biomass Flows in the Twentieth Century: A Comprehensive Assessment of Available Data." Global Environmental Change 25: 41–52.

Kumar, S., I. Dev, and N. Kumar. 2023. "Forestry and Agroforestry." In Trajectory of 75 Years of Indian Agriculture after Independence, edited by P. K. Ghosh, A. Das, R. Saxena, K. Banerjee, G. Kar, and D. Vijay, 497–519. Singapore: Springer. https://doi.org/10.1007/978-981-19-7997-2_19.

Kurz, W. A., C. C. Dymond, G. Stinson, G. J. Rampley, et al. 2008. "Mountain Pine Beetle and Forest Carbon Feedback to Climate Change." Nature 452, no. 7190: 987–90.

Langley, A. 2014. "Using Results from Global Change Experiments to Inform Land Model Development and Calibration." Energies 6, no. 9: 4607–38.

Larson, A. M., and G. R. Dahal. 2012. "Forest Tenure Reform: New Resource Rights for Forest-Based Communities?" Conservation and Society 10, no. 2: 77–90.

Lasco, Rodel D., F. Pulhin, P. Sanchez, G. B. Villamor, and K. A. L. Villegas. 2008. "Climate Change and Forest Ecosystems in the Philippines: Vulnerability, Adaptation and Mitigation." Journal of Environmental Science and Management 11, no. 1: 1–14.

Lewis, S. L., P. M. Brando, O. L. Phillips, G. M. van der Heijden, and D. Nepstad. 2011. "The 2010 Amazon Drought." Science 331, no. 6017: 554.

Lindner, M., M. Maroschek, S. Netherer, A. Kremer, et al. 2014. "Climate Change and European Forests: What Do We Know, What Are the Uncertainties, and What Are the Implications for Forest Management?" Journal of Environmental Management 146: 69–83.

Luyssaert, S., I. Inglima, M. Jung, A. D. Richardson, M. Reichstein, D. Papale, et al. 2007. "CO_2 Balance of Boreal, Temperate, and Tropical Forests Derived from a Global Database." Global Change Biology 13, no. 12: 2509–37.

Maathai, W. 2006. Unbowed: A Memoir. https://en.wikipedia.org/wiki/Unbowed:_A_Memoir.

Malhi, Y., D. D. Baldocchi, and P. G. Jarvis. 1999. "The Carbon Balance of Tropical, Temperate, and Boreal Forests." Plant, Cell & Environment 22, no. 6: 715–40. https://doi.org/10.1046/j.1365-3040.1999.00453.x.

Milder, J. C., S. J. Scherr, and C. Bracer. 2010. "Trends and Future Potential of Payment for Ecosystem Services to Alleviate Rural Poverty in Developing Countries." Ecology and Society 15, no. 2: 4. https://doi.org/10.5751/ES-03098-150204.

Moore, B., and G. Allard. 2008. Climate Change Impacts on Forest Health. Forest Health and Biosecurity Working Papers, FBS/34E. Rome: Food and Agriculture Organization.

Nair, P. K. R. 2004. Agroforestry Systems: A Classification for Research and Development. Food and Agriculture Organization of the United Nations. http://www.fao.org/3/y1579e/y1579e00.htm.

Nair, P. K. R. 2012. "Carbon Sequestration Studies in Agroforestry Systems: A Reality Check." Agroforestry Systems 86, no. 2: 243–53.

Nair, P. K. R., B. M. Kumar, and V. Nair. 2021. "Definition and Concepts of Agroforestry." In An Introduction to Agroforestry, edited by P. K. R. Nair, 1–23. Cham: Springer. https://doi.org/10.1007/978-3-030-75358-0_2.

Nair, P. R., V. D. Nair, B. M. Kumar, and J. M. Showalter. 2010. "Carbon Sequestration in Agroforestry Systems." Advances in Agronomy 108: 237–307.

Nepstad, D., S. Schwartzman, B. Bamberger, et al. 2014. "Inhibition of Amazon Deforestation and Fire by Parks and Indigenous Lands." Conservation Biology 20, no. 1: 65–73.

Ojha, H. R., J. Cameron, and C. Kumar. 2009. "Deliberation or Symbolic Violence? The Governance of Community Forestry in Nepal." Forest Policy and Economics 11, no. 5–6: 365–74.

Pan, Y., R. A. Birdsey, J. Fang, R. Houghton, P. E. Kauppi, et al. 2011. "A Large and Persistent Carbon Sink in the World's Forests." Science 333, no. 6045: 988–93.

Place, F., and K. Otsuka. 2001. "Tenure, Agricultural Investment, and Productivity in the Customary Tenure Sector of Malawi." Economic Development and Cultural Change 50, no. 1: 77–99.

Plieninger, Tobias, et al. 2020. "Agroforestry for Sustainable Landscape Management." Sustainability Science 15, no. 5: 1401–15.

Poorter, L., F. Bongers, T. M. Aide, A. M. Almeyda Zambrano, et al. 2016. "Biomass Resilience of Neotropical Secondary Forests." Nature 530, no. 7589: 211–14.

Porras, I., D. N. Barton, M. Miranda, and A. Chacon-Cascante. 2013. Learning from 20 Years of Payments for Ecosystem Services in Costa Rica. London: International Institute for Environment and Development.

Putz, F. E., P. Sist, T. Fredericksen, and D. Dykstra. 2008. "Reduced-Impact Logging: Challenges and Opportunities." Forest Ecology and Management 256, no. 7: 1427–33.

Rice, J. 2010. "Carbon Stock Conditions and How Climate and Disturbance May Influence Carbon Dynamics on the Shoshone National Forest, Wyoming." IPCC: 1–16.

Rizvi, R. H., A. K. Handa, K. B. Sridhar, A. Kumar, S. Bhaskar, et al. 2020. Mapping Agroforestry and Trees Outside Forest. Jhansi and New Delhi: ICAR, Central Agroforestry Research Institute (CAFRI) and World Agroforestry (ICRAF), South Asia Regional Programme.

Rocheleau, D., and D. Edmunds. 1997. "Women, Men and Trees: Gender, Power and Property in Forest and Agrarian Landscapes." World Development 25, no. 8: 1351–71. https://doi.org/10.1016/S0305-750X(97)00036-3.

Schwartzman, S., B. Zimmerman, A. Parkhurst, and C. A. Peres. 2019. "The Kayapo: Indigenous Conservationists." Science Advances 5, no. 12: eaax6657.

Sharma, P., R. Rai, A. K. Singh, R. Singh, and A. K. Tiwari. 2020. "Impact of Agroforestry on Soil Carbon Sequestration and Agricultural Productivity in the Deccan Plateau, India." Journal of Environmental Management 268: 110783.

Shikakura, Y., T. Oguchi, X. Yu, M. Ohtani, T. Demura, A. Kikuchi, and K. N. Watanabe. 2022. "Transgenic Poplar Trees Overexpressing AtGolS2, a Stress-Responsive Galactinol Synthase Gene Derived from Arabidopsis thaliana, Improved Drought Tolerance in a Confined Field." Transgenic Research 31, no. 4–5: 579–91. https://doi.org/10.1007/s11248-022-00321-x.

Streck, C., O. Payne, D. Nepstad, and C. Maurer. 2019. "Forest Governance and the Politics of Conservation in Brazil and Indonesia: A Comparative Analysis." Global Environmental Change 55: 130–41.

Tacconi, L., S. Mahanty, and H. Suich. 2013. "The Livelihood Impacts of Payments for Environmental Services and Implications for REDD+." Society & Natural Resources 26, no. 6: 733–44. https://doi.org/10.1080/08941920.2012.724151.

Tavernier, H., L. Borremans, J. Bracke, B. Reubens, and E. Wauters. 2024. "Assessing the Potential of Different Economic Incentives for Stimulating Temperate Agroforestry: A Study in Flanders, Belgium." Agroforestry Systems 98: 1873–89. https://doi.org/10.1007/s10457-024-00996-7.

Tranchina, M., B. Reubens, M. Frey, M. Mele, and A. Mantino. 2024. "What Challenges Impede the Adoption of Agroforestry Practices? A Global Perspective through a Systematic Literature Review." Agroforestry Systems 98: 1817–37. https://doi.org/10.1007/s10457-024-00993-w.

9 Significance of Functional Traits in Nutrient Cycling and Soil Interaction

Abhishek Srivastava, Hariom Singh, Ajay, Rohit Kumar Singh, Suresh Choudhary, Lhingneivah Chongloi, and Deepak Kumar Jaiswal

9.1 INTRODUCTION

The development of trait-based ecology originated in the early 20th century but gained substantial momentum in the late 20th and early 21st centuries. Traits were driven by a growing need to understand the importance of biodiversity in ecosystem functioning, especially in the context of global environmental changes (Dawson et al., 2021). Functional traits refer to the Organisms' biological traits that influence their survival, growth, reproduction, and overall interactions with the environment. These traits include a wide range of attributes, such as morphological features (e.g., leaf shape, root architecture), physiological processes (e.g., photosynthesis, nutrient uptake efficiency), and phenological patterns (e.g., the timing of flowering, leaf senescence) (Pérez-Harguindeguy et al., 2013). Functional traits have become essential in ecological context studies to help understand how organisms engage with their environment and contribute to essential ecosystems, particularly nutrient cycling and soil health (Lavorel & Grigulis, 2012). It offers a unifying framework linking individual organisms to broader ecosystem processes, allowing for predictions about how ecosystems react to shifts in biodiversity, climate, and land use (Díaz et al., 2006; McGill et al., 2006).

Nutrient cycling is a critical ecosystem process involving the flow and transformation of key elements such as carbon (C), nitrogen (N), and phosphorus (P) through different biotic and abiotic components. The functional traits of plants, microbes, and soil fauna are pivotal in regulating the rates and pathways of nutrient cycling, influencing ecosystem productivity, stability, and resilience (Schmitz et al., 2015). Microbes, e.g., bacteria, fungi, and archaea, are the primary decomposers, decomposing organic matter and converting nutrients into forms that plants can readily use (Fierer et al., 2009). Microbial functional traits, such as enzyme production and substrate utilization patterns, determine the efficiency of nutrient cycling in soil (Schimel & Schaeffer, 2012). For example, enzymes produced by microbes degrade complex organic molecules into simple compounds that can be mineralized into inorganic nutrients, with the efficiency of these processes depending on the functional traits and combination of the microbial community (Sinsabaugh et al., 2008).

 DOI: 10.1201/9781003588580-9

Conversely, plants obtain nutrients from the soil and atmosphere, making their functional traits crucial in determining efficiency and the flow of nutrients through ecosystems (Wright et al., 2001). Important plant traits influencing nutrient cycling include root architecture, mycorrhizal associations, leaf nutrient content, and litter quality.

The interaction among plant and microbial functional traits is essential in determining the efficiency and sustainability of nutrient cycling (Bever et al., 1997). For example, plants that produce high-quality litter promote microbial communities that rapidly decompose organic matter, leading to quick nutrient mineralization and availability (Wardle et al., 2004). Conversely, plants that produce recalcitrant litter foster microbial communities that slow decomposition, resulting in soil nutrient immobilization and organic matter accumulation (Cornwell et al., 2008). These interactions can also be synergistic, where the functional traits of different organisms complement each other, or antagonistic, where they compete for resources or inhibit each other's activity (Pezzani et al., 2019). For example, the interaction between plant roots and mycorrhizal fungi can boost nutrient uptake and soil structure, leading to improved soil fertility and plant growth (Smith & Read, 2010).

By studying the functional traits of microbes and plants, scientists can develop strategies to promote sustainable nutrient cycling, reduce the impacts of climate change, and enhance ecosystem resilience and services. Thus, this chapter focuses on an in-depth exploration of the functional traits of plants and microbes, their relationships with soil, and their role in nutrient cycling, which aims to promote resilience in the ecosystem.

9.1.1 Plant Functional Traits

Plant functional traits are specific attributes of plants that directly affect their growth, reproduction, and survival, and they could be morphological, physiological, or phenological traits (Masarovičová et al., 2015). Understanding how plants interact with their environment is crucial for grasping broader ecological and agricultural processes. Plants' different forms and growth strategies, mainly concerning size, shape, and reproductive tactics, strongly affect nutrient cycling and soil interactions (Freschet et al., 2021b). Plants have varying growth forms that affect nutrient uptake, biomass production, and litter dynamics (Table 9.1). For example, the leaves of annual plants often exhibit high turnover rates, adding substantial amounts of litter to the soil within a relatively short period (Song et al., 2023). Such rapid input of organic matter in soil may increase nutrient cycling and general fertility (Coonan et al., 2020). Inversely, Perennial plants and woody species typically produce litter at a slower rate, contributing to the gradual accumulation of organic matter and supporting soil structure over time (Kuyper, 2005). There are reproductive strategies where plants devoting resources to seed production can reduce local soil nutrients. In contrast, plants investing in vegetative growth may increase the stability of soils and their organic matter content (Soto-Gómez & Pérez-Rodríguez, 2022). With the added stress of climate change, these functional traits are increasingly being tested, impacting forest structure, plant physiology, and ecological interactions.

TABLE 9.1
Major Functional Traits of Plants and Their Effects on Plant Performance and Ecosystem Functioning

Functional Trait	Description	Effects on Plant	Effects on Ecosystem	References
		Aboveground Traits		
Leaf area	Surface area of leaves	Larger leaves permit high light capture, but it may increase water loss	Influences photosynthesis, water relation, and microclimate regulation	Aparecido et al. (2017)
Specific leaf area	Leaf area per unit of dry mass	High SLA indicates fast growth and vice versa	It affects carbon sequestration and nutrient cycling	Anten and Selaya (2011)
Leaf N_2 content	Concentration of N_2 in leaves	Higher N_2 in leaves indicates higher photosynthesis	Influences nutrient cycling	Jia et al. (2021)
Plant height	The height of the plant	Taller plants may capture light but are susceptible to wind and heat stress	Creates competition for light, influencing species dynamics in changing climates	Valladares and Niinemets (2007)
Leaf lifespan	The duration of time the leaf remains functional	Longer leaf lifespan may enhance resource use efficiency	It affects litter quality and decomposition rates	Giweta (2020)
Wood density	Density reflecting structural and storage capacity	High wood density provides structural support in extreme weather but may reduce growth rates.	Contributes to long-term carbon storage, affecting ecosystem carbon balance	Hardiman et al. (2013)
Reproductive strategy	Method and frequency of reproduction	Perennials may be more climate-resilient	Affect ecosystem stability and species diversity under climate change	Kumar et al. (2019)
Photosynthetic pathway	Biochemical pathway for photosynthesis	Water use efficiency is higher in C4 and CAM	Alters carbon cycling and water use efficiency	Sage (2014)
Drought tolerance	Ability to survive under water-limited conditions	Improves survival in drier conditions	Enhances ecosystem resilience to climate variability in arid conditions	Norton et al. (2016)
Phenology	Life cycle events	Altered phenology can affect reproductive success	It affects species interactions and ecosystem dynamics	Yang and Rudolf (2010)
		Belowground Traits		
Root depth	Vertical extent of the root system	Deeper roots can access water from deeper soil layers	Influence groundwater recharge and drought resilience of ecosystems	Kühnhammer et al. (2023)

(Continued)

TABLE 9.1 ***(Continued)***
Major Functional Traits of Plants and Their Effects on Plant Performance and Ecosystem Functioning

Functional Trait	Description	Effects on Plant	Effects on Ecosystem	References
Root biomass	Total mass of roots in a given volume of soil	Improve nutrient and water uptake	Enhances soil structure and carbon sequestration and stabilizes soil against erosion	Reubens et al. (2007)
Root density	Root length/ unit volume of soil	Improves soil exploration for nutrients and water,	Enhances nutrient cycling and soil aeration	Ning et al. (2019)
Root exudation	Organic compounds are produced from roots.	Enhance nutrient availability, supporting plant resilience to nutrient limitations.	Affects soil microbial activity, nutrient cycling, and soil carbon storage	Canarini et al. (2019)
Root architecture	Spatial configuration of roots within the soil	Improve water and nutrient uptake efficiency,	Affects soil structure, water infiltration, and nutrient dynamics	Pierret et al. (2007)

9.1.1.1 Architecture of Roots

The root architecture of plants, including root length, branching patterns, and exudation of organic compounds, plays a crucial role in shaping soil structure. Roots create channels and pores in the soil, which improve aeration and water infiltration (Stuart Chapin et al., 2012) and are important in nutrient cycling and soil interactions (Awasthi & Laxmi, 2021). A deep-rooted plant can access water and nutrients from deeper soil layers that are not available to shallow-rooted species, which are pretty much stuck on surface soils (Freschet et al., 2021a). These plants penetrate the soil to anchor themselves during drought periods, which helps reduce erosion by holding particles together within a solid mass of aggregated materials or organic matter present in rocks like clay minerals (Jafari et al., 2022; Joly, 2021). This may lead to improved water infiltration, oxygen exchange, more carbon storage, and increased mineralization rates (Rumpel & Kögel-Knabner, 2011; Solomon et al., 2024) to support crop growth for good yields and healthy soil quality (Fageria & Moreira, 2011). By influencing water and nutrient cycling, root traits also play a crucial role in buffering ecosystems against the effects of climate change. Hence, understanding and managing these traits and their role in nutrients and water cycling is crucial for boosting ecosystem resilience, especially in the face of global climate change.

9.1.1.2 Leaf Traits

The rate of leaf decomposition, along with other leaf traits, is key to understanding nutrient cycling and soil function (Zhang et al., 2019). Leaf litter decomposition is an essential determinant of how rapidly nutrients are released in the soil (Ge et al., 2013). Some leaves have more nitrogen content and tissues with less lignification, usually decomposing quickly (Taylor et al., 1989), releasing their contents as plant food to the soil, enhancing their fertility and assisting plants' growth (Leghari et al., 2016). This may result in increased microbial activities in the soil, which further increases the rate at which these elements become available (Prescott, 2010). On the other hand, lignin-rich leaves decompose slowly due to their high lignin content, contributing to the long-term build-up of organic matter and the formation of humus in the soil (Rahman et al., 2013). This process improves soil structure and increases water retention capacity and nutrient-holding ability (Guo et al., 2019). Consequently, understanding leaf traits offers insights into how plants manipulate soil processes, nutrient cycles, and ecosystem functioning (Freschet et al., 2021b).

9.1.1.3 Effect of Climate Change on Plants' Functional Traits

With the shift in climatic factors and their impact on the ecosystem, traits face new selection pressure, which can significantly alter forest structure and function (Ahrens et al., 2020). Increased temperatures, shifts in precipitation patterns, and prolonged droughts place considerable stress on plant functional traits, which affects the delicate balance of forest ecosystems (Lloret & Batllori, 2021). Changes in root architecture, leaf traits, and growth forms may affect how plants interact with the environment and respond to changing conditions (Maurel & Nacry, 2020). Functional characteristics that adapt to climate change vary significantly among biome ecosystems, like shrublands, and may respond more to temperature changes than forested ecosystems (Boonman et al., 2022).

Climate change directly impacts trees by altering physiological mechanisms of photosynthesis and water transport and indirectly impacts through increased biotic stressors, such as insect infestations (Teshome et al., 2020). Trees must effectively utilize carbohydrates to maintain hydraulic function for growth, adjust osmoregulation at the cellular level, and produce defense-related compounds (Huang et al., 2023). In many instances, efficient control of transpiration is required for water homeostasis, which often results in reduced stomatal conductance and a smaller leaf area, which has cascading effects on photosynthesis and ultimately leads to reduced carbon uptake (López et al., 2021; McDowell et al., 2022). Carbon starvation is a primary mechanism underlying tree mortality under warmer and drier conditions (Camarero, 2021).

Climate change also disrupts the food web by influencing plant productivity, fruit and flower allocation, and plant secondary chemistry, which affects herbivores and other animals that depend on these plants (Kessler & Chautá, 2020). Drought and heat stress directly affect tree physiology through carbon starvation processes and are further complicated by biotic interactions, such as bark-boring beetles, emphasizing the interaction between climate stressors and biotic factors in driving tree mortality (Hartmann et al., 2022).

9.1.2 Microbial Functional Traits

Microbial functional traits encompass a broad array of characteristics that define the roles and behaviors of microorganisms within various ecosystems. These traits are essential for understanding microbial ecology, environmental processes, and

TABLE 9.2
Microbial Functional Traits and Their Effects on Ecosystem Functioning

Functional Trait	Microbes	Effects	Ecosystem Function	References
Nutrient cycling	*Nitrosomonas*, *Nitrobacter*	Nitrification	Enhances nitrogen availability to plants and soil fertility	Hayatsu et al. (2008)
Carbon sequestration	*Cyanobacteria*, Algae	Photosynthesis and carbon fixation	Reduces atmospheric CO_2 and contributes to carbon cycling	Singh and Ahluwalia (2013)
Decomposition	*Bacillus*, *Pseudomonas*	Breaking down of organic matter	Enhances soil health	Rao et al. (2019)
Symbiotic nitrogen fixation	*Rhizobium*, *Frankia*	Converts atmospheric nitrogen into available forms	It enhances plant growth, improves soil fertility, and is crucial for nutrient cycling	Singh et al. (2022)
Phosphorus solubilization	*Pseudomonas*, *Bacillus*	Converts insoluble phosphorus to available forms for plants	Enhances phosphorus availability to plants and maintains soil fertility	Arif et al. (2017)
Sulfate reduction	*Desulfovibrio*	Reduces sulfate to hydrogen sulfide	It influences sulfur cycling, metal corrosion, and soil and water chemistry	Hao et al. (1996)
Bioremediation	*Pseudomonas*, *Deinococcus*	Degradation of pollutants like heavy metals and hydrocarbons	Cleans contaminated environments, restores ecosystem health, supports biodiversity	Tripathi and Yadav (2016)
Antibiotic production	*Streptomyces*	Produces substances that kill other microbes	Controls microbial population dynamics influences soil and plant microbiomes	Hou and Kolodkin-Gal (2020)
Lignin degradation	White-rot fungi (e.g., *Phanerochaete*)	Breaks down lignin in plants	It aids in the decomposition of complex organic matter and plays a role in carbon cycling	Singh and Chen (2008)
Quorum sensing	*Aliivibrio fischeri*	Cell-to-cell communication for regulating population-based behaviors	Coordinates microbial activities, influences biofilm formation, impacts nutrient cycling	Wang et al. (2020)

the interaction between microorganisms and their environments (Yang, 2021). Microorganisms, including bacteria, fungi, archaea, and viruses, exhibit various functional traits that influence their behavior, interactions, and ecological functions (Table 9.2). These traits determine how microbes contribute to nutrient cycling, decomposition of organic matter, disease suppression, and other critical processes in ecosystems. Microbes possess diverse metabolic pathways that enable them to utilize different energy sources and carbon sources. For instance, some microorganisms are capable of photosynthesis, using sunlight as an energy source, while others rely on organic matter or inorganic compounds. The metabolic versatility of microbes allows them to thrive in various environments and contribute to different biogeochemical cycles.

9.1.2.1 Metabolic Versatility

Microorganisms exhibit a remarkable range of metabolic capabilities, utilizing various energy and carbon sources. For example, some bacteria can perform anaerobic respiration, using compounds like nitrate or sulfate as electron donors, while others can use photosynthesis or chemosynthesis (Madigan et al., 2019). These metabolic processes further highlight the versatility and adaptability of microorganisms in diverse environments.

9.1.2.2 Enzyme Production

Enzymes break down complex organic molecules, facilitating essential ecosystem processes. For instance, bacteria and fungi produce specific enzymes like cellulases to degrade cellulose in plant materials and proteases to break down proteins (Van Der Heijden et al., 2008). This enzymatic activity is fundamental to the decomposition of organic matter. It plays a vital role in nutrient cycling, ensuring that essential nutrients are returned to the environment for use by other organisms (Sahu et al., 2017).

9.1.2.3 Stress Tolerance

Microorganisms possess various characteristics that enable their survival across different environmental conditions. Extremophiles, such as thermophiles and acidophiles, are specially adapted to thrive in high-temperature and acidic environments (Canganella & Wiegel, 2011). These adaptations ensure their survival and expand our understanding of the limits within which life can exist (D'Amico et al., 2006).

9.1.2.4 Symbiotic Interactions

Many microorganisms establish symbiotic relationships with plants and animals, playing key roles in maintaining ecological balance. For example, nitrogen-fixing bacteria in the root nodules of legumes convert atmospheric nitrogen into forms that plants can absorb, thereby improving soil fertility (Pankievicz et al., 2019).

9.1.2.5 Pathogenicity and Defense Mechanisms

Some microorganisms have evolved traits that enable them to cause diseases in plants, animals, or humans. These include producing virulence factors and mechanisms to evade host immune responses (Finlay & Cossart, 1997).

9.1.2.6 Climate Change Effect on Microbial Traits

Climate change significantly influences microbial traits, altering their activity, composition, and interactions within ecosystems (Waldrop & Firestone, 2006). Shifts in temperature, precipitation, and atmospheric carbon dioxide levels modify microbial community structure and function, impacting processes like nutrient cycling, organic matter decomposition, and soil carbon storage (Meena et al., 2023). For instance, warming temperatures may accelerate microbial respiration, leading to increased carbon dioxide emissions from soils, while changes in moisture levels can affect microbial biomes and enzymatic functions, ultimately influencing soil fertility and plant productivity. Changes in microbial traits due to climate change can significantly impact ecosystem resilience and stability, potentially intensifying the effects of climate change on global carbon and nutrient cycles (Schimel et al., 2007).

9.1.2.7 Biofilm Formation

Many microorganisms form biofilms, complex communities of cells that adhere to surfaces. Biofilms protect microorganisms against environmental stressors and antibiotics, which are essential in ecological and clinical settings (Costerton et al., 1999).

9.1.2.8 Bioremediation and Degradation

Certain microorganisms can degrade pollutants and toxins, making them valuable for environmental cleanup processes. This includes the breakdown of hydrocarbons in oil spills and the detoxification of hazardous substances (Salari et al., 2022). Understanding microbial functional traits provides valuable insights into the ecological function of microorganisms and their usage in different fields. e.g., agriculture, medicine, and environmental management (Faucon et al., 2017).

9.1.3 Root-Soil-Microbial Interaction

As a determinant factor for plant development and nutrient availability, the relationships between soil, microorganisms, and plants may be influenced by several variables such as nutrient levels, plant species, and plant density (Adomako et al., 2022).

9.1.4 Root Exudates and Their Influence on Microbial Activity

Root exudates enable interaction between plants and soil microorganisms to impact plant health by stimulating beneficial microorganisms (Upadhyay et al., 2022). In agricultural fields, the development of roots is influenced by soil strength, and deeper roots use macropores to avoid mechanical resistance (Atkinson et al., 2020). Roots can decrease and increase the size of soil aggregates depending on the status of growth (Bai et al., 2021). In case large aggregates exist, roots grow amongst them, reducing the mean size of aggregates. Root growth increases aggregate size if the aggregates are small (Vanhees et al., 2021). Pioneer roots enhance the number of bio-pore channels when decomposing (Lee et al., 2020). Microorganisms in the rhizosphere are crucial for plant health, yield, and overall fitness. Microorganisms in the rhizosphere are crucial for plant health, yield, and overall fitness (Dlamini et al., 2022). Under environmental stress, roots secrete certain compounds into the soil environment, alter soil properties, and attract beneficial microbes that impact plant behavior (Vocciante et al., 2022).

Over the life cycle of a plant, changes in root exudation are known to affect the composition of the rhizosphere microbiome (Steinauer et al., 2023). Low nutrient conditions of soil promote a fungal food web to develop, while high nutrient conditions promote bacterial food (Zhao et al., 2021). It is postulated that root exudation at various stages of development affects the quantity of nutrients available in soil and the microbial populations that further impact plant growth (Tian et al., 2020). The critical point in the priming effect of forest soils is that understory grasses function similarly to canopy trees (Michel et al., 2023). Grasses allocate a more significant percentage of photosynthates to belowground organs, leading to rapid responses of soils and higher microbial activity in grasslands than forest lands (Zhou et al., 2022).

9.1.5 Resource Allocation, Decomposition, and Nutrient Cycling

Microbial competition for resources is another critical interaction mechanism in soil. Different microbial species compete for nutrients, water, and space, which can influence microbial community composition and activity (Bier et al., 2015). This contest can be fierce, particularly in soils lacking nutrients where microbes need to adjust to scarce resources. Resource allocation tactics, like the production of antimicrobial compounds by certain microorganisms, can influence the competitive relationships among soil microbes (Malik et al., 2020).

Microbes are central to the breakdown of organic matter and nutrient cycling in soils. Decomposers, such as bacteria and fungi, break down plant residues and other organic materials, releasing essential nutrients into the soil (Rashid et al., 2016). This process is vital for sustaining soil fertility and promoting plant growth. The efficiency of decomposition and nutrient release depends on microbial diversity and activity, affected by environmental factors, temperature, and moisture (Jiaying et al., 2024).

9.1.6 Functional Traits' Impact on Carbon Storage of Soil

Carbon storage is essential for maintaining ecosystem resilience and regulating the climate. The amount and stability of carbon stored in soils are significantly affected by the functional traits of plants, soil organisms, and microbes (Trivedi et al., 2018). These characteristics affect many processes that lead to long-term sequestration of carbon, the breakdown of organic things, the creation of soil structure, and the stabilization of carbon within soil matrices (Bronick & Lal, 2005). Carbon Input Plant traits, including root architecture & biomass allocation and litter quality, are fundamental to determining how much carbon is introduced into the soil and how it is stored (De Deyn et al., 2008; Tameirão et al., 2021).

For instance, root characteristics that encourage deep carbon allocation have been demonstrated to improve soil carbon storage, particularly in habitats like grasslands and forests. As roots deteriorate, Root biomass contributes organic matter to the soil, enhancing the soil's carbon content (Pardon et al., 2017). Likewise, the amount of lignin, cellulose, and nitrogen in plant litter affects its chemical composition, affecting how quickly it breaks down and releases or stores carbon in the soil (Findlay, 2021). Plant species high in lignin create litter that breaks down more slowly, increasing the organic matter that builds up in the soil and improving carbon storage (Prescott & Vesterdal, 2021). Furthermore, stable organic matter (SOM), a vital carbon source in soils, can form from litter rich in resistant compounds (Gul & Whalen, 2022).

As aforementioned, in soil ecosystems, Microorganisms like bacteria and fungi are also essential for breaking down organic matter and cycling carbon (Bhattacharyya et al., 2022). With high diversity and functional redundancy, microbial communities break down organic materials more efficiently, which, depending on the environment, results in either the quick emission of carbon as carbon dioxide or its incorporation into stable forms of organic matter (Hu et al., 2022). Moreover, the other factor affecting soil carbon dynamics is microbial carbon utilization efficiency, measured as the ratio of carbon assimilated into biomass to total carbon taken up (Spohn et al., 2016). With high Carbon Use Efficiency, more carbon goes into microbial biomass, an eventual route to soil organic matter (Tao et al., 2023). In contrast, low Carbon Use Efficiency reduces this potential for long-term carbon storage by being associated with increased carbon dioxide emissions from soil respiration (Sauvadet et al., 2018).

9.1.7 Implications for Management and Conservation

9.1.7.1 Agricultural Management

In agriculture, functional trait refers to the heritable-morphological, physiological, and phonological characteristics of a crop, livestock, or any organism that impact their performance, productivity, and interaction with the environment, which can be used to improve resource use efficiency like water, nutrients, or land use pattern (Yadav et al., 2017; Zanklan et al., 2018). These traits are vital for understanding how plants respond to environmental conditions and for breeding and managing crops to address food security challenges in a changing climate. The selection and enhancement of functional traits in crops can improve resilience to stress, optimize resource use, and increase yield stability. The changing climate system and its adverse effect on the growth of crop varieties that are more resilient to environmental stresses, with higher yields, and requiring fewer inputs, is a prime focus (Schröder et al., 2019). Morphological and anatomical traits like longer root length or deep root system, volume of root, dry weight of root, thickness of root, phenological traits like early maturity, late flowering, seedling vitality, photosensitivity, weed aggression, and physiological and biochemical characteristics like stomatal conductance, epicuticular wax, heat shock proteins, water potential, and water usage are critical for developing climate resilient varieties (Ranjith & Srinivasa Rao, 2021). Root architecture is the most promising characteristic for drought avoidance in breeding. For instance, wheat varieties with longer root systems can tap into water from lower soil layers, making them more resilient to dry conditions (Ober et al., 2021). Functional traits of the leaf, e.g., area, specific area, dry matter content, and turgor loss point, also predict drought resistance in plants. Thus, Low leaf area values provide resistance to water deficits in plant tissues (Villagra et al., 2023). Likewise, the high dry matter content in leaves can help maintain leaf structure despite turgor loss by ensuring a lower ratio of water to structural components (Markesteijn et al., 2011).

Concerning pest and disease management, different groups of natural enemies may react differently to crop plant traits due to their distinct biology and the specific ecological roles they occupy (Peterson et al., 2016), which may include plant traits like toxin production and plant nutrients. For instance, Gossypol in cotton leaves and seeds deters a wide variety of lepidopteran pests (Ismail, 2020). An indirect relationship between plant nutrients and natural enemies happens when nutrient-induced delays in herbivore growth and development prolong the time herbivores are susceptible to predation (Peterson et al., 2016). The natural enemy the spined soldier bug, *Podisus maculiventris*, controlled *Epilachna varivestis* on crop varieties that reduce the herbivore growth rate (Sahayaraj & Hassan, 2023).

Understanding the functional traits of a particular crop will give a better idea of how effective management practices can be adopted in agriculture. For instance, weed management practices require understanding seed dormancy and germination, seed dispersal mechanisms, phenology, drought tolerance, herbicide resistance, allelopathy, etc., with their interaction with different cropping systems (Gaba et al., 2014). Tilling of soil inhibits the establishment of chamaephyte phanerophytes, and most hemicryptophyte weeds, while its impact on therophytes and perennial species (such as geophytes with bulbs, tubers, or rhizomes) is less predictable (Alhammad et al., 2023). In the case of a no-tillage system, Perennial and biennial weed species are often favored by inconsistencies in soil tillage owing to diverse weeds in the cropping system (Coleman et al., 2024). Likewise, pollination plays a crucial role in regulating ecosystem services in nature. Functional traits of pollinators accompanied by farming systems improved crop production. Studies have shown that Pollinator abundance, species richness, and bee diversity were notably higher in agroforestry systems compared to arable systems (Jha & Vandermeer, 2010). Intercropping of plants with complementary interaction and breeding approaches that focus on functional traits of plants with particular growing environments or soil conditions could supplement successful crop management.

9.1.7.2 Ecosystem Restoration

Ecosystems offer essential services like nutrient cycling, water purification, climate regulation, and cultural benefits. However, both human-induced and natural changes are significantly impacting these ecosystems, potentially diminishing their capacity to provide these services (Fauzi et al., 2023). As such, ecologists have adopted several approaches to assess the components of biodiversity accurately, the factors affecting it, and its impact on ecosystem structure and function. Restoring degraded ecosystems represents one of the current century's pre-eminent environmental challenges. A globally coordinated ecological restoration policy has catalyzed ambitious initiatives to address biodiversity and ecosystem services declines. It also aims at restoring the species composition of communities to their pre-disturbance state (Triatmojo, 2019) either by reintroducing native species or by encouraging spontaneous re-colonization on degraded sites (Holmes et al., 2020; Watson, 2009). Recently, a new approach that focuses on the diversity of an organism's functional traits has evolved. This approach enhances ecological understanding by focusing on characteristics that define how organisms interact with their surrounding physical, chemical, and biological environments (Freschet et al., 2021b). Applying a functional trait-based framework in ecological restoration is gaining global attention. However, the limited understanding of functional traits and their connection to ecosystem services poses a significant challenge to effectively implementing this approach.

Selecting plants and their species is an essential step in restoring ecology. Pioneer and climax species typically exhibit contrasting traits such as wood density, mass of seed, mass of leaf per area, and leaf density, with pioneers generally having reduced values and climaxes having higher values for these characteristics (Jones et al., 2022; Triatmojo, 2019). In restoration projects, species are often chosen along a succession timeline for their life history strategies. They are planted in combinations that aim to replicate natural forest recovery more quickly (Geneletti, 2019).

Species mixes often consist of fast-growing pioneer species and slow-growing climax species. The idea is to have the pioneer species quickly establish canopy cover or outcompete native grasses and weeds, while the climax species gradually replace the pioneers over time.

An effective strategy should address the high mortality rates of pioneer species, which can delay canopy closure or impede succession. Conversely, while climax species may have higher survival rates, they might hinder the speed of canopy closure and site capture (Petrokas, 2020). Functional traits of plant species like wood density, seed mass, and maximum tree height impacted seedling growth and survival in restoration. Alternatively, particular trait values may consistently confer a more significant competitive advantage, irrespective of trait similarity. For instance, taller plant species may routinely outcompete their shorter neighbors for light, or those with higher wood density and specific leaf area may exert stronger competitive influences on neighbors with lower values for these traits (Kunstler et al., 2012).

Wood density was found to be a highly effective indicator of seedling survival. Species with higher wood density experienced greater survival rates than others (Visser et al., 2016). Previous reports have demonstrated that small-seeded animal-dispersed species with low wood density are more appropriate for restoration plantings as they have higher development rates (Sangsupan et al., 2018). Species with large seeds that are dispersed by animals often adopt slow life history strategies, taking decades or even centuries to reach their full carbon storage potential. In contrast, small-seeded species usually achieve this much more rapidly (Hodgson et al., 2020). Higher plant growth also suppresses the weed species that often dominate during the early restoration period. In restoration, more excellent and reliable seed supply and faster growth rates help reduce the costs related to nursery seedling production and site maintenance in restoration projects (Patil et al., 2020).

The diversity of microbial life, including archaea, fungi, bacteria, and protists, is fundamental to the key ecological processes and overall biodiversity within an ecosystem (George et al., 2019). Microorganisms play a critical role in strengthening ecosystems against biotic and abiotic stressors, thereby enhancing resilience. Research highlights the mutual effects of plant and microbial biodiversity on essential ecosystem processes like nutrient cycling and soil organic matter dynamics, including in agricultural landscapes and systems that support plant growth. The stability and functionality of ecosystems are closely linked to the diversity of microbial life (Varma & Choudhary, 2019). Additionally, the potential of microorganisms in bioremediation is exemplified by species like *Pseudomonas putida* and *Aspergillus niger*, which are effective in environmental clean-up. Plant functional traits, such as the composition of root exudates, play a crucial role in soil carbon sequestration and nutrient availability by regulating carbon assimilation from the soil through various processes (Lori et al., 2020).

Recent studies have explored how the functional characteristics of plants influence soil microbial communities and ecosystem functions. Fast-growing plant species promote bacterial communities and accelerate nutrient cycling, while slow-growing species exert distinct effects. Successful ecological restoration necessitates considering the comprehensive array of plant functional traits supporting human and ecosystem well-being.

9.1.8 Bottlenecks and Future Directions

The exploration of functional traits in nutrient cycling and soil interactions is hindered by several critical bottlenecks, reflecting the complexity of ecological systems and the limitations of current scientific approaches. One of the foremost challenges is the incomplete understanding of trait variability and its ecological consequences. Functional traits exhibit significant variability across species, populations, and environmental conditions, which complicates efforts to predict how changes in trait composition might influence nutrient cycling and soil interactions, especially under changing environmental factors such as climate change (Cizauskas et al., 2017). Another challenge is the scarcity of comprehensive datasets that effectively link specific functional traits to nutrient cycling processes. The absence of robust, large-scale datasets that integrate functional traits with nutrient cycling metrics limits the ability to develop predictive models essential for understanding how shifts in trait distributions—due to factors like species invasions, extinctions, or environmental changes—impact ecosystem functions (Renault et al., 2022).

Moreover, many studies are conducted over short timescales, failing to capture the long-term trends in nutrient cycling and soil interactions in forestland or grasslands where nutrient cycling processes unfold over extended periods (Jo et al., 2017). Additionally, the context-dependence of functional traits poses a significant challenge. The influence of a particular trait on nutrient cycling can vary widely depending on the environmental context. For instance, one trait may promote nutrient retention in one context but contribute to nutrient loss in another, making it challenging to generalize findings across different ecosystems.

Addressing these bottlenecks will require a concerted effort to integrate functional traits more effectively into ecosystem models. Future research should focus on developing comprehensive, trait-based models that link specific traits to nutrient-cycling processes across different environmental contexts. This will involve expanding current datasets to include more species, traits, and ecological conditions and improving the temporal and spatial scales of studies. Technology and data analytics advances offer promising avenues for overcoming some of the current limitations. For instance, remote sensing technologies and advanced statistical models can bridge the gap between trait data and ecosystem processes, allowing for more accurate predictions of how changes in trait composition will influence nutrient cycling and soil relationships (Funk et al., 2017). Additionally, interdisciplinary studies and collaborative research efforts that integrate ecologists, soil scientists, and data scientists will be crucial in developing the tools and models required to deepen our understanding of the role of functional traits in nutrient cycling and soil interactions. This interdisciplinary approach will help build the required predictive capacity to manage ecosystems in the face of rapid environmental change.

9.2 CONCLUSION

In conclusion, functional traits are crucial in nutrient cycling and soil interactions, linking plant physiology, microbial activity, and ecosystem processes. These traits, encompassing characteristics like root depth, nutrient uptake efficiency, and microbial decomposition rates, directly influence how ecosystems manage nutrient availability and maintain soil fertility. The diversity of functional traits within an ecosystem enhances nutrient cycling efficiency and resilience. For instance, plants with deep root systems or efficient mycorrhizal associations improve nutrient acquisition, while microbial traits regulate decomposition and mineralization. This interaction creates a dynamic system that supports sustained plant growth and ecosystem stability. Understanding and utilizing functional traits in agriculture can optimize nutrient use efficiency, reduce reliance on fertilizers, and minimize environmental impacts. A diverse range of functional characteristics in natural ecosystems contributes to resilience, enabling ecosystems to adapt to disturbances and maintain essential processes. Ultimately, studying functional traits offers a valuable understanding of the mechanisms of nutrient cycling and soil interactions, offering a foundation for sustainable ecosystem management in the face of global environmental challenges.

REFERENCES

Adomako, M. O., Roiloa, S., & Yu, F. H. (2022). Potential roles of soil microorganisms in regulating the effect of soil nutrient heterogeneity on plant performance. *Microorganisms*. https://doi.org/10.3390/microorganisms10122399

Ahrens, C. W., Andrew, M. E., Mazanec, R. A., Ruthrof, K. X., Challis, A., Hardy, G., Byrne, M., Tissue, D. T., & Rymer, P. D. (2020). Plant functional traits differ in adaptability and are predicted to be differentially affected by climate change. *Ecology and Evolution*. https://doi.org/10.1002/ece3.5890

Alhammad, B. A., Roy, D. K., Ranjan, S., Padhan, S. R., Sow, S., Nath, D., Seleiman, M. F., & Gitari, H. (2023). Conservation tillage and weed management influencing weed dynamics, crop performance, soil properties, and profitability in a rice–wheat–greengram system in the eastern Indo-Gangetic Plain. *Agronomy*. https://doi.org/10.3390/agronomy13071953

Anten, N. P. R., & Selaya, N. G. (2011). *Ecophysiology of secondary succession in tropical moist forest: Scaling from individual traits to whole-plant performance*. https://doi.org/10.1007/978-94-007-1242-3_16

Aparecido, L. M. T., Miller, G. R., Cahill, A. T., & Moore, G. W. (2017). Leaf surface traits and water storage retention affect photosynthetic responses to leaf surface wetness among wet tropical forest and semiarid savanna plants. *Tree Physiology*. https://doi.org/10.1093/treephys/tpx092

Arif, M. S., Shahzad, S. M., Yasmeen, T., Riaz, M., Ashraf, M., Ashraf, M. A., Mubarik, M. S., & Kausar, R. (2017). Improving plant phosphorus (p) acquisition by phosphate-solubilizing bacteria. In *Essential plant nutrients: Uptake, use efficiency, and management*. https://doi.org/10.1007/978-3-319-58841-4_21

Atkinson, J. A., Hawkesford, M. J., Whalley, W. R., Zhou, H., & Mooney, S. J. (2020). Soil strength influences wheat root interactions with soil macropores. *Plant Cell and Environment*. https://doi.org/10.1111/pce.13659

Awasthi, P., & Laxmi, A. (2021). *Root architectural plasticity in changing nutrient availability*. https://doi.org/10.1007/978-3-030-84985-6_2

Bai, T., Wang, P., Ye, C., & Hu, S. (2021). Form of nitrogen input dominates n effects on root growth and soil aggregation: A meta-analysis. *Soil Biology and Biochemistry*. https://doi.org/10.1016/j.soilbio.2021.108251

Bever, J. D., Westover, K. M., & Antonovics, J. (1997). Incorporating the soil community into plant population dynamics: The utility of the feedback approach. *Journal of Ecology*, *85*(5), 561–573.

Bhattacharyya, S. S., Ros, G. H., Furtak, K., Iqbal, H. M. N., & Parra-Saldívar, R. (2022). Soil carbon sequestration – An interplay between soil microbial community and soil organic matter dynamics. *Science of the Total Environment, 815*. https://doi.org/10.1016/j.scitotenv.2022.152928

Bier, R. L., Bernhardt, E. S., Boot, C. M., Graham, E. B., Hall, E. K., Lennon, J. T., Nemergut, D. R., Osborne, B. B., Ruiz-González, C., Schimel, J. P., Waldrop, M. P., & Wallenstein, M. D. (2015). Linking microbial community structure and microbial processes: An empirical and conceptual overview. *FEMS Microbiology Ecology*. https://doi.org/10.1093/femsec/fiv113

Boonman, C. C. F., Huijbregts, M. A. J., Benítez-López, A., Schipper, A. M., Thuiller, W., & Santini, L. (2022). Trait-based projections of climate change effects on global biome distributions. *Diversity and Distributions*. https://doi.org/10.1111/ddi.13431

Bronick, C. J., & Lal, R. (2005). Soil structure and management: A review. *Geoderma*. https://doi.org/10.1016/j.geoderma.2004.03.005

Camarero, J. J. (2021). The drought–dieback–death conundrum in trees and forests. *Plant Ecology and Diversity*. https://doi.org/10.1080/17550874.2021.1961172

Canarini, A., Kaiser, C., Merchant, A., Richter, A., & Wanek, W. (2019). Root exudation of primary metabolites: Mechanisms and their roles in plant responses to environmental stimuli. *Frontiers in Plant Science*. https://doi.org/10.3389/fpls.2019.00157

Canganella, F., & Wiegel, J. (2011). Extremophiles: From abyssal to terrestrial ecosystems and possibly beyond. *Naturwissenschaften*. https://doi.org/10.1007/s00114-011-0775-2

Cizauskas, C. A., Carlson, C. J., Burgio, K. R., Clements, C. F., Dougherty, E. R., Harris, N. C., & Phillips, A. J. (2017). Parasite vulnerability to climate change: An evidence-based functional trait approach. *Royal Society Open Science*. https://doi.org/10.1098/rsos.160535

Coleman, M. J., Kristiansen, P., Sindel, B. M., & Fyfe, C. (2024). Imperatives for integrated weed management in vegetable production: Evaluating research and adoption. *Weed Biology and Management*. https://doi.org/10.1111/wbm.12285

Coonan, E. C., Kirkby, C. A., Kirkegaard, J. A., Amidy, M. R., Strong, C. L., & Richardson, A. E. (2020). Microorganisms and nutrient stoichiometry as mediators of soil organic matter dynamics. *Nutrient Cycling in Agroecosystems*. https://doi.org/10.1007/s10705-020-10076-8

Cornwell, W. K., Cornelissen, J. H. C., Amatangelo, K., Dorrepaal, E., Eviner, V. T., Godoy, O., Hobbie, S. E., Hoorens, B., Kurokawa, H., Pérez-Harguindeguy, N., Quested, H. M., Santiago, L. S., Wardle, D. A., Wright, I. J., Aerts, R., Allison, S. D., Van Bodegom, P., Brovkin, V., Chatain, A., … Westoby, M. (2008). Plant species traits are the predominant control on litter decomposition rates within biomes worldwide. *Ecology Letters*. https://doi.org/10.1111/j.1461-0248.2008.01219.x

Costerton, J. W., Stewart, P. S., & Greenberg, E. P. (1999). Bacterial biofilms: A common cause of persistent infections. *Science*. https://doi.org/10.1126/science.284.5418.1318

Dawson, N. M., Coolsaet, B., Sterling, E. J., Loveridge, R., Gross-Camp, N. D., Wongbusarakum, S., Sangha, K. K., Scherl, L. M., Phan, H. P., Zafra-Calvo, N., Lavey, W. G., Byakagaba, P., Idrobo, C. J., Chenet, A., Bennett, N. J., Mansourian, S., & Rosado-May, F. J. (2021). The role of indigenous peoples and local communities in effective and equitable conservation. *Ecology and Society, 26*(3), 19. https://doi.org/10.5751/ES-12625-260319

D'Amico, S., Collins, T., Marx, J. C., Feller, G., Gerday, C., & Gerday, C. (2006). Psychrophilic microorganisms: Challenges for life. *EMBO reports, 7*(4), 385–389. https://www.embopress.org/doi/full/10.1038/sj.embor.7400662

De Deyn, G. B., Cornelissen, J. H. C., & Bardgett, R. D. (2008). Plant functional traits and soil carbon sequestration in contrasting biomes. *Ecology Letters*. https://doi.org/10.1111/j.1461-0248.2008.01164.x

Díaz, S., Tilman, D., Fargione, J., Chapin, F. S., III, Dirzo, R., Kitzberger, T., Gemmill, B., Zobel, M., Vilà, M., Mitchell, C., Wilby, A, Daily, G. C., Galetti, M., Laurance, W. F.,

Pretty, J., Naylor, R., Power, A, Harvell, A, Potts, S., ... Eardley, C. (2006). Biodiversity regulation of ecosystem services (pp. 297–329). In *Ecosystems and human well-being: Current state and trends.*

Dlamini, S. P., Akanmu, A. O., & Babalola, O. O. (2022). Rhizospheric microorganisms: The gateway to a sustainable plant health. *Frontiers in Sustainable Food Systems.* https://doi.org/10.3389/fsufs.2022.925802

Fageria, N. K., & Moreira, A. (2011). Chapter four – The role of mineral nutrition on root growth of crop plants. In D. L. Sparks (Ed.), *Advances in Agronomy* (Vol. 110, pp. 251–331). Academic Press. https://doi.org/10.1016/B978-0-12-385531-2.00004-9

Faucon, M. P., Houben, D., & Lambers, H. (2017). Plant functional traits: Soil and ecosystem services. *Trends in Plant Science.* https://doi.org/10.1016/j.tplants.2017.01.005

Fauzi, A. I., Azizah, N., Yati, E., Atmojo, A. T., Rohman, A., Putra, R., Rahadianto, M. A. E., Ramadhanti, D., Ardani, N. H., Robbani, B. F., Nuha, M. U., Perdana, A. M. P., Sakti, A. D., Aufaristama, M., & Wikantika, K. (2023). Potential loss of ecosystem service value due to vessel activity expansion in Indonesian marine protected areas. *ISPRS International Journal of Geo-Information.* https://doi.org/10.3390/ijgi12020075

Fierer, N., Strickland, M. S., Liptzin, D., Bradford, M. A., & Cleveland, C. C. (2009). Global patterns in belowground communities. *Ecology Letters.* https://doi.org/10.1111/j.1461-0248.2009.01360.x

Findlay, S. E. G. (2021). Chapter 4 – Organic matter decomposition. In K. C. Weathers, D. L. Strayer, & G. E. Likens (Eds.), *Fundamentals of ecosystem science* (2nd ed., pp. 81–102). Academic Press. https://doi.org/10.1016/B978-0-12-812762-9.00004-6

Finlay, B. B., & Cossart, P. (1997). Exploitation of mammalian host cell functions by bacterial pathogens. *Science.* https://doi.org/10.1126/science.276.5313.718

Freschet, G. T., Pagès, L., Iversen, C. M., Comas, L. H., Rewald, B., Roumet, C., Klimešová, J., Zadworny, M., Poorter, H., Postma, J. A., Adams, T. S., Bagniewska-Zadworna, A., Bengough, A. G., Blancaflor, E. B., Brunner, I., Cornelissen, J. H. C., Garnier, E., Gessler, A., Hobbie, S. E., ... McCormack, M. L. (2021a). A starting guide to root ecology: Strengthening ecological concepts and standardising root classification, sampling, processing and trait measurements. *New Phytologist.* https://doi.org/10.1111/nph.17572

Freschet, G. T., Roumet, C., Comas, L. H., Weemstra, M., Bengough, A. G., Rewald, B., Bardgett, R. D., De Deyn, G. B., Johnson, D., Klimešová, J., Lukac, M., McCormack, M. L., Meier, I. C., Pagès, L., Poorter, H., Prieto, I., Wurzburger, N., Zadworny, M., Bagniewska-Zadworna, A., ... Stokes, A. (2021b). Root traits as drivers of plant and ecosystem functioning: Current understanding, pitfalls and future research needs. *New Phytologist.* https://doi.org/10.1111/nph.17072

Funk, J. L., Larson, J. E., Ames, G. M., Butterfield, B. J., Cavender-Bares, J., Firn, J., Laughlin, D. C., Sutton-Grier, A. E., Williams, L., & Wright, J. (2017). Revisiting the Holy Grail: Using plant functional traits to understand ecological processes. *Biological Reviews.* https://doi.org/10.1111/brv.12275

Gaba, S., Fried, G., Kazakou, E., Chauvel, B., & Navas, M. L. (2014). Agroecological weed control using a functional approach: A review of cropping systems diversity. *Agronomy for Sustainable Development.* https://doi.org/10.1007/s13593-013-0166-5

Ge, X., Zeng, L., Xiao, W., Huang, Z., Geng, X., & Tan, B. (2013). Effect of litter substrate quality and soil nutrients on forest litter decomposition: A review. *Acta Ecologica Sinica.* https://doi.org/10.1016/j.chnaes.2013.01.006

Geneletti, D. (2019). Forest Landscape Restoration. In *Multicriteria analysis for environmental decision-making.* https://doi.org/10.2307/j.ctvhhhg9x.12

George, P. B. L., Lallias, D., Creer, S., Seaton, F. M., Kenny, J. G., Eccles, R. M., Griffiths, R. I., Lebron, I., Emmett, B. A., Robinson, D. A., & Jones, D. L. (2019). Divergent national-scale trends of microbial and animal biodiversity revealed across diverse temperate soil ecosystems. *Nature Communications.* https://doi.org/10.1038/s41467-019-09031-1

Giweta, M. (2020). Role of litter production and its decomposition, and factors affecting the processes in a tropical forest ecosystem: A review. *Journal of Ecology and Environment*. https://doi.org/10.1186/s41610-020-0151-2

Gul, S., & Whalen, J. K. (2022). Chapter six – Perspectives and strategies to increase the microbial-derived soil organic matter that persists in agroecosystems. In D. L. Sparks (Ed.), *Advances in Agronomy* (Vol. 175, pp. 347–401). Academic Press. https://doi.org/10.1016/bs.agron.2022.04.004

Guo, X-x., Liu, H-t., & Wu, S.-b. (2019). Humic substances developed during organic waste composting: Formation mechanisms, structural properties, and agronomic functions. *Science of the Total Environment*. https://doi.org/10.1016/j.scitotenv.2019.01.137

Hao, O. J., Chen, J. M., Huang, L., & Buglass, R. L. (1996). Sulfate-reducing bacteria. *Journal of Environmental Science and Health – Part C Environmental Carcinogenesis and Ecotoxicology Reviews*. https://doi.org/10.5956/jriet.18.229

Hardiman, B. S., Gough, C. M., Halperin, A., Hofmeister, K. L., Nave, L. E., Bohrer, G., & Curtis, P. S. (2013). Maintaining high rates of carbon storage in old forests: A mechanism linking canopy structure to forest function. *Forest Ecology and Management*. https://doi.org/10.1016/j.foreco.2013.02.031

Hartmann, H., Bastos, A., Das, A. J., Esquivel-Muelbert, A., Hammond, W. M., Martínez-Vilalta, J., Mcdowell, N. G., Powers, J. S., Pugh, T. A. M., Ruthrof, K. X., & Allen, C. D. (2022). Climate change risks to global forest health: Emergence of unexpected events of elevated tree mortality worldwide. *Annual Review of Plant Biology*. https://doi.org/10.1146/annurev-arplant-102820-012804

Hayatsu, M., Tago, K., & Saito, M. (2008). Various players in the nitrogen cycle: Diversity and functions of the microorganisms involved in nitrification and denitrification. *Soil Science and Plant Nutrition*. https://doi.org/10.1111/j.1747-0765.2007.00195.x

Hodgson, J. G., Montserrat Marti, G., Šerá, B., Jones, G., Bogaard, A., Charles, M., Font, X., Ater, M., Taleb, A., Santini, B. A., Hmimsa, Y., Palmer, C., Wilson, P. J., Band, S. R., Styring, A., Diffey, C., Green, L., Nitsch, E., Stroud, E., & Warham, G. (2020). Seed size, number and strategies in annual plants: A comparative functional analysis and synthesis. *Annals of Botany*. https://doi.org/10.1093/aob/mcaa151

Holmes, P. M., Esler, K. J., van Wilgen, B. W., & Richardson, D. M. (2020). Ecological restoration of ecosystems degraded by invasive alien plants in South African Fynbos: Is spontaneous succession a viable strategy? *Transactions of the Royal Society of South Africa*. https://doi.org/10.1080/0035919X.2020.1781291

Hou, Q., & Kolodkin-Gal, I. (2020). Harvesting the complex pathways of antibiotic production and resistance of soil bacilli for optimizing plant microbiome. In *FEMS microbiology ecology*. https://doi.org/10.1093/femsec/fiaa142

Hu, A., Jang, K. S., Meng, F., Stegen, J., Tanentzap, A. J., Choi, M., Lennon, J. T., Soininen, J., & Wang, J. (2022). Microbial and environmental processes shape the link between organic matter functional traits and composition. *Environmental Science and Technology*. https://doi.org/10.1021/acs.est.2c01432

Huang, J., Hartmann, H., Ogaya, R., Schöning, I., Reichelt, M., Gershenzon, J., & Peñuelas, J. (2023). Hormone and carbohydrate regulation of defense secondary metabolites in a Mediterranean forest during drought. *Environmental and Experimental Botany*. https://doi.org/10.1016/j.envexpbot.2023.105298

Ismail, S. (2020). Gossypol as a natural insecticide in cotton plants against cotton thrips and pink bollworm. *Progress in Chemical and Biochemical Research*, *4*(December 2020), 68–79. https://doi.org/10.22034/pcbr.2021.120448

Jafari, M., Tahmoures, M., Ehteram, M., Ghorbani, M., & Panahi, F. (2022). Soil erosion control in drylands. In *Soil erosion control in drylands*. https://doi.org/10.1007/978-3-031-04859-3

Jha, S., & Vandermeer, J. H. (2010). Impacts of coffee agroforestry management on tropical bee communities. *Biological Conservation*. https://doi.org/10.1016/j.biocon.2010.03.017

Jia, M., Colombo, R., Rossini, M., Celesti, M., Zhu, J., Cogliati, S., Cheng, T., Tian, Y., Zhu, Y., Cao, W., & Yao, X. (2021). Estimation of leaf nitrogen content and photosynthetic nitrogen use efficiency in wheat using sun-induced chlorophyll fluorescence at the leaf and canopy scales. *European Journal of Agronomy*. https://doi.org/10.1016/j.eja.2020.126192

Jiaying, L., Changjun, D., Chao, T., Weixi, Z., Xiaohua, S., & Wenxu, Z. (2024). Impacts of litter microbial community on litter decomposition in the absence of soil microorganisms. *Applied and Environmental Microbiology*, *90*(4), e00239–24. https://doi.org/10.1128/aem.00239-24

Jo, I., Fridley, J. D., & Frank, D. A. (2017). Invasive plants accelerate nitrogen cycling: Evidence from experimental woody monocultures. *Journal of Ecology*. https://doi.org/10.1111/1365-2745.12732

Joly, F. (2021). Water falling onto soil and the effects it produces. In *Mankind and deserts 2*. https://doi.org/10.1002/9781119808275.ch1

Jones, J., Ellison, D., Ferraz, S., Lara, A., Wei, X., & Zhang, Z. (2022). Forest restoration and hydrology. *Forest Ecology and Management*. https://doi.org/10.1016/j.foreco.2022.120342

Kessler, A., & Chautá, A. (2020). The ecological consequences of herbivore-induced plant responses on plant-pollinator interactions. *Emerging Topics in Life Sciences*. https://doi.org/10.1042/ETLS20190121

Kühnhammer, K., van Haren, J., Kübert, A., Bailey, K., Dubbert, M., Hu, J., Ladd, S. N., Meredith, L. K., Werner, C., & Beyer, M. (2023). Deep roots mitigate drought impacts on tropical trees despite limited quantitative contribution to transpiration. *Science of the Total Environment*. https://doi.org/10.1016/j.scitotenv.2023.164763

Kumar, S., Saxena, S., Rai, A., Radhakrishna, A., & Kaushal, P. (2019). Ecological, genetic, and reproductive features of Cenchrus species indicate evolutionary superiority of apomixis under environmental stresses. *Ecological Indicators*. https://doi.org/10.1016/j.ecolind.2019.05.036

Kunstler, G., Lavergne, S., Courbaud, B., Thuiller, W., Vieilledent, G., Zimmermann, N. E., Kattge, J., & Coomes, D. A. (2012). Competitive interactions between forest trees are driven by species' trait hierarchy, not phylogenetic or functional similarity: Implications for forest community assembly. *Ecology Letters*, *15*(8), 831–840. https://doi.org/10.1111/j.1461-0248.2012.01803.x

Kuyper, T. W. (2005). Book review: B. Berg and C. McClaugherty (2003). Plant litter–decomposition, humus formation, carbon sequestration. Springer Verlag, 286 pp., 76 figs, 64 tabs. Hardcover. ISBN 3-540-44329-0, price €129.95. *Plant Ecology*. https://doi.org/10.1007/s11258-005-2822-6

Lavorel, S., & Grigulis, K. (2012). How fundamental plant functional trait relationships scale-up to trade-offs and synergies in ecosystem services. *Journal of Ecology*. https://doi.org/10.1111/j.1365-2745.2011.01914.x

Lee, J. T., Chu, M. Y., Lin, Y. S., Kung, K. N., Lin, W. C., & Lee, M. J. (2020). Root traits and biomechanical properties of three tropical pioneer tree species for forest restoration in landslide areas. *Forests*. https://doi.org/10.3390/f11020179

Leghari, S. J., Wahocho, N. A., Laghari, G. M., HafeezLaghari, A., MustafaBhabhan, G., HussainTalpur, K., Bhutto, T. A., Wahocho, S. A., & Lashari, A. A. (2016). Role of nitrogen for plant growth and development : A review. *Advances in Environmental Biology*, *10*(9), 209–219.

Lloret, F., & Batllori, E. (2021). *Climate-Induced Global Forest Shifts due to Heatwave-Drought*. https://doi.org/10.1007/978-3-030-71330-0_7

López, R., Cano, F. J., Martin-StPaul, N. K., Cochard, H., & Choat, B. (2021). Coordination of stem and leaf traits define different strategies to regulate water loss and tolerance ranges to aridity. *New Phytologist*. https://doi.org/10.1111/nph.17185

Lori, M., Piton, G., Symanczik, S., Legay, N., Brussaard, L., Jaenicke, S., Nascimento, E., Reis, F., Sousa, J. P., Mäder, P., Gattinger, A., Clément, J. C., & Foulquier, A. (2020).

Compared to conventional, ecological intensive management promotes beneficial proteolytic soil microbial communities for agro-ecosystem functioning under climate change-induced rain regimes. *Scientific Reports*, *10*(1), 1–16. https://doi.org/10.1038/s41598-020-64279-8

Madigan, M., Bender, K., Buckley, D., Sattley, W. M., & Stahl, D. (2019). *Brock biology of microorganisms* (15th ed.). Pearson.

Malik, A. A., Martiny, J. B. H., Brodie, E. L., Martiny, A. C., Treseder, K. K., & Allison, S. D. (2020). Defining trait-based microbial strategies with consequences for soil carbon cycling under climate change. *ISME Journal*. https://doi.org/10.1038/s41396-019-0510-0

Markesteijn, L., Poorter, L., Paz, H., Sack, L., & Bongers, F. (2011). Ecological differentiation in xylem cavitation resistance is associated with stem and leaf structural traits. *Plant, Cell and Environment*. https://doi.org/10.1111/j.1365-3040.2010.02231.x

Masarovičová, E., Májeková, M., & Vykouková, I. (2015). Functional traits and plasticity of plants in ecological research and education. *Chemistry-Didactics-Ecology-Metrology*. https://doi.org/10.1515/cdem-2015-0006

Maurel, C., & Nacry, P. (2020). Root architecture and hydraulics converge for acclimation to changing water availability. *Nature Plants*. https://doi.org/10.1038/s41477-020-0684-5

McDowell, N. G., Sapes, G., Pivovaroff, A., Adams, H. D., Allen, C. D., Anderegg, W. R. L., Arend, M., Breshears, D. D., Brodribb, T., Choat, B., Cochard, H., De Cáceres, M., De Kauwe, M. G., Grossiord, C., Hammond, W. M., Hartmann, H., Hoch, G., Kahmen, A., Klein, T., … Xu, C. (2022). Mechanisms of woody-plant mortality under rising drought, CO_2 and vapour pressure deficit. *Nature Reviews Earth and Environment*. https://doi.org/10.1038/s43017-022-00272-1

McGill, B. J., Enquist, B. J., Weiher, E., & Westoby, M. (2006). Rebuilding community ecology from functional traits. *Trends in Ecology and Evolution*. https://doi.org/10.1016/j.tree.2006.02.002

Meena, M., Yadav, G., Sonigra, P., Nagda, A., Mehta, T., Swapnil, P., Harish, Marwal, A., & Kumar, S. (2023). Multifarious responses of forest soil microbial community toward climate change. *Microbial Ecology*. https://doi.org/10.1007/s00248-022-02051-3

Michel, J., Hartley, I. P., Buckeridge, K. M., van Meegen, C., Broyd, R. C., Reinelt, L., Ccahuana Quispe, A. J., & Whitaker, J. (2023). Preferential substrate use decreases priming effects in contrasting treeline soils. *Biogeochemistry*. https://doi.org/10.1007/s10533-022-00996-8

Ning, S., Chen, C., Zhou, B., & Wang, Q. (2019). Evaluation of normalized root length density distribution models. *Field Crops Research*. https://doi.org/10.1016/j.fcr.2019.107604

Norton, M. R., Malinowski, D. P., & Volaire, F. (2016). Plant drought survival under climate change and strategies to improve perennial grasses. A review. *Agronomy for Sustainable Development*. https://doi.org/10.1007/s13593-016-0362-1

Ober, E. S., Alahmad, S., Cockram, J., Forestan, C., Hickey, L. T., Kant, J., Maccaferri, M., Marr, E., Milner, M., Pinto, F., Rambla, C., Reynolds, M., Salvi, S., Sciara, G., Snowdon, R. J., Thomelin, P., Tuberosa, R., Uauy, C., Voss-Fels, K. P., … Watt, M. (2021). Wheat root systems as a breeding target for climate resilience. *Theoretical and Applied Genetics*. https://doi.org/10.1007/s00122-021-03819-w

Pankievicz, V. C. S., Irving, T. B., Maia, L. G. S., & Ané, J. M. (2019). Are we there yet? The long walk towards the development of efficient symbiotic associations between nitrogen-fixing bacteria and non-leguminous crops. *BMC Biology*. https://doi.org/10.1186/s12915-019-0710-0

Pardon, P., Reubens, B., Reheul, D., Mertens, J., De Frenne, P., Coussement, T., Janssens, P., & Verheyen, K. (2017). Trees increase soil organic carbon and nutrient availability in temperate agroforestry systems. *Agriculture, Ecosystems and Environment*. https://doi.org/10.1016/j.agee.2017.06.018

Patil, A. G., Neelambika, M., Nagendra, K., M, R., & Singh, V. P. (2020). A review of weed management in different crops. *International Journal of Chemical Studies*. https://doi.org/10.22271/chemi.2020.v8.i4ay.10276

Pérez-Harguindeguy, N., Díaz, S., Garnier, E., Lavorel, S., Poorter, H., Jaureguiberry, P., Bret-Harte, M. S., Cornwell, W. K., Craine, J. M., Gurvich, D. E., Urcelay, C., Veneklaas, E. J., Reich, P. B., Poorter, L., Wright, I. J., Ray, P., Enrico, L., Pausas, J. G., De Vos, A. C., … Cornelissen, J. H. C. (2013). New handbook for standardised measurement of plant functional traits worldwide. *Australian Journal of Botany*. https://doi.org/10.1071/BT12225

Peterson, J. A., Ode, P. J., Oliveira-Hofman, C., & Harwood, J. D. (2016). Integration of plant defense traits with biological control of arthropod pests: Challenges and opportunities. *Frontiers in Plant Science*. https://doi.org/10.3389/fpls.2016.01794

Petrokas, R. (2020). Forest climax phenomenon: An invariance of scale. *Forests*. https://doi.org/10.3390/f11010056

Pezzani, R., Salehi, B., Vitalini, S., Iriti, M., Zuñiga, F. A., Sharifi-Rad, J., Martorell, M., & Martins, N. (2019). Synergistic effects of plant derivatives and conventional chemotherapeutic agents: An update on the cancer perspective. *Medicina (Lithuania)*. https://doi.org/10.3390/medicina55040110

Pierret, A., Doussan, C., Capowiez, Y., Bastardie, F., & Pagès, L. (2007). Root functional architecture: A framework for modeling the interplay between roots and soil. *Vadose Zone Journal*. https://doi.org/10.2136/vzj2006.0067

Prescott, C. E. (2010). Litter decomposition: What controls it and how can we alter it to sequester more carbon in forest soils? *Biogeochemistry*. https://doi.org/10.1007/s10533-010-9439-0

Prescott, C. E., & Vesterdal, L. (2021). Decomposition and transformations along the continuum from litter to soil organic matter in forest soils. *Forest Ecology and Management*. https://doi.org/10.1016/j.foreco.2021.119522

Rahman, M. M., Tsukamoto, J., Rahman, M. M., Yoneyama, A., & Mostafa, K. M. (2013). Lignin and its effects on litter decomposition in forest ecosystems. *Chemistry and Ecology*. https://doi.org/10.1080/02757540.2013.790380

Ranjith, P., & Srinivasa Rao, M. (2021). Breeding for drought resistance. In *Plant breeding – Current and future views*. https://doi.org/10.5772/intechopen.97276

Rao, D., Aparna, K., & Mohanty, S. R. (2019). Microbiology and biochemistry of soil organic matter, carbon sequestration and soil health. *Indian Journal of Fertilisers*, *15*(2), 124–138.

Rashid, M. I., Mujawar, L. H., Shahzad, T., Almeelbi, T., Ismail, I. M. I., & Oves, M. (2016). Bacteria and fungi can contribute to nutrients bioavailability and aggregate formation in degraded soils. *Microbiological Research*. https://doi.org/10.1016/j.micres.2015.11.007

Renault, D., Hess, M. C. M., Braschi, J., Cuthbert, R. N., Sperandii, M. G., Bazzichetto, M., Chabrerie, O., Thiébaut, G., Buisson, E., Grandjean, F., Bittebiere, A. K., Mouchet, M., & Massol, F. (2022). Advancing biological invasion hypothesis testing using functional diversity indices. *Science of the Total Environment*. https://doi.org/10.1016/j.scitotenv.2022.155102

Reubens, B., Poesen, J., Danjon, F., Geudens, G., & Muys, B. (2007). The role of fine and coarse roots in shallow slope stability and soil erosion control with a focus on root system architecture: A review. *Trees – Structure and Function*. https://doi.org/10.1007/s00468-007-0132-4

Rumpel, C., & Kögel-Knabner, I. (2011). Deep soil organic matter-a key but poorly understood component of terrestrial C cycle. *Plant and Soil*. https://doi.org/10.1007/s11104-010-0391-5

Sage, R. F. (2014). Preface: Photosynthetic efficiency and carbon concentration in terrestrial plants: The C4 and CAM solutions. *Journal of Experimental Botany*. https://doi.org/10.1093/jxb/eru262

Sahayaraj, K., & Hassan, E. (2023). Bioefficacy of insect predators under laboratory. In *Worldwide predatory insects in agroecosystems*. https://doi.org/10.1007/978-981-99-1000-7_16

Sahu, N., Vasu, D., Sahu, A., Lal, N., & Singh, S. K. (2017). Strength of microbes in nutrient cycling: A key to soil health. In *Agriculturally important microbes for sustainable agriculture*. https://doi.org/10.1007/978-981-10-5589-8_4

Salari, M., Rahmanian, V., Hashemi, S. A., Chiang, W. H., Lai, C. W., Mousavi, S. M., & Gholami, A. (2022). Bioremediation treatment of polyaromatic hydrocarbons for environmental sustainability. *Water (Switzerland)*. https://doi.org/10.3390/w14233980

Sangsupan, H. A., Hibbs, D. E., Withrow-Robinson, B. A., & Elliott, S. (2018). Seed and microsite limitations of large-seeded, zoochorous trees in tropical forest restoration plantations in northern Thailand. *Forest Ecology and Management*. https://doi.org/10.1016/j.foreco.2018.03.021

Sauvadet, M., Lashermes, G., Alavoine, G., Recous, S., Chauvat, M., Maron, P. A., & Bertrand, I. (2018). High carbon use efficiency and low priming effect promote soil C stabilization under reduced tillage. *Soil Biology and Biochemistry*. https://doi.org/10.1016/j.soilbio.2018.04.026

Schimel, J., Balser, T. C., & Wallenstein, M. (2007). Microbial stress-response physiology and its implications for ecosystem function. *Ecology*. https://doi.org/10.1890/06-0219

Schimel, J. P., & Schaeffer, S. M. (2012). Microbial control over carbon cycling in soil. *Frontiers in Microbiology*. https://doi.org/10.3389/fmicb.2012.00348

Schmitz, O. J., Buchkowski, R. W., Burghardt, K. T., & Donihue, C. M. (2015). Functional traits and trait-mediated interactions. Connecting community-level interactions with ecosystem functioning. *Advances in Ecological Research*. https://doi.org/10.1016/bs.aecr.2015.01.003

Schröder, P., Sauvêtre, A., Gnädinger, F., Pesaresi, P., Chmeliková, L., Doğan, N., Gerl, G., Gökçe, A., Hamel, C., Millan, R., Persson, T., Ravnskov, S., Rutkowska, B., Schmid, T., Szulc, W., Teodosiu, C., & Terzi, V. (2019). Discussion paper: Sustainable increase of crop production through improved technical strategies, breeding and adapted management – A European perspective. *Science of the Total Environment*. https://doi.org/10.1016/j.scitotenv.2019.04.212

Singh, A., Hidangmayum, A., Tiwari, P., kumar, V., Singh, B. N., & Dwivedi, P. (2022). How the soil nitrogen nutrient promotes plant growth—A critical assessment. In *New and future developments in microbial biotechnology and bioengineering: Sustainable agriculture: Revisiting green chemicals*. https://doi.org/10.1016/B978-0-323-85581-5.00018-5

Singh, D., & Chen, S. (2008). The white-rot fungus *Phanerochaete chrysosporium*: Conditions for the production of lignin-degrading enzymes. *Applied Microbiology and Biotechnology*. https://doi.org/10.1007/s00253-008-1706-9

Singh, U. B., & Ahluwalia, A. S. (2013). Microalgae: A promising tool for carbon sequestration. *Mitigation and Adaptation Strategies for Global Change*. https://doi.org/10.1007/s11027-012-9393-3

Sinsabaugh, R. L., Lauber, C. L., Weintraub, M. N., Ahmed, B., Allison, S. D., Crenshaw, C., Contosta, A. R., Cusack, D., Frey, S., Gallo, M. E., Gartner, T. B., Hobbie, S. E., Holland, K., Keeler, B. L., Powers, J. S., Stursova, M., Takacs-Vesbach, C., Waldrop, M. P., Wallenstein, M. D., … Zeglin, L. H. (2008). Stoichiometry of soil enzyme activity at global scale. *Ecology Letters*. https://doi.org/10.1111/j.1461-0248.2008.01245.x

Smith, S. E., & Read, D. J. (2010). *Mycorrhizal symbiosis* (3rd ed.). Academic Press.

Solomon, W., Janda, T., & Molnár, Z. (2024). Unveiling the significance of rhizosphere: Implications for plant growth, stress response, and sustainable agriculture. *Plant Physiology and Biochemistry*. https://doi.org/10.1016/j.plaphy.2023.108290

Song, Y., Yu, Y., Li, Y., & Du, M. (2023). Leaf litter chemistry and its effects on soil microorganisms in different ages of *Zanthoxylum planispinum* var. *Dintanensis*. *BMC Plant Biology*. https://doi.org/10.1186/s12870-023-04274-z

Soto-Gómez, D., & Pérez-Rodríguez, P. (2022). Sustainable agriculture through perennial grains: Wheat, rice, maize, and other species. A review. *Agriculture, Ecosystems and Environment*. https://doi.org/10.1016/j.agee.2021.107747

Spohn, M., Pötsch, E. M., Eichorst, S. A., Woebken, D., Wanek, W., & Richter, A. (2016). Soil microbial carbon use efficiency and biomass turnover in a long-term fertilization experiment in a temperate grassland. *Soil Biology and Biochemistry*. https://doi.org/10.1016/j.soilbio.2016.03.008

Steinauer, K., Thakur, M. P., Emilia Hannula, S., Weinhold, A., Uthe, H., van Dam, N. M., & Martijn Bezemer, T. (2023). Root exudates and rhizosphere microbiomes jointly determine temporal shifts in plant-soil feedbacks. *Plant Cell and Environment*. https://doi.org/10.1111/pce.14570

Stuart Chapin, F., Matson, P. A., & Vitousek, P. M. (2012). Principles of terrestrial ecosystem ecology. In *Principles of terrestrial ecosystem ecology*. https://doi.org/10.1007/978-1-4419-9504-9

Tameirão, L. B. S., Caminha-Paiva, D., Negreiros, D., Veloso, M. D. D. M., Berbara, R. L. L., Dias, L. E., Pierce, S., & Fernandes, G. W. (2021). Role of environmental filtering and functional traits for species coexistence in a harsh tropical montane ecosystem. *Biological Journal of the Linnean Society*. https://doi.org/10.1093/biolinnean/blaa181

Tao, F., Huang, Y., Hungate, B. A., Manzoni, S., Frey, S. D., Schmidt, M. W. I., Reichstein, M., Carvalhais, N., Ciais, P., Jiang, L., Lehmann, J., Wang, Y. P., Houlton, B. Z., Ahrens, B., Mishra, U., Hugelius, G., Hocking, T. D., Lu, X., Shi, Z., … Luo, Y. (2023). Microbial carbon use efficiency promotes global soil carbon storage. *Nature*. https://doi.org/10.1038/s41586-023-06042-3

Taylor S. E., & Lobel M. (1989). Social comparison activity under threat: Downward evaluation and upward contacts. *Psychological Review*, *96*(4), 569–575. https://doi.org/10.1037/0033-295x.96.4.569.

Teshome, D. T., Zharare, G. E., & Naidoo, S. (2020). The threat of the combined effect of biotic and abiotic stress factors in forestry under a changing climate. *Frontiers in Plant Science*. https://doi.org/10.3389/fpls.2020.601009

Tian, P., Razavi, B. S., Zhang, X., Wang, Q., & Blagodatskaya, E. (2020). Microbial growth and enzyme kinetics in rhizosphere hotspots are modulated by soil organics and nutrient availability. *Soil Biology and Biochemistry*. https://doi.org/10.1016/j.soilbio.2019.107662

Triatmojo, A. N. (2019). Optimasi polyvynil alchol dan gliserin sediaan masker peel off ekstrak daunsirsak (*Annona muricata* Linn) aplikasi desain faktrorial. *Journal of Wind Engineering and Industrial Aerodynamics*.

Tripathi, S., Yadav, A., & Tripathi, D. M. (2016). Plastic waste: Environmental pollution, health hazards and biodegradation strategies. In *Bioremediation of industrial waste for environmental safety* (pp. 99–133). Springer Nature.

Trivedi, P., Wallenstein, M. D., Delgado-Baquerizo, M., & Singh, B. K. (2018). Microbial modulators and mechanisms of soil carbon storage. In *Soil carbon storage: Modulators, mechanisms and modeling*. https://doi.org/10.1016/B978-0-12-812766-7.00003-2

Upadhyay, S. K., Srivastava, A. K., Rajput, V. D., Chauhan, P. K., Bhojiya, A. A., Jain, D., Chaubey, G., Dwivedi, P., Sharma, B., & Minkina, T. (2022). Root exudates: Mechanistic insight of plant growth promoting rhizobacteria for sustainable crop production. *Frontiers in Microbiology*. https://doi.org/10.3389/fmicb.2022.916488

Valladares, F., & Niinemets, U. (2007). The architecture of plant crowns: From design rules to light capture and performance. In *Functional plant ecology* (2nd ed.). https://doi.org/10.1201/9781420007626-4

Van Der Heijden, M. G. A., Bardgett, R. D., & Van Straalen, N. M. (2008). The unseen majority: Soil microbes as drivers of plant diversity and productivity in terrestrial ecosystems. *Ecology Letters*. https://doi.org/10.1111/j.1461-0248.2007.01139.x

Vanhees, D. J., Loades, K. W., Bengough, A. G., Mooney, S. J., & Lynch, J. P. (2021). The ability of maize roots to grow through compacted soil is not dependent on the amount of roots formed. *Field Crops Research*. https://doi.org/10.1016/j.fcr.2020.108013

Varma, A., & Choudhary, D. K. (2019). Mycorrhizosphere and pedogenesis. *Mycorrhizosphere and Pedogenesis*, 1–420. https://doi.org/10.1007/978-981-13-6480-8

Villagra, M., Di Francescantonio, D., Munaretto, N., & Campanello, P. I. (2023). Yerba mate (*Ilex paraguariensis*) agroforestry systems: Intraspecific differences in water relations and hydraulic architecture. *Functional Plant Biology*. https://doi.org/10.1071/FP22300

Visser, M. D., Bruijning, M., Wright, S. J., Muller-Landau, H. C., Jongejans, E., Comita, L. S., & de Kroon, H. (2016). Functional traits as predictors of vital rates across the life cycle of tropical trees. *Functional Ecology*. https://doi.org/10.1111/1365-2435.12621

Vocciante, M., Grifoni, M., Fusini, D., Petruzzelli, G., & Franchi, E. (2022). The role of plant growth-promoting rhizobacteria (PGPR) in mitigating plant's environmental stresses. *Applied Sciences (Switzerland)*. https://doi.org/10.3390/app12031231

Waldrop, M. P., & Firestone, M. K. (2006). Response of microbial community composition and function to soil climate change. *Microbial Ecology*. https://doi.org/10.1007/s00248-006-9103-3

Wang, S., Payne, G. F., & Bentley, W. E. (2020). Quorum sensing communication: Molecularly connecting cells, their neighbors, and even devices. *Annual Review of Chemical and Biomolecular Engineering*. https://doi.org/10.1146/annurev-chembioeng-101519-124728

Wardle, D. A., Bardgett, R. D., Klironomos, J. N., Setälä, H., Van Der Putten, W. H., & Wall, D. H. (2004). Ecological linkages between aboveground and belowground biota. *Science*. https://doi.org/10.1126/science.1094875

Watson, J. E. M. (2009). Ecological restoration: Principles, values, and structure of an emerging profession. *Pacific Conservation Biology*. https://doi.org/10.1071/pc090277

Wright, I. J., Reich, P. B., & Westoby, M. (2001). Strategy shifts in leaf physiology, structure and nutrient content between species of high- and low-rainfall and high- and low-nutrient habitats. *Functional Ecology*. https://doi.org/10.1046/j.0269-8463.2001.00542.x

Yadav, M. R., Kumar, R., Parihar, C. M., Yadav, R. K., Jat, S. L., Ram, H., Meena, R. K., Singh, M. B., Verma, A. P., Ghoshand, A., & Jat, M. L. (2017). Strategies for improving nitrogen use efficiency: A review. *Agricultural Reviews*. https://doi.org/10.18805/ag.v0iof.7306

Yang, L. H., & Rudolf, V. H. W. (2010). Phenology, ontogeny and the effects of climate change on the timing of species interactions. *Ecology Letters*. https://doi.org/10.1111/j.1461-0248.2009.01402.x

Yang, Y. (2021). Emerging patterns of microbial functional traits. *Trends in Microbiology*. https://doi.org/10.1016/j.tim.2021.04.004

Zanklan, A. S., Becker, H. C., Sørensen, M., Pawelzik, E., & Grüneberg, W. J. (2018). Genetic diversity in cultivated yam bean (*Pachyrhizus* spp.) Evaluated through multivariate analysis of morphological and agronomic traits. *Genetic Resources and Crop Evolution*. https://doi.org/10.1007/s10722-017-0582-5

Zhang, M., Cheng, X., Geng, Q., Shi, Z., Luo, Y., & Xu, X. (2019). Leaf litter traits predominantly control litter decomposition in streams worldwide. *Global Ecology and Biogeography*. https://doi.org/10.1111/geb.12966

Zhao, M., Zhao, J., Yuan, J., Hale, L., Wen, T., Huang, Q., Vivanco, J. M., Zhou, J., Kowalchuk, G. A., & Shen, Q. (2021). Root exudates drive soil-microbe-nutrient feedbacks in response to plant growth. *Plant Cell and Environment*. https://doi.org/10.1111/pce.13928

Zhou, J., Li, Z., Shi, L., Kuzyakov, Y., & Pausch, J. (2022). Microbial utilization of photosynthesized carbon depends on land-use. *Geoderma*. https://doi.org/10.1016/j.geoderma.2022.116160

10 Adaptation and Mitigation to Climate Change through Agroforestry

Suraj Goswami, Vipul Sharma, Avantika Adhruj, Ritika Verma, and Yalal Mallesh

10.1 INTRODUCTION

Trees have been an important component of the farming system since ancient times and have continuously served many services to humanity. The growing of trees on agricultural land is an age-old practice and popularly known as agroforestry. Land fragmentation, the rise of mechanization, and the intensification of agriculture have marginalized this practice (Gibbons et al., 2008). Agroforestry practices are getting worldwide attention as they offer a vital strategy for global climate targets and resolving many societal problems (Prajapati et al., 2024).

Agroforestry encompasses various systems, including home gardens, taungya cultivation, shelterbelts, windbreaks, protein banks, and woody hedgerows (Nair et al., 2021). These systems incorporate multiple components that are grown concurrently in diverse arrangements within a shared environment, wherein at least one component is a perennial species. These perennial species are often referred to as "companion species" because they provide multiple benefits to the main crop, including shade and nitrogen requirement (Gassner et al., 2022). Consequently, agroforestry functions as a unifying framework that amalgamates tree cultivation with annual crops, fostering a cohesive relationship between agriculture, forestry, and communities. Trees on the farmland not only serve as a source of additional income to the farmers but also provide a wide range of services such as water regulation, carbon sequestration, and soil protection (Muchane et al., 2020). Thus, agroforestry has massive potential to serve as an important strategy for mitigating the impacts of climate change.

Conventional farming systems are increasingly threatened by climate change, which changes the environmental conditions for crop cultivation and livestock production (Altieri et al., 2015; Goswami et al., 2022). Increasing temperatures put high heat stress on plants and animals, resulting in low agricultural production (Dash et al., 2016; Hatfield et al., 2011). Additionally, higher frequency and concentration of extreme weather events (drought, floods, and tsunamis) due to climate change, resulting in soil degradation and disrupting physical properties (Das, 2005). Recently, the shifting of weather patterns has been observed, which makes

DOI: 10.1201/9781003588580-10

it difficult for irrigation strategies and farmers' ability to safeguard irrigation, and poses a danger to food security and the rural economy (Kotir, 2011). The interaction of these factors emphasizes the requirement for effective strategies that can increase resilience within conventional farming practices, as farmers fight to steer a gradually unstable climate landscape (De Pinto et al., 2019; Mhagama, 2020). The impact of climate change extends beyond direct effects on arable productivity, augmenting current problems like outbreaks of pests and diseases. High temperatures can allow pests to spread their terrestrial range, resulting in new infestations that conventional farming practices may be tough to control (Skendžić et al., 2021). Besides, climate pattern change can affect insect life cycle timing, making it more challenging for growers to suppress and address crop threats. These new patterns necessitate the adoption of different plans and technologies by growers, who often need knowledge and capital support that small and marginal farmers lack (Lamichhane et al., 2015). Due to the high pressure on traditional farming systems, the impacts triggered by climate change underscore the vital need for a detailed reconsideration of agricultural practices. To steer the intricacies of altering arable land, it is vital to adopt modern ways that increase sustainability and resilience. By encompassing adoptive practices that contemplate both financial and environmental aspects, growers can better prepare themselves to address the current problems faced by climate change, eventually shielding food security and the rural economy.

10.1.1 Role of Agroforestry in Climate Resilience

Agroforestry has become one of the most familiar adaptation and mitigation approaches globally, especially in developing nations. It offers you multiple benefits (Figure 10.1), including carbon sequestration, improving soil, conserving water, improving biodiversity, and providing many ecosystem services. Agroforestry is a better approach to sustainable agriculture (Sahoo et al., 2020; Singh et al., 2024; Wilson & Lovell, 2016). The tree component adds leaf litter to the surface layer, improving soil fertility and water retention capacity (Bot & Benites, 2005).

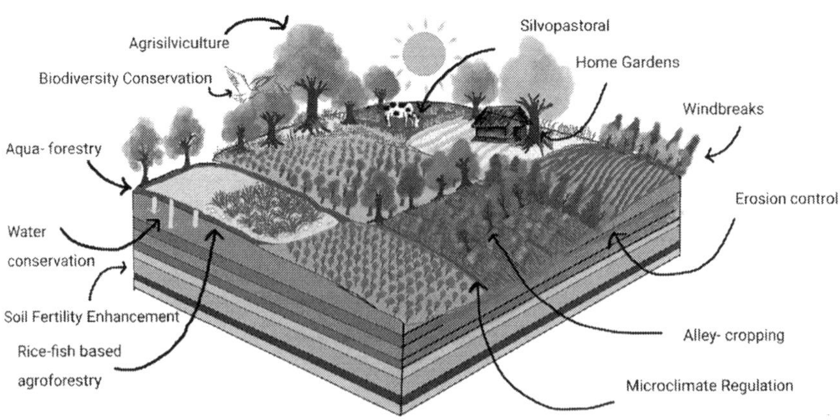

FIGURE 10.1 Different agroforestry practices and their benefits.

Therefore, it demonstrates higher resilience to soil degradation and environmental stressors like drought and flood (Dobhal et al., 2024). Additionally, trees provide shade to cover crops, making a unique microclimate for better conservation and economically viable system (Jose, 2012). Trees store atmospheric CO_2 in their biomass and soil (Abbas et al., 2017; Ghale et al., 2022; Ramachandran Nair et al., 2009). These benefits make it a perfect strategy for adaptation and mitigation of climate change (Mbow et al., 2014; Mulatu & Hunde, 2019; Verma et al., 2021), and currently, many countries are adopting it (Rosenstock et al., 2019). By establishing more stable microclimates, agroforestry enhances local water cycles, moderates temperature extremes, and promotes resilience to climate variability (Lasco et al., 2014; Pancholi et al., 2023). Agroforestry offers many direct benefits apart from ecological preservation, including better food security and additional livelihood sources for rural farmers (Keprate et al., 2024). Multiple components like trees, crops, and animals can mitigate the risks related to monocropping farming. This method not only progresses nutritional diversity and food security but also boosts rural economies through the sale of extra goods in markets. Additionally, agroforestry allows farmers to adopt sustainable agriculture practices (Kumari et al., 2024; Satish et al., 2024).

10.1.2 Carbon Sequestration: Agroforestry's Role in Capturing and Storing Carbon

The combination of trees into agricultural land has notably contributed to better agricultural productivity and better living standards of both developed and developing countries (Murthy et al., 2013). The carbon sequestration is an attractive ability of agroforestry systems, which has heightened worldwide interest in adaptation and mitigation of climate change. Agroforestry practices have been endorsed as a method for sequestering carbon in soils within the frameworks of afforestation and reforestation initiatives, as well as under the Kyoto Protocol's Clean Development Mechanisms (Sharma et al., 2016). Current research and many initiatives globally are focused on examining and improving the carbon sequestration potential of agroforestry. Table 10.1 shows the carbon sequestration potential of different agroforestry systems across the world, and Figure 10.2 illustrates the Indian agroforestry system's potential for sequestering carbon.

10.1.2.1 Soil Health and Fertility: How Agroforestry Improves Soil Structure and Nutrient Content

Agroforestry's potential to improve soil quality has been widely recognized as a major benefit since its inception as a scientifically recognized discipline and practice (Nair, 2011; Young, 1989). Collectively, agroforestry can (1) enrich soil organic carbon (SOC) better than monocropping systems, (2) improve soil nutrient availability and soil fertility due to the presence of trees in the system, and (3) enhance soil microbial dynamics, which would positively influence soil health. The soil organic matter (OM) is a function of the rate of decomposition and replacement of the OM in the soil. The incorporation of trees in agroforestry enhances the soil OM by adding litter both above and below ground. Soil OM serves as an energy source for

TABLE 10.1
Above and below Ground Carbon Sequestration Potential of Major Agroforestry Systems across the Globe

S. No.	Continent	Promising Agroforestry System	Country	Carbon Sequestration Potential (Mg ha^{-1} year^{-1})	Reference
1.	Asia	Indonesian homegarden	Sumatra	8.00	Nair et al. (2010)
2.	Africa	Parkland agroforestry	Ethiopia	8.34	Ghale et al. (2022)
3.	North America	Silvopasture	USA	6.10	Udawatta and Jose (2012)
4.	South America	Alley cropping	Peru	1.60	Oelbermann et al. (2004)
5.	Europe	Hedgerow	Germany	3.80	Golicz et al. (2021)

soil organisms and impacts both soil biodiversity and related biological processes. Consequently, SOC is a key indicator in evaluating soil health. Noiha Noumi et al. (2018) compared the global C storage of Eucalyptus afforested plots with traditional Savannah in Cameroon. They observed that Eucalyptus stands stored more C than traditional Savannahs, especially the oldest Eucalyptus stands.

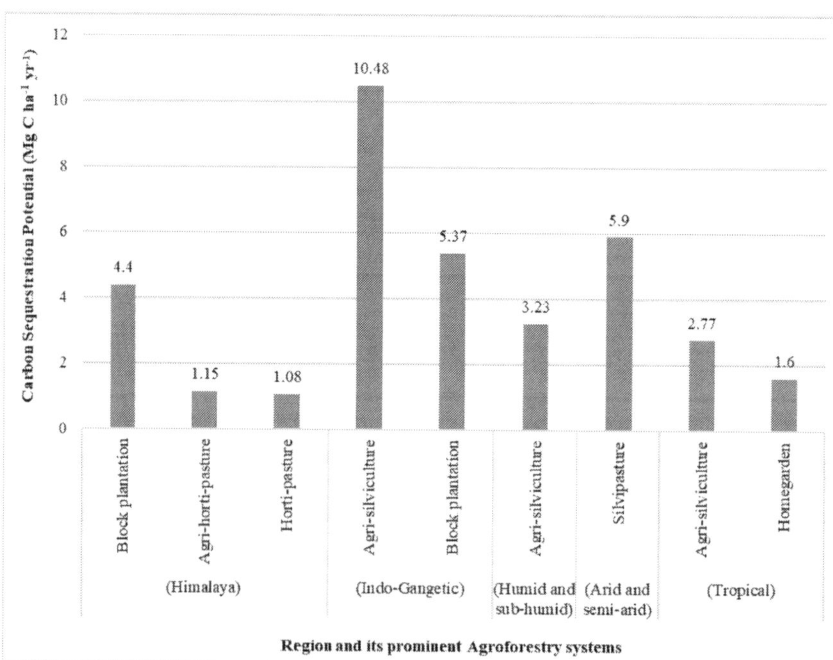

FIGURE 10.2 Different agroforestry systems and their potential carbon sequestration in different regions of India. (Modified from Dhyan et al. 2016.)

The tree leaf decomposition and mineralization give ample nutrients in agroforestry fields and thus better agricultural yields. Yengwe et al. (2018) examined the nutritive potential of *Faidherbia albida* with maize crop in Zambia and found that leaf litter input from *F. albida* could give more than 18 kg N ha^{-1} $year^{-1}$ and accelerate the microbial activity. Partey et al. (2018) compared the N-supply potential of tree leaves from *Acacia auriculiformis*, *Baphia nitida*, *Albizia zygia*, *Azadirachta indica*, *Senna siamea*, *Senna spectabilis*, *Tithonia diversifolia*, *Gliricidia sepium*, and *Leucaena leucocephala* for maize fertilization in Ghana. They estimated that tree leaves could supply up to 93 mg N kg^{-1}. The nutritive contents and mineralization rates of the leaves differed between the tree species, with *T. diversifolia* and *G. sepium* improving maize yield the best.

Soil health and ecosystem sustainability are heavily dependent on the biota, which plays a crucial role in breaking down soil OM, cycling nutrients, and influencing soil chemical and physical characteristics. These processes ultimately determine soil fertility and long-term sustainability. Research by Hailemariam et al. (2018) examined the impact of arbuscular mycorrhizal fungi (AMF) inoculation and water stress on *F. albida* seedling survival in Ethiopia. Their findings revealed that AMF inoculation enhanced seedling growth, regardless of the inoculum's origin or water conditions. The design of agroforestry systems can affect AMF diversity, as demonstrated by Shukla et al. (2018). Their study showed how shade influences the effectiveness of bioinoculants composed of rhizobacteria, phosphate-solubilizing bacteria, and AMF.

10.1.2.2 Water Conservation and Management: Benefits of Agroforestry for Water Retention and Reduction of Runoff

Agroforestry is a viable solution that is more efficient to combat the increasing challenges of water management and conservation as it is the integration of trees, shrubs, and crops on the same unit of land that not only reduces soil erosion, improves groundwater recharge but also plays a major role in conserving biodiversity (Altieri, 1999), which is a way forward toward sustainability. In comparison to traditional practices, the large canopies and root systems of agroforestry systems improve the microclimate as a result of which evaporation decreases, water retention increases, and the runoff reduces up to 50% (Jose, 2009; Kaushal et al., 2021; Lal, 2009). Various agroforestry practices such as silvopastoral systems, bioengineering structures, home gardens, street crops, and landscaped gardens can improve soil structure by facilitating the porosity and water infiltration rate (Kaur et al., 2023).

10.1.2.3 Biodiversity and Ecosystem Services: Enhancing Biodiversity and Providing Ecosystem Services

Agroforestry not only supports large-scale agricultural diversification, offering ecological benefits for environmental improvement (Chaturvedi & Krishnan, 2009) but is also an economically viable option in contrast to monocropping (Anshiso et al., 2017). Furthermore, it also improves the cultural, aesthetic, and recreational value, nitrogen retention, and water quality improvement (Jose, 2009; Pimm et al., 1995). Additionally, it ensures food security, reduces poverty, and improves soil fertility and water quality through the strategic interaction of vegetation components (Jose, 2011; Pimm, 2008; Pimm et al., 1995). Also, as a key approach to mitigating biodiversity loss in human-dominated ecosystems, agroforestry fosters diverse habitats, thereby conserving biodiversity while enhancing the sustainability of agricultural landscapes (Jose, 2011).

10.1.3 Challenges and Limitations

10.1.3.1 Economic and Financial Barriers: Initial Costs and Financial Constraints for Farmers

The modern world needs to integrate the three fundamental pillars of sustainability (economic, social, and environmental) for the fulfillment of present needs (Fischer et al., 2002). Most of the farmers in developing countries are extremely vulnerable to climate change (Masud et al., 2017; Verchot et al., 2007). Economic and financial factors are one of the major barriers for marginal and small farmers to adopting a sustainable practice to combat climate change (Grigorieva et al., 2023). Upgrading the technology and adopting climate-resilient crops often entails substantial upfront costs, particularly among small and marginal farmers (Ziro et al., 2023). These farmers do not have enough investment for better irrigation infrastructure, renewable energy resources, and modern soil management practices. Furthermore, access to capital is mostly difficult for small rural farmers in agricultural businesses. Due to low productivity in rural areas, which makes them financially vulnerable, agriculture financing is important to reduce these gaps (Balana et al., 2022; Maulu et al., 2021).

Uncertain market prices and unpredictability of weather events result in lower income (FAO et al., 2011), which makes it difficult to adopt sustainable practices that require capital investment, and this leads to continued reliance on monoculture methods, making farmers more vulnerable to climate change impacts. This kind of situation underlines the demand for financial interventions and policy support that can promote a smooth transition to sustainable agroforestry practices.

10.1.3.2 Knowledge and Technical Gaps: Lack of Expertise and Technical Support

Recently, agroforestry is recognized as a climate change adaptation strategy (Swamy & Tewari, 2017; Verchot et al., 2007); yet the full potential of agroforestry remains limited in many regions due to huge technological and knowledge gaps (Mbow et al., 2014; Tranchina et al., 2024). Many farmers lack the expertise in species compatibility, soil management, and landscape planning needed to implement these systems effectively. This knowledge gap is a major barrier to adoption, particularly in vulnerable communities, and addressing these gaps through targeted education and resource sharing is crucial for maximizing the benefits of agroforestry.

The absence of modern technological support further obscures the application of agroforestry (Scherr, 1991). Although few areas have government and NGO programs that endorse agroforestry, these programs mostly have low funding, skilled staff, or experts to provide continuous direction to farmers (Ajayi & Place, 2012). For instance, training programs on tree selection, planting techniques, pest management, and soil health monitoring are often limited in scope and availability. Without tailored technical support, farmers may struggle to achieve the full benefits of agroforestry, resulting in systems that are poorly adapted to local climate and soil conditions. Additionally, the absence of technical infrastructure for monitoring and evaluating agroforestry systems restricts the ability to track their effectiveness in mitigating climate impacts and improving ecosystem health.

Addressing these gaps requires a coordinated approach that involves collaboration among governments, research institutions, and development organizations (Buck, 1995; Mbow et al., 2014). Investing in capacity-building programs to train local experts and agricultural extension workers is critical to bridging the knowledge gap and providing on-the-ground technical support. Furthermore, partnerships with universities and research institutions can facilitate the development of region-specific best practices, allowing agroforestry to be tailored to diverse environments and cultural contexts. By strengthening technical expertise and support networks, agroforestry can more effectively contribute to climate resilience, creating sustainable, adaptive agricultural landscapes that respond to the challenges posed by a changing climate (Akanbi, 2024; Bourne et al., 2020; Fones-Sundell & Teklehaimanot, 2006; Johansson, 2015).

10.1.3.3 Policy and Market Constraints: Insufficient Policies and Market Support

Agroforestry initiatives face similar challenges, including financial and labor costs, as well as institutional barriers. Nevertheless, integrated approaches utilizing financial incentives and farmer participation have shown promise. Recent IPCC reports highlight agroforestry's role in enhancing climate resilience through diverse benefits (IPCC, 2023). Recognition began with the Kyoto Protocol and has expanded in national policies under the UNFCCC Paris Agreement, with many NDCs citing agroforestry as a key strategy. However, it is often underrepresented in carbon sink assessments and adaptation evaluations.

Despite growing acknowledgment of National Adaptation Plans (NAPs), a need remains for comprehensive analyses to set clear adaptation targets (Morgan et al., 2019). Existing policies frequently overlook agroforestry's benefits, and regulatory frameworks favor traditional monocultures, discouraging farmers from adopting sustainable practices. Market constraints further hinder agroforestry implementation (VijayKumar et al., 2024). Farmers struggle to access markets for diverse agroforestry products, facing inconsistent demand and prioritization of large-scale monocultures (Ibrahim et al., 2019). Addressing these intertwined challenges requires targeted policies to enhance market access, provide financial incentives, and integrate agroforestry into national strategies (Keprate et al., 2024), ultimately unlocking its potential to mitigate climate change and promote sustainable agriculture.

Gender discrimination in participatory approaches can pose a significant constraint to agroforestry. According to the FAO, women make up 43% of global agricultural labor, while facing enormous discrimination in the ownership of land and livestock, being significantly disadvantaged compared to men regarding land ownership, management, transfer, and economic rights (Karia, 2018). Also, unclear or insecure land ownership discourages women from investing in agroforestry since they lack long-term assurance of benefiting from their trees and crops. Future research should place greater emphasis on ensuring gender representation in participatory studies focused on the adoption of agroforestry as a sustainable practice. This approach would help to fairly capture the perspectives of all farmers or particular subgroups, supporting equity within agriculture. Highlighting the distinct challenges faced by women in agriculture could encourage their involvement and facilitate their transition into leadership roles.

10.1.3.4 Physical and Environmental Limitations: Land Availability and Climate Suitability

Agroforestry systems offer a promising solution for the biological enhancement and restoration of degraded soils, including dryland soils susceptible to wind and water erosion, and acidic sulfate soils in low-pH, humid regions with limited resources (Laskar et al., 2021; Sileshi et al., 2014). It also addresses soil nutrient deficiencies by overcoming the environmental and physical constraints, particularly climate suitability and land availability, that present challenges to their implementation. For instance, in the Eastern Himalayas, approximately 26% of the land within 5 km of roads or water bodies is suitable for agroforestry, increasing to 54% within 10 km, but significant portions of the region face severe soil fertility issues (Nath et al., 2021). Land Suitability Analysis (LSA) is a scientific methodology employed to evaluate land appropriateness for specific purposes, integrating factors such as ecological, climatic, soil, socio-economic and topographical (Ahmad et al., 2019; Collins et al., 2001; Hopkins, 1977; Malczewski, 2004; Nath et al., 2021; Singha & Swain, 2016). By assessing suitability, agroforestry practices can be carried out to improve land productivity and ecosystem services through strategic tree-crop combinations that optimize resource use and soil health (Ranjitkar et al., 2016). Therefore, both land and climate factors are crucial determinants of agroforestry system viability.

10.1.4 Policy Support and Future Directions

10.1.4.1 Current Policies and Support Systems: Existing Frameworks Promoting Agroforestry

Agroforestry has emerged as an important land management practice with much potential to increase climate resilience, enhance biodiversity, and promote rural development (Mosquera-Losada et al., 2018; Santiago-Freijanes et al., 2021). Some countries such as India and Nepal have already formulated strategies promoting agroforestry as it is considered to be one of the solutions to climate change, food security, and soil conservation (Dhyani et al., 2021). Agroforestry is mentioned in the Nationally Determined Contributions documents of many countries that are parties to the Paris Agreement, including the pledges to combat climate change and its adverse effects through tree planting and rehabilitation of degraded areas.

International organizations like the Global Environment Facility (GEF) and the Green Climate Fund (GCF) also focus on agroforestry in climate resilience programs, funding systems to implement and scale up in vulnerable places (Barnard et al., 2014). Working in synergy with NGOs and local authorities allows for agroforestry to become a reality, supporting biodiversity and the improvement of soil and water quality. Programs such as the AFR100 (African Forest Landscape Restoration Initiative) and the Bonn Challenge further advance agroforestry as landscape restoration in the context of international awareness and demand.

At the local level, agricultural extension programs and research institutes are the main drivers of agroforestry. Extension services give technical support and training to farmers; research centers develop regional models of agroforestry and spread information about sustainable practices. Community-based programs engage and manage agroforestry systems together and support farmers' exchange of knowledge. All these policy frameworks and supporting systems together establish the right conditions for agroforestry to be adopted as an option for sustainable, climate-friendly agriculture.

10.1.4.2 Recommendations for Policy Improvements: Suggestions for Strengthening Agroforestry Policies

To maximize agroforestry's potential as a strategy for climate adaptation and mitigation, it is essential to strengthen existing policy frameworks. Main improvements should concentrate on capital support for growers, including subsidies, allowances, and low-interest loans for tree growers, along with tax relief on agroforestry goods, and along with creating payments for ecosystem services. Simple land tenure procedure and safeguarding future land rights inspire growers to participate in agroforestry practices. Circulating new information services and collaborating with agriculture institutes can reduce knowledge gaps.

Making comprehensive frameworks that vigorously involve local communities is critical for better agroforestry legislation. Making crop insurance Schemes that cover agroforestry practices can protect growers from unpredictable losses, providing them with security. Participatory decision-making processes ensure that the voices of farmers and indigenous groups are heard, enhancing the relevance of policies. Recognizing and integrating traditional agroforestry practices can further bolster policy effectiveness. Incorporating agroforestry into wider national agricultural, environmental, and rural development plans to establish it as a core aspect of sustainable land use planning. Moreover, improving market access through better infrastructure and cooperative networks can strengthen rural economies. Creating a market system that values ecosystem services and offers compensation for positive contributions could support the shift toward adopting multifunctional agroforestry, making it a more financially sustainable and profitable option. Developing certification programs for agroforestry products enables farmers to access premium markets. Encouraging partnerships and collaborations with private agencies and NGOs as a part of corporate social responsibility programs can enhance the benefits for the stakeholders. By adopting these recommendations, policymakers can position agroforestry as a resilient and impactful solution for climate adaptation and sustainable rural development.

10.1.4.3 Future Research Needs: Areas Requiring Further Study for Effective Agroforestry Integration

The previous and current research only focus on major agroforestry practices and create a knowledge gap; thus, future research should prioritize the integration of science and technology, focusing on applied studies that build on foundational research to develop practical agroforestry technologies for farm, regional, and global applications. Additionally, utilized the previously unnoticed parts, such as the use of indigenous fruit-bearing trees and the agronomic components of agroforestry (Jamnadass et al., 2011; Muschler, 2016; Nair, 1998) while also addressing broader societal challenges.

There is a need for cost-effective inclusive assessment methodologies for agroforestry that cover economic, social, and environmental aspects along with suitable policy to facilitate its adoption (Amare & Darr, 2020). There should be developed region-specific agroforestry models for better adoption based on climate, soil, and socioeconomic factors (Nair, 1998). Additionally, understanding the interaction between different tree species and crops through agroecological assessments and agroforestry system modeling can enhance food production and maintain biodiversity (Malézieux et al., 2009), and long-term trial-based assessment of agroforestry using modern techniques for better understanding and generating reliable data.

Market potential and economic aspects of agroforestry-related products are the other key research areas (Polglase et al., 2008), and economic modeling should be used to compare with conventional farming systems and also accounting for tangible services like carbon sequestration for tackling climate change issues (Montagnini & Nair, 2004). New incentive programs and robust policies should promote better adaptation strategies to fully utilize the potential of agroforestry (Devi et al., 2024; Goswami et al., 2024). By addressing these interconnected research needs, scientists can generate the data necessary to advocate for the widespread adoption of agroforestry, ultimately fostering sustainable farming practices and enhancing resilience to climate change.

10.2 CONCLUSION

Agroforestry plays a crucial role in climate change adaptation by optimizing resource utilization, influencing local rainfall patterns, restoring tree cover, and regulating tree-crop interactions in response to climatic variability (Pancholi et al., 2023). The adverse impacts of climate change, such as disruptions to food and water security, pose significant risks to vulnerable populations, particularly low-income farmers who rely heavily on natural resources (UNEP, 2009). Moreover, global warming is expected to increase the frequency and severity of extreme weather events, including hurricanes, droughts, and floods, further exacerbating these challenges (Goswami et al., 2006). These climatic changes threaten crop productivity, food security, and ecosystem stability, as shifts in baseline conditions such as temperature and precipitation disrupt ecological balances (Fagre et al., 2009; Williams & Jackson, 2007). These problems could be overcome by effective agroforestry management, which relies on robust tree adaptation research, including geographic data analysis and tree-site matching, to understand tree responses to climate variability. By integrating trees with crops, agroforestry systems enhance ecosystem services and bolster landscape resilience against climate fluctuations, providing a sustainable strategy for adaptation (Van Noordwijk et al., 2021). Given the urgency of predicted climate impacts, coordinated and comprehensive global action is essential. Resources must be used efficiently, and adaptation efforts should be aligned with mitigation strategies to maximize effectiveness. Government planners and project developers should aim to minimize trade-offs, ensuring that progress in one area does not undermine outcomes in another. By deepening our understanding of the interactions between natural and human-managed systems, opportunities for synergizing climate adaptation and mitigation efforts can be expanded, fostering more cohesive and impactful responses to climate change.

REFERENCES

Abbas, F., Hammad, H. M., Fahad, S., Cerdà, A., Rizwan, M., Farhad, W., Ehsan, S., & Bakhat, H. F. (2017). Agroforestry: A sustainable environmental practice for carbon sequestration under the climate change scenarios—A review. *Environmental Science and Pollution Research*, *24*(12), 11177–11191. https://doi.org/10.1007/s11356-017-8687-0

Ahmad, F., Uddin, M. M., & Goparaju, L. (2019). Agroforestry suitability mapping of India: Geospatial approach based on FAO guidelines. *Agroforestry Systems*, *93*, 1319–1336.

Ajayi, O. C., & Place, F. (2012). Policy support for large-scale adoption of agroforestry practices: Experience from Africa and Asia. In *Agroforestry – The future of global land use* (pp. 175–201). Springer Netherlands.

Akanbi, M. T. (2024). Policy innovations for advancing sustainable agriculture in developing countries: Integrating educational programs, resource management strategies, and community empowerment for long-term growth. *Tensorgate Journal of Sustainable Technology and Infrastructure for Developing Countries*, *7*(6), 41–61.

Altieri, M. A. (1999). The ecological role of biodiversity in agroecosystems. In *Invertebrate biodiversity as bioindicators of sustainable landscapes* (pp. 19–31). Elsevier.

Altieri, M. A., Nicholls, C. I., Henao, A., & Lana, M. A. (2015). Agroecology and the design of climate change-resilient farming systems. *Agronomy for Sustainable Development*, *35*(3), 869–890.

Amare, D., & Darr, D. (2020). Agroforestry adoption as a systems concept: A review. *Forest Policy and Economics*, *120*, 102299.

Anshiso, A., Woldeamanuel, T., & Asfaw, Z. (2017). Financial analysis of fruit tree based agroforestry practice in Hadero Tunto Zuria Woreda, Kembata Tembaro Zone, South Ethiopia. *Research Journal of Finance and Accounting*, *8*(3), 72–80.

Balana, B. B., Mekonnen, D., Haile, B., Hagos, F., Yimam, S., & Ringler, C. (2022). Demand and supply constraints of credit in smallholder farming: Evidence from Ethiopia and Tanzania. *World Development*, *159*, 106033.

Barnard, S., Watson, C., Greenhill, R., Caravani, A., Trujillo, N. C., Hedger, M., & Whitley, S. (2014). *Climate finance: Is It making a difference?*. Overseas Development Institute.

Bot, A., & Benites, J. (2005). *The importance of soil organic matter: Key to drought-resistant soil and sustained food production* (no. 80). Food & Agriculture Org.

Bourne, M., Chesterman, S., & Wardell, D. A. (2020). *Capacity needs assessment of CIFOR, ICRAF and their partners for the implementation of the CGIAR research program on forestry, trees and agroforestry (FTA): Phase II, 2017–2021*. CIFOR.

Buck, L. E. (1995). Agroforestry policy issues and research directions in the US and less developed countries: Insights and challenges from recent experience. *Agroforestry Systems*, *30*, 57–73.

Chaturvedi, O. P., & Krishnan, P. R. (2009). Ecosystem development through agroforestry measures. In *Agroforestry: Natural resources sustainability, livelihood and climate moderation* (pp. 55–70). Satish Serial Publishing House.

Collins, M. G., Steiner, F. R., & Rushman, M. J. (2001). Land-use suitability analysis in the United States: Historical development and promising technological achievements. *Environmental Management*, *28*, 611–621.

Das, H. P. (2005). Agrometeorological impact assessment of natural disasters and extreme events and agricultural strategies adopted in areas with high weather risks. In *Natural disasters and extreme events in agriculture: Impacts and mitigation* (pp. 93–118). Springer Berlin Heidelberg.

Dash, S., Chakravarty, A. K., Singh, A., Upadhyay, A., Singh, M., & Yousuf, S. (2016). Effect of heat stress on reproductive performances of dairy cattle and buffaloes: A review. *Veterinary World*, *9*(3), 235.

De Pinto, A., Bryan, E., Ringler, C., & Cenacchi, N. (2019). *Adapting the global food system to new climate realities: Guiding principles and priorities.* International Food Policy Research Institute (IFPRI).

Devi, S., Goswami, S., Bisaria, J., & Sinha, B. (2024). Approaches for land restoration: An alternative strategy for climate change mitigation. In *Forests and climate change: Biological perspectives on impact, adaptation, and mitigation strategies* (pp. 535–551). Springer Nature Singapore.

Dhyan, S. K., Ram, A., & Dev, I. (2016). Potential of agroforestry systems in carbon sequestration in India. 1103–1112. National Agricultural Science Fund (NASF), ICAR.

Dhyani, S., Murthy, I. K., Kadaverugu, R., Dasgupta, R., Kumar, M., & Adesh Gadpayle, K. (2021). Agroforestry to achieve global climate adaptation and mitigation targets: Are South Asian countries sufficiently prepared?. *Forests*, *12*(3), 303.

Dobhal, S., Chavan, S., Upadhyay, K., Kumar, M., Lal, P., Chichaghare, A. R., & Kumar, R. (2024). Role of agroforestry in moderating extreme temperature conditions under climate change scenarios. In *Agroforestry solutions for climate change and environmental restoration* (pp. 85–102). Springer Nature Singapore.

Fagre, D. B., Charles, C. W., Allen, C. D., Birkeland, C., Chapin, F. S., Groffman, P. M., Guntenspergen, G. R., Knapp, A. K., McGuire, A. D., Mulholland, P. J., Peters, D. P. C., Roby, D. D., & Sugihara, G. (2009). *Thresholds of climate change in ecosystems.* U.S. Geological Survey.

FAO, IFAD, IMF, OECD, UNCTAD, WFP, World Bank, WTO, IFPRI, & United Nations High Level Task Force on Global Food and Nutrition. (2011). *Price volatility in food and agricultural markets: Policy responses.* World Bank.

Fischer, G., Shah, M. M., & Van Velthuizen, H. T. (2002). *Climate change and agricultural vulnerability.* International Institute for Applied Systems Analysis (IIASA).

Fones-Sundell, M., & Teklehaimanot, Z. (2006). *Mobilising agroforestry capacity for development. Final evaluation of the African Network for Agriculture, Agroforestry and Natural Resources Education (ANAFE) and Zambia Agroforestry Project (ZAP).* Swedish International Development Cooperation Agency (Sida).

Gassner, A., Cornelius, J. P., Dobie, P., Mercado, A., Mukuralinda, A., Okia, C. A., Pinon C, Somarriba E, & Thorne, P. (2022). *Components of agroforestry systems.* In Gassner A and Dobie P. eds. *Agroforestry: A primer. Design and management principles for people and the environment.* Bogor, Indonesia: Center for International Forestry Research (CIFOR) and Nairobi: World Agroforestry (ICRAF) 24–49.

Ghale, B., Mitra, E., Sodhi, H. S., Verma, A. K., & Kumar, S. (2022). Carbon sequestration potential of agroforestry systems and its potential in climate change mitigation. *Water, Air, & Soil Pollution*, *233*(7), 228.

Gibbons, P., Lindenmayer, D. B., Fischer, J., Manning, A. D., Weinberg, A., Seddon, J., & Barrett, G. (2008). The future of scattered trees in agricultural landscapes. *Conservation Biology*, *22*(5), 1309–1319.

Golicz, K., Ghazaryan, G., Niether, W., Wartenberg, A. C., Breuer, L., Gattinger, A., Jacobs, S., Kleinebecker, T., Weckenbrock, P., & Große-Stoltenberg, A. (2021). The role of small woody landscape features and agroforestry systems for national carbon budgeting in Germany. *Land*, *10*(10), 1028.

Goswami, B. N., Venugopal, V., Sengupta, D., Madhusoodanan, M. S., & Xavier, P. K. (2006). Increasing trend of extreme rain events over India in a warming environment. *Science*, *314*(5804), 1442–1445.

Goswami, S., Kumar, K., & Yadav, S. (2022). *Chapter 6. Indian agriculture and climate change*. In D. Dongre, S. Rahangdale, A. K. Shah, & A. Barela (Eds.), *Recent advances in agriculture and forestry* (pp. 69–76). AkiNik Publications.

Goswami, S., Reddy, B. V., Yadav, S., Adhruj, A., Dash, U., & Rathore, A. (2024). Rice–fish-based agroforestry system: A climate smart way to reconcile sustainable livelihood options. In *Agroforestry to combat global challenges: Current prospects and future challenges* (pp. 551–568). Springer Nature Singapore.

Grigorieva, E., Livenets, A., & Stelmakh, E. (2023). Adaptation of agriculture to climate change: A scoping review. *Climate*, *11*(10), 202.

Hailemariam, M., Birhane, E., Gebresamuel, G., Gebrekiros, A., Desta, Y., Alemayehu, A., Muruts, H., Araya, T., & Norgrove, L. (2018). Arbuscular mycorrhiza effects on *Faidherbia albida* (Del.) A. Chev. growth under varying soil water and phosphorus levels in Northern Ethiopia. *Agroforestry Systems*, *92*, 485–498.

Hatfield, J. L., Boote, K. J., Kimball, B. A., Ziska, L. H., Izaurralde, R. C., Ort, D., Thomson, A. M., & Wolfe, D. (2011). Climate impacts on agriculture: Implications for crop production. *Agronomy Journal*, *103*(2), 351–370.

Hopkins, L. D. (1977). Methods for generating land suitability maps: A comparative evaluation. *Journal of the American Institute of Planners*, *43*(4), 386–400.

Ibrahim, A. O., Adedeji, A. S., & Meduna, P. N. (2019). Constraints facing agroforestry practices among farmers in New Bussa, Nigeria. *Journal of Research in Forestry, Wildlife and Environment*, *11*(3), 133–141.

Jamnadass, R. H., Dawson, I. K., Franzel, S., Leakey, R. R. B., Mithöfer, D., Akinnifesi, F. K., & Tchoundjeu, Z. (2011). Improving livelihoods and nutrition in sub-Saharan Africa through the promotion of indigenous and exotic fruit production in smallholders' agroforestry systems: A review. *International Forestry Review*, *13*(3), 338–354.

Johansson, K. E. (2015). *Barriers and bridges for introducing agroforestry and community-based forestry among food insecure households in eastern Africa*. School for Forest Management, Swedish University of Agricultural Sciences.

Jose, S. (2009). *Agroforestry for ecosystem services and environmental benefits: An overview* (pp. 1–10). Springer.

Jose, S. (2011). Managing native and non-native plants in agroforestry systems. *Agroforestry Systems*, *83*, 101–105.

Jose, S. (2012). Agroforestry for conserving and enhancing biodiversity. *Agroforestry Systems*, *85*, 1–8.

Karia, S. (2018). *Thegendergapinlandrights*. FAO. https://openknowledge.fao.org/handle/20.500.14283/i8796en

Kaur, A., Paruchuri, R. G., Nayak, P., Devi, K. B., Upadhyay, L., Kumar, A., Pancholi, R., & Yousuf, M. (2023). The role of agroforestry in soil conservation and sustainable crop production: A comprehensive review. *International Journal of Environment and Climate Change*, *13*(11), 3089–3095.

Kaushal, R., Mandal, D., Panwar, P., Kumar, P., Tomar, J. M. S., & Mehta, H. (2021). Soil and water conservation benefits of agroforestry. In *Forest resources resilience and conflicts* (pp. 259–275). Elsevier.

Keprate, A., Sharma, V., Bhatnagar, S., Thakur, R., Abbas, G., Bhardwaj, D. R., & Sharma, P. (2024). Economic studies in agroforestry for livelihood security. In A. Raj, M. K. Jhariya, A. Banerjee, R. K. Jha, & K. P. Singh (Eds.), *Agroforestry for sustainable agriculture: Trends and innovations* (Chapter 15). Wiley.

Kotir, J. H. (2011). Climate change and variability in sub-Saharan Africa: A review of current and future trends and impacts on agriculture and food security. *Environment, Development and Sustainability*, *13*, 587–605.

Kumari, Y., Garima, Kumari, N., Mehta, A., Walia, K., Devi, M., Kumari, R., Rana, D., & Verma, S. (2024). Empowering climate change resilience through forests and agroforestry. In

Forests and climate change: Biological perspectives on impact, adaptation, and mitigation strategies (pp. 489–512). Springer Nature Singapore.

Lal, R. (2009). Soil degradation as a reason for inadequate human nutrition. *Food Security, 1*, 45–57.

Lamichhane, J. R., Barzman, M., Booij, K., Boonekamp, P., Desneux, N., Huber, L., Kudsk, P., Langrell, S., Ratnadass, A., Ricci, P., Sarah, J.-L., & Messéan, A. (2015). Robust cropping systems to tackle pests under climate change. A review. *Agronomy for Sustainable Development, 35*, 443–459.

Lasco, R. D., Delfino, R. J. P., & Espaldon, M. L. O. (2014). Agroforestry systems: Helping smallholders adapt to climate risks while mitigating climate change. *Wiley Interdisciplinary Reviews: Climate Change, 5*(6), 825–833.

Laskar, S. Y., Sileshi, G. W., Pathak, K., Debnath, N., Nath, A. J., Laskar, K. Y., Singnar, P., & Das, A. K. (2021). Variations in soil organic carbon content with chronosequence, soil depth and aggregate size under shifting cultivation. *Science of the Total Environment, 762*, 143114.

Malczewski, J. (2004). GIS-based land-use suitability analysis: A critical overview. *Progress in Planning, 62*(1), 3–65.

Malézieux, E., Crozat, Y., Dupraz, C., Laurans, M., Makowski, D., Ozier-Lafontaine, H., Rapidel, B., de Tourdonnet, S., & Valantin-Morison, M. (2009). Mixing plant species in cropping systems: Concepts, tools and models: A review. *Sustainable Agriculture, 29*(1), 329–353.

Masud, M. M., Azam, M. N., Mohiuddin, M., Banna, H., Akhtar, R., Alam, A. F., & Begum, H. (2017). Adaptation barriers and strategies towards climate change: Challenges in the agricultural sector. *Journal of Cleaner Production, 156*, 698–706.

Maulu, S., Hasimuna, O. J., Mutale, B., Mphande, J., & Siankwilimba, E. (2021). Enhancing the role of rural agricultural extension programs in poverty alleviation: A review. *Cogent Food & Agriculture, 7*(1), 1886663.

Mbow, C., van Noordwijk, M., Prabhu, R., & Simons, T. (2014). Knowledge gaps and research needs concerning agroforestry's contribution to sustainable development goals in Africa. *Current Opinion in Environmental Sustainability, 6*, 162–170.

Mhagama, M. L. (2020). *Vulnerability and resilience of dry land irrigation farming schemes against climate change: A case of Ruvuma Basin, South Eastern Tanzania* [Doctoral dissertation, Sokoine University of Agriculture].

Montagnini, F., & Nair, P. R. (2004). Carbon sequestration: An underexploited environmental benefit of agroforestry systems. In *New vistas in agroforestry: A compendium for 1st world congress of agroforestry, 2004* (pp. 281–295). Springer Netherlands.

Morgan, E. A., Nalau, J., & Mackey, B. (2019). Assessing the alignment of national-level adaptation plans to the Paris Agreement. *Environmental Science & Policy, 93*, 208–220.

Mosquera-Losada, M. R., Santiago-Freijanes, J. J., Rois-DíAz, M., Moreno, G., den Herder, M., Aldrey-Vázquez, J. A., Ferreiro-Domínguez, N., Pantera, A., Pisanelli, A., & Rigueiro-Rodríguez, A. (2018). Agroforestry in Europe: A land management policy tool to combat climate change. *Land Use Policy, 78*, 603–613.

Muchane, M. N., Sileshi, G. W., Gripenberg, S., Jonsson, M., Pumariño, L., & Barrios, E. (2020). Agroforestry boosts soil health in the humid and sub-humid tropics: A meta-analysis. *Agriculture, Ecosystems & Environment, 295*, 106899.

Mulatu, K., & Hunde, D. (2019). Agroforestry: A supplementary tool for biodiversity conservation and climate change mitigation and adaptation. *System, 9*(19).

Murthy, I. K., Gupta, M., Tomar, S., Munsi, M., Tiwari, R., Hegde, G. T., & Ravindranath, N. H. (2013). Carbon sequestration potential of agroforestry systems in India. *Journal of Earth Science and Climatic Change, 4*(1), 1–7.

Muschler, R. G. (2016). Agroforestry: Essential for sustainable and climate-smart land use. *Tropical Forestry Handbook, 2*, 2013–2116.

Nair, P. K. R. (1998). Directions in tropical agroforestry research: past, present, and future. In *Directions in Tropical Agroforestry Research: Adapted from selected papers presented to a symposium on Tropical Agroforestry organized in connection with the annual meetings of the American Society of Agronomy, 5 November 1996, Indianapolis, Indiana, USA* (pp. 223–245). Springer Netherlands.

Nair, P. R. (2011). Agroforestry systems and environmental quality: Introduction. *Journal of Environmental Quality, 40*(3), 784–790.

Nair, P. R., Kumar, B. M., Nair, V. D., Nair, P. R., Kumar, B. M., & Nair, V. D. (2021). Classification of agroforestry systems. *An introduction to agroforestry: Four decades of scientific developments* (pp. 29–44). Springer.

Nair, P. R., Nair, V. D., Kumar, B. M., & Showalter, J. M. (2010). Carbon sequestration in agroforestry systems. *Advances in Agronomy, 108*, 237–307.

Nath, A. J., Kumar, R., Devi, N. B., Rocky, P., Giri, K., Sahoo, U. K., Bajpai, R. K., Sahu, N., & Pandey, R. (2021). Agroforestry land suitability analysis in the Eastern Indian Himalayan region. *Environmental Challenges, 4*, 100199.

Noiha Noumi, V., Zapfack, L., Hamadou, M. R., Awe Djongmo, V., Witanou, N., Nyeck, B., Ngossomo, J., Hamadou, M., Chimi, C., & Mapongmetsem, P. M. (2018). Floristic diversity, carbon storage and ecological services of eucalyptus agrosystems in Cameroon. *Agroforestry Systems, 92*, 239–250.

Oelbermann, M., Voroney, R. P., & Gordon, A. M. (2004). Carbon sequestration in tropical and temperate agroforestry systems: A review with examples from Costa Rica and southern Canada. *Agriculture, Ecosystems & Environment, 104*(3), 359–377.

Pancholi, R., Yadav, R., Gupta, H., Vasure, N., Choudhary, S., Singh, M. N., & Rastogi, M. (2023). The role of agroforestry systems in enhancing climate resilience and sustainability – A review. *International Journal of Environment and Climate Change, 13*(11), 4342–4353.

Partey, S. T., Thevathasan, N. V., Zougmoré, R. B., & Preziosi, R. F. (2018). Improving maize production through nitrogen supply from ten rarely-used organic resources in Ghana. *Agroforestry Systems, 92*, 375–387.

Pimm, S. L. (2008). Biodiversity: Climate change or habitat loss—Which will kill more species?. *Current Biology, 18*(3), R117–R119.

Pimm, S. L., Russell, G. J., Gittleman, J. L., & Brooks, T. M. (1995). The future of biodiversity. *Science, 269*(5222), 347–350.

Polglase, P., Paul, K., Hawkins, C., Siggins, A., Turner, J., Booth, T., Crawford, D., Jovanovic, T., Hobbs, T., Opie, K., Almeida, A., & Carter, J. (2008). *Regional opportunities for agroforestry systems in Australia*. Rural Industries Research and Development Corporation, *Canberra. Journal of Sustainable Forestry, 36*(3), 277–288.

Prajapati, R., Singh, S., Anand, A., Mallesh, Y., & Gulaiya, S. (2024). Green solutions: The impact of agroforestry on climate change resilience – A review. *International Journal of Environment and Climate Change, 14*(11), 1–13.

Ramachandran Nair, P. K., Mohan Kumar, B., & Nair, V. D. (2009). Agroforestry as a strategy for carbon sequestration. *Journal of Plant Nutrition and Soil Science, 172*(1), 10–23.

Ranjitkar, S., Sujakhu, N. M., Lu, Y., Wang, Q., Wang, M., He, J., Mortimer, P., Xu, J., Kindt, R., & Zomer, R. J. (2016). Climate modelling for agroforestry species selection in Yunnan Province, China. *Environmental Modelling & Software, 75*, 263–272.

Rosenstock, T. S., Wilkes, A., Jallo, C., Namoi, N., Bulusu, M., Suber, M., Mboi, D., Mulia, R., Simelton, E., Richards, M., Gurwick, N., & Wollenberg, E. (2019). Making trees count: Measurement and reporting of agroforestry in UNFCCC national communications of non-Annex I countries. *Agriculture, Ecosystems & Environment, 284*, 106569.

Sahoo, G., Wani, A., Sharma, A., & Rout, S. (2020). Agroforestry for forest and landscape restoration. *International Journal of Advance Study and Research Work, 9*, 536–542.

Santiago-Freijanes, J. J., Mosquera-Losada, M. R., Rois-Díaz, M., Ferreiro-Domínguez, N., Pantera, A., Aldrey, J. A., & Rigueiro-Rodríguez, A. (2021). Global and European policies to foster agricultural sustainability: Agroforestry. *Agroforestry Systems*, *95*(5), 775–790.

Satish, P., Madiwalar, A. F., Lallawmkimi, M. C., Reddy, K. J., Parveen, S., Ashoka, P., Thejavath, L., Karikalan, M. K., & Anand, G. (2024). Agroforestry: Multifunctional benefits and implementation strategies. *Journal of Geography, Environment and Earth Science International*, *28*(10), 1–12.

Scherr, S. J. (1991). On-farm research: The challenges of agroforestry. *Agroforestry Systems*, *15*, 95–110.

Sharma, R., Chauhan, S. K., & Tripathi, A. M. (2016). Carbon sequestration potential in agroforestry system in India: An analysis for carbon project. *Agroforestry Systems*, *90*, 631–644.

Shukla, A., Kumar, A., Chaturvedi, O. P., Nagori, T., Kumar, N., & Gupta, A. (2018). Efficacy of rhizobial and phosphate-solubilizing bacteria and arbuscular mycorrhizal fungi to ameliorate shade response on six pulse crops. *Agroforestry Systems*, *92*, 499–509.

Sileshi, G. W., Mafongoya, P., Akinnifesi, F. K., Phiri, E., Chirwa, P., Beedy, T., Makumba, W., Nyamadzawo, G., Njoloma, J., Wuta, M., Nyamugafata, P., & Jiri, O. (2014). *Agroforestry: Fertilizer trees.In Encyclopedia of agriculture and food systems (Vol.* 1, pp. 222–234). Elsevier

Singh, N. R., Singh, A., Devi, N. P., Kumar, Y. B., Sangma, R. H. C., Philanim, W. S., Devi, M., & Bhutia, P. L. (2024). Agroforestry for soil health. In A. Raj, M. K. Jhariya, A. Banerjee, R. K. Jha, & K. P. Singh (Eds.), *Agroforestry* (pp. 255–283). Wiley.

Singha, C., & Swain, K. C. (2016). Land suitability evaluation criteria for agricultural crop selection: A review. *Agricultural Reviews*, *37*(2), 125–132.

Skendžić, S., Zovko, M., Živković, I. P., Lešić, V., & Lemić, D. (2021). The impact of climate change on agricultural insect pests. *Insects*, *12*(5), 440.

Swamy, S. L., & Tewari, V. P. (2017). Mitigation and adaptation strategies to climate change through agroforestry practices in the tropics. In J. C. Dagar & V. P. Tewari (Eds.), *Agroforestry* (pp. 725–738). Springer.

Tranchina, M., Reubens, B., Frey, M., Mele, M., & Mantino, A. (2024). What challenges impede the adoption of agroforestry practices? A global perspective through a systematic literature review. *Agroforestry Systems*, *98*, 1817–1837.

Udawatta, R. P., & Jose, S. (2012). Agroforestry strategies to sequester carbon in temperate North America. *Agroforestry Systems*, *86*, 225–242.

UNEP. (2009). The environmental food crisis – The environment's role in averting future food crises. In: C. Nellemann, M. MacDevette, T. Manders, B. Eickhout, B Svihus, A.G. Prins, and B.P. Kaltenborn (Eds.), *A UNEP rapid response assessment*. United Nations Environment Programme, GRIDArendal, Norway, www.grida.no. Feb 2009

Van Noordwijk, M., Coe, R., Sinclair, F. L., Luedeling, E., Bayala, J., Muthuri, C., Cooper, P., Kindt, R., Duguma, L., Lamanna, C., & Minang, P. A. (2021). Climate change adaptation in and through agroforestry: Four decades of research initiated by Peter Huxley. *Mitigation and Adaptation Strategies for Global Change*, *26*(5), 18.

Verchot, L. V., Van Noordwijk, M., Kandji, S., Tomich, T., Ong, C., Albrecht, A., Mackensen, J., Bantilan, C., Anupama, K. V., & Palm, C. (2007). Climate change: Linking adaptation and mitigation through agroforestry. *Mitigation and Adaptation Strategies for Global Change*, *12*, 901–918.

Verma, K., Sharma, P., Kumar, D., Vishwakarma, S. P., & Meena, N. K. (2021). Strategies sustainable management of agroforestry in climate change mitigation and adaptation. *International Journal of Current Microbiology and Applied Sciences*, *10*(01), 2439–2449.

VijayKumar, R., Lallawmkimi, M. C., Kumar Gautam, S., Tiwari, P., KS, A., Sekhar, M., & Upadhyay, L. (2024). Agroforestry practices as a keystone for biodiversity conservation: A review. *Journal of Experimental Agriculture International*, *46*(9), 61–76. https://doi.org/10.9734/jeai/2024/v46i92805

Williams, J. W., & Jackson, S. T. (2007). Novel climates, no-analog communities, and ecological surprises. *Frontiers in Ecology and the Environment*, *5*, 475–482.

Wilson, M. H., & Lovell, S. T. (2016). Agroforestry—The next step in sustainable and resilient agriculture. *Sustainability*, *8*(6), 574.

Yengwe, J., Gebremikael, M. T., Buchan, D., Lungu, O., & De Neve, S. (2018). Effects of *Faidherbia albida* canopy and leaf litter on soil microbial communities and nitrogen mineralization in selected Zambian soils. *Agroforestry Systems*, *92*, 349–363.

Young, A. (1989). *Agroforestry for soil conservation*. C.A.B. International & International Council for Research in Agroforestry.

Ziro, J. S., Kichamu-Wachira, E., Ross, H., & Palaniappan, G. (2023). Adoption of climate resilient agricultural practices among the Giriama community in South East Kenya: Implications for conceptual frameworks. *Frontiers in Climate*, *5*, 1032780.

11 Forest Adaptation Potential to Maintain the Flow of Ecosystem Services

Abhishek Kumar, Aakash Kumar, Vatsala, Shilpa Kumari, Mukesh Kumar, Darshita, Kunal Biswas, Yashaswi Singh, Swati Singh, Shivansh Negi, Komal Rani, and Sanjay Singh

11.1 INTRODUCTION

11.1.1 Importance of Forests in Providing Ecosystem Services

Forests are crucial in providing ecosystem services (ESs) globally (Jenkins & Schaap, 2018). Forests are also important to determine majority of conservation decisions and promote sustainable living (Pfund et al., 2011). They balance the human need for goods and services and also provide habitat for a plethora of flora and fauna. Forest ecosystems provide a wide range of vital services, i.e., ESs which is essential for environmental sustainability and human well-being (Butler & Oluoch-Kosura, 2006). Regardless of knowing the fact that ecosystems linked to forests are extremely important, they still require a better and broader understanding (Kanta et al., 2024). These services include the production of ecosystem goods such as timber, non-timber forest products, and medicinal resources. Forests help to regulate climate by sequestering carbon by storing them and participate in the processes of soil formation which also prevents soil loss and its degradation (Kumar et al., 2024). Forests also support a large variety of plant and animal species and help in pollination, which is critical for many plant species. Forests also provide natural pest control, which decreases the overuse of chemical pesticides. Seed dispersal by forest animals ensures the propagation of various plant species. Forests add aesthetic beauty and are a source of spiritual stimulation that increases human mental health and cultural value (Hall & Knuth, 2019). Such ESs are not evaluated on monetary terms and are invaluable. Their preservation is essential to preserve ecological harmony and sustain life on Earth.

The ES that forests provide influence a wide range of ecological, political, economic, social, and cultural issues and activities in addition to helping people make sensible decisions regarding the climate (Singh et al., 2022). One of the earlier studies on ES valuation, Costanza et al. (2017), calculated that wetlands and forests

DOI: 10.1201/9781003588580-11

account for 38% of the overall ES value globally. Later, Millennium Ecosystem Assessment System (Assesment, 2005) classified ESs in four categories, i.e., cultural, regulating, provisioning, and supporting. Costanza (1996) stated that ES encompass the functions of ecosystems, including their habitat, biological, and system properties or processes. These functions give rise to ecosystem goods, such as food, and services, like waste assimilation, which humans benefit from directly or indirectly. Borma et al. (2022) in their study focused on the provision of several ESs that would be enhanced by managing forests to preserve huge trees, create canopy gaps, and increase structural variation. These findings demonstrate how forest management may promote multifunctional forests and imply that a coordinated approach at the landscape size may be able to lessen trade-offs in areas that are controlled by humans.

11.1.2 Overview of the Concept of Forest Adaptation

Forests are the second most vital resource that stores carbon and acts as a stabilizing force for regulating the climate (Funk et al., 2019). Forests act as a carbon sink and source and have the potential to regulate the climate patterns and are the best carbon capture technology available till date (Lewis et al., 2019). The forests mitigate climate change through the absorption of carbon dioxide from the atmosphere and the storage. This is a direct contribution because too much carbon dioxide in the atmosphere is one of the primary causes of climate change. Besides that, there are various indirect benefits offered by the forests that will help society cope with impacts of climate change. Healthy forests are very essential in managing the rainfall pattern (Keleş, 2019). Forests soak up the rain and hold back it from becoming floods, thus preserving the integrity of the soil and avoiding damage to the quality of water. Moreover, the temperature variation between urban and rural areas is controlled by forests. This reduces the energy needed for human population sustenance and, consequently, aligns our efforts towards sustainable development goals. Besides, forests are natural purifiers of air and water and filter out many harmful airborne pollutants that are exacerbating the climate change process and help clean up waterborne pollutants (Sadhu et al., 2018).

11.1.3 How Climate Regulation Is Done by Forests

Forests stabilize the climate and offer livelihood, preserve biodiversity, conserve ecosystems, play a major role in the carbon cycle, and even facilitate sustainable development (Swamy et al., 2018). Forests regulate temperature within both the urban and rural environments. The climate considerably impacts the development, process, and conditions of the forests. For this reason, the forests may be directly influenced by climate changes such as temperature and rainfall (Keenan, 2015). The impact can vary, which depends upon the species as well as the region of the forest. While some impacts may be positive, others are going to be negative. Forests are very long-lived and resistant by nature to climate changes. Not only that, there are other things which mankind can do to make forests vulnerable to climate change.

11.1.4 Forest Ecosystems Behave, Conserve Soil, and Stabilize Stream Flows and Water Runoff

Forests are natural sponges and their role in the control of Earth's water cycle prevents floods and ensures water quality (Nguyen et al., 2019). Thus, the ability of forests to retain water and stabilize the soil, besides building cloud space, maintains hydrological balance at the worldwide level. Forests do not merely represent passive receivers but very active contributors to the hydrological cycle. The difference in forest cover can have a significant bearing on the hydrological dynamics of an entire region. Furthermore, because forests contain high content of organic material, forest soils absorb and retain moisture just like a sponge (Ghazoul & Sheil, 2010). In this regard, contaminants as well as extraneous nutrients are flushed out of water passing through the soil layers while preventing erosion.

11.1.5 Forests Help Societies Adapt to Climate Variability and Change

Forests and their products provide a direct source of income for millions of indigenous forest dwellers (Hegde et al., 1996). More generally, woods provide a well-known array of benefits that enhance human well-being. Therefore, it is crucial that forests adapt to climate change (Locatelli, 2010). Trees ability to sequester carbon, how forests respond to climate change, and the forest services that local populations receive are all likely to be impacted locally by forest management and silviculture. In this case, adaptation and mitigation must coexist (Locatelli et al., 2015). Forests play an important role in maintaining the hydrological ESs as they regulate stream flow and equalize the water potential underneath for fresh ground. Forest cover in upland watersheds improves infiltration of rainwater into it and reduces surface runoffs, hence avoiding soil erosion leading to floods, factors that are otherwise destructive to the affected areas (Hamilton, 2019). Forests also provide storage for runoff, thus acting as natural recharge zones for streams, hence providing an important service in face of risks rising with increased drought. The economic and welfare implications for the poorer communities in developing and least developed countries rely on forests for food, fodder, fuelwood, medicines, and all these other essential ESs (Chand et al., 2021). Challenges such as water scarcity and irregular rainfalls, coupled with population growth and land degradation, put additional stress on forest ecosystems and their capacity to sustain livelihoods in the local community. Under such conditions, community forestry promoted in most developing regions may strengthen adaptive capacity by moving decision-making power to those who feel the impacts of climate change most intensely and by injecting indigenous knowledge into sustainable forest management (SFM) practices.

11.1.6 Forest Ecosystem Services and Sustainable Development

Specific impacts of climate change on the health, growth, distribution, and composition of certain forest ecosystems are unknown. There are very few local projections for future climates, and the complex interactions between biotic and abiotic factors in these systems introduce much uncertainty. The industrialized countries are possibly more prepared for the immediate effects of climate change and certainly less vulnerable to direct short-term consequences – they stand to suffer much fewer economic and welfare effects from climate change than many of the poorer communities found throughout much of the developing and least developed world (Costello et al., 2009). For these communities, forests are among the most important sources of food, fodder, fuelwood, medicines, and other ESs. Pressures on forest ecosystems are heightened by the challenges of water scarcity, unpredictable rainfall, population growth, and land degradation; meanwhile, local livelihood needs remain further unmet by forests' productivity potential (Brauch et al., 2011). Community forestry approaches initiated in these regions could lift community adaptive capacity by engaging communities that are most likely to be experiencing early climate change effects and facilitate the embedding of traditional knowledge within SFM practice (Fernandez-Gimenez et al., 2008).

11.1.7 Threats to Ecosystem Services

Forests are essential for regulating climate of the Earth and providing ESs that are important for survival of human beings. Climate change also poses threats to forest ecosystems, affecting their health, composition, and ability to provide services like carbon sequestration, water cycle regulation, and biodiversity conservation. As global temperatures continue to rise, forests become vulnerable to a range of stressors, all of which are being worsened by these changing climatic conditions (Trumbore et al., 2015). Since pre-industrial times, the global temperatures have risen by approximately 1.2°C, and projections suggest that this trend could reach 2°C or more by the end of the 21st century (IPCC, 2021). These temperature changes already have significant impact on forest ecosystems in many ways. For example, tree growth and productivity are being affected. In temperate and boreal regions, warmer temperatures may extend the growing season and temporarily boost tree productivity. While in tropical and arid regions, many tree species are already near their physiological limits, and the added heat stress can stunt tree growth, impair photosynthesis, and disrupt nutrient cycling (Allen et al., 2010).

Due to the increase in global temperature, the tree species tends to shift their geographical range, usually moving towards higher altitudes. Boreal forests, for example, may expand into tundra regions, while temperate forests could encroach upon boreal zones (Thuiller et al., 2005). But not all species can migrate quickly enough to keep up with these changes, leading to habitat mismatches, local extinctions, biodiversity loss, and shifts in forest composition. Climate change also has its impact on the frequency and intensity of wildfires (Jones et al., 2020). Extended periods of drought due to higher temperatures affect vegetation which creates conditions for fire outbreaks. Wildfires cause destruction to large areas of forest and release sequestered carbon back into the atmosphere, leading to global warming (Bowman et al., 2009). This feedback loop further threatens forest resilience and the ESs they provide. Another critical impact of climate change on forests comes from altered precipitation patterns. Climate change is reshaping global rainfall distribution, which directly affects water availability for forest ecosystems (Grimm et al., 2013). Many regions in the world now experience more frequent and severe droughts, affecting the water table. This decreases their ability to photosynthesize and take up nutrients, leading to tree death and forest structure changes (Anderegg et al., 2013). Conversely, increased rainfall in some areas leads to soil erosion, and root system architecture gets disturbed affecting forest health (Bonan, 2008). The changing climate affects the composition of tree species because various species have different tolerances to changes in precipitation. Drought-tolerant species may become dominant in areas with less rainfall, while moisture-dependent species may decline, changing the dynamics and biodiversity of forests. For instance, tropical rainforests, which depend on high precipitation, are highly susceptible to changes in rainfall patterns and may lead to shifts in species dominance and forest structure (Clark et al., 2010). Climate change has exacerbated extreme weather events like storms, hurricanes, and floods, which leads to devastating impacts on forests. Strong winds can destroy a significant amount of tree canopy and uproot them, and flooding can take away nutrient-rich soils from an ecosystem, damage tree roots, and thus, alter the ecosystems in the long run (Lugo, 2000; Tockner et al., 2010). The inherent resilience of a forest is thus often outweighed by continuing stressors such as deforestation, fragmentation of habitats, and climate change itself. Forests also play vital roles as carbon sinks, annually removing up to 30% of global carbon dioxide emissions, and climate change can reduce their ability to sequester carbon, creating a positive feedback loop that accelerates global warming (Brando et al., 2014). Degradation and deforestation lead to global carbon emissions as the carbon stored in trees and soil is released into the atmosphere.

To prevent this, one approach is to adjust forest management practices in order to respond to changing climatic conditions. This can involve using climate-resilient tree species, enhancing forest connectivity to enable species to migrate, and developing a fire management approach to diminish the risks of wildfires (Millar et al., 2007). Restoring degraded forests and expanding replanting efforts promote carbon sequestration as well as enhance forest resistance to climatic stressors, making them vital strategies as well (Figure 11.1). More emphasis on native species planting should be given in plantation projects while focusing on forming multi-layered diverse forest which is more resistant towards climate change. Continuous survey and research is equally important, as this will help identify forest health and emerging potential threats in advance. Emerging technologies by involving advanced remote sensing and shift of focus on long-term ecological studies can help policymakers and forest managers to make informed decisions for protection of forests.

11.1.8 Deforestation and Land-Use Change

Forests are an integral part to sustain global biodiversity and regulate climate supporting human livelihoods. They, however, face threats from various anthropogenic and natural factors like deforestation and land-use changes, followed by pests and diseases. Critical challenges, along with unsustainable human activities in logging and agricultural expansion, reduce the forest cover and compromise various ESs that support biodiversity. The most serious threats to forest ecosystems include loss and land-use change, the latter of which is a result of agricultural land demand, infrastructural development, and urbanization. Millions of hectares of forests are lost or

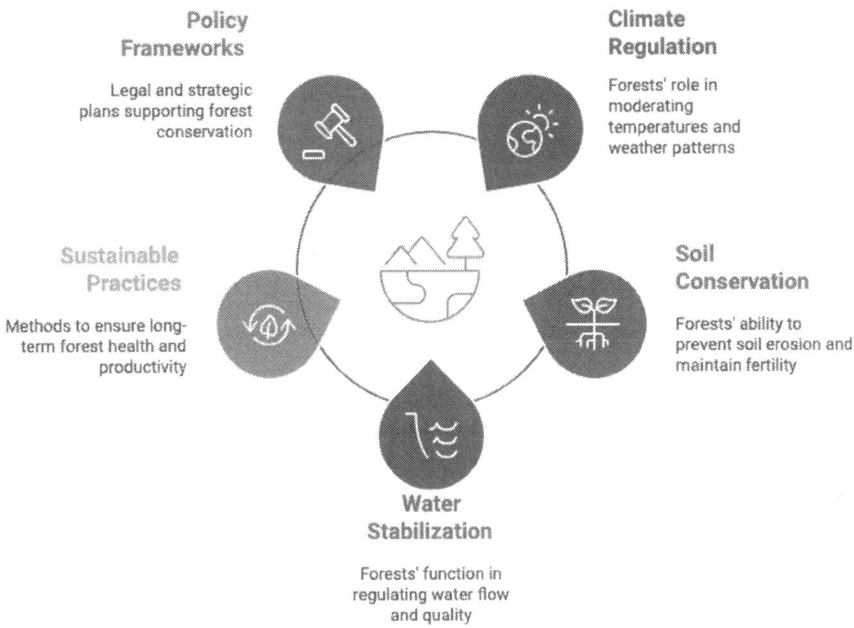

FIGURE 11.1 Main factors involved in forest adaptation.

degraded each year, with the majority coming from tropical regions, which host the highest rates of deforestation. The changes in the forest structure exert far-reaching consequences that range from local to global systems.

11.1.9 Drivers of Deforestation

It has been found that agricultural development is one of the biggest reasons for deforestation. Due to the rising demands on food production caused by the population of the world, farmlands are now coming into place where forests have begun to be cleared out. Because of industrialization of agriculture and focusing more on the short-term gain in yield, clearing the land is often more paramount than practising sustainable land-use that leads to a large amount of forest being lost (Hosonuma et al., 2012). The other key cause for forests loss is the construction of infrastructure like roads, mines, and dams. Roads encroach on the forests, fragment ecosystems, and make isolated forest patches to be used for harvesting and farming. Deforestation is also largely caused by urbanization, especially in emerging nations where fast economic development has resulted in growing cities and the conversion of forest land for residential and commercial uses. Forest ecosystem directly faces threats from urbanization which leads to habitat destruction and a decline in biodiversity (Geist & Lambin, 2002). Forests are essential for controlling water levels through the distribution of rainfall and evapotranspiration. Clearance of forests results in less rainfall, changes stream flow patterns, and reduces water availability for human populations. Deforestation can lead to greater chance of devastation in form of landslides and floods because forests lose the ability to stabilize soil and control excess water flow (Ellison et al., 2017).

Ecosystems become more vulnerable to environmental changes due to habitat destruction, species extinctions, and reduced genetic diversity caused by deforestation (Sodhi et al., 2010). Over 80% of terrestrial species live in forests, and many are highly specialized and cannot survive outside their natural forest habitat. Because species loss can affect ecological interactions such as pollination, seed distribution, and nutrient cycling, losing biodiversity further threatens the ability of forests to provide ESs. Around 10% of global emissions from greenhouse gases are due to deforestation, indicating how critical forest protection is for the slowing down of climate change (Pan et al., 2011). In the absence of a forest cover, soils tend to get exposed to the forces of water and wind; this accelerates erosion while depleting nutrients within the soil and lowers its fertility. Beyond land productivity implications, soil degradation fuels more universal environmental problems like desertification and sedimentation of lakes and rivers. Soil degradation in tropical areas is a common occurrence which leads to deforestation. These degraded soil patches when left exposed further fasten the cycle of land degradation by making it challenging to restore these degraded areas (Lal, 2003; Singh et al., 2024).

11.1.9.1 Land-Use Planning and Management

There is a need for efficient land-use planning and management to prevent deforestation and its associated repercussions. Sustainable land-use planning can help balance the need to preserve forests and the need for development. Agroforestry is one of the strategies involving the integration of trees in agricultural systems that can help the farmer economically while conserving part of the environmental functions that the forests provide. Agroforestry remains a good strategy for decreasing the pressures of deforestation, as it can enhance biodiversity, increase soil fertility, and trap carbon (Jose, 2009). Moreover, afforestation and reforestation need to be undertaken to reclaim degraded landscapes and enhance the carbon sequestration capabilities (Hudson, 2023).

11.1.9.2 Invasive Species

Invasive species are non-native organisms which are introduced to new ecological realm and have the capacity to outcompete native species having broad ecological amplitude (Kumar et al., 2024). Some of the plant species like *Lantana camara*, *Parthenium hysterophorus*, and *Ageratum conyzoides* significantly harm the forest ecosystems and disrupt the ecological balance. Climate change is facilitating the dispersal of invasive species through the creation of favourable conditions for their existence (Kumar & Singh, 2019). Similarly, invasive species of animals like beetles and insects are also disrupting the ecological balance. Liebhold et al. (2017) note that the rising temperatures along with altered precipitation patterns may enhance the expansion of pests in new areas where the native species may not be as resistant to their attacks. An increase in bark beetle population in many places in North America and Europe reduces tree mortality in large proportions. Millions of hectares of forest are invaded by bark beetles due to warm winters and longer growth periods. In the same way, a large population of frogs in forest ecosystems has been killed by fungal infections caused by chytrid fungus (Cudmore et al., 2010).

11.1.9.3 Forest Vulnerability and Mitigation Strategies

With changes in environmental conditions, certain species of trees are more prone to diseases and pests. For instance, pest outbreaks are more likely in monoculture plantations compared to mixed-species forests due to a lack of species diversity. Some tree species will likely become more vulnerable to pests due to the impacts of climate change on temperature and precipitation, threatening biodiversity and forest resilience. Because such key tree species are important habitats for many species and crucial for carbon storage and nutrient cycling, their loss is bound to trigger a domino effect within the entire ecosystem (Boyd et al., 2013).

Integrated pest management (IPM) gives a holistic approach integrating chemical, mechanical, and biological control methods to halt the spread of diseases and pests. Biological controls, such as introducing viruses or natural predators that can target pests, may also be helpful in keeping pest populations under control without jeopardizing the ecosystem. Preserving species variety and planting pest-resistant tree species may also make a forest more resistant to outbreaks of insects (Liebhold et al., 2017). More importantly, monitoring and discovering the early spread of diseases and pests is very important. In combination with field surveys, remote sensing technology can be utilized to monitor changes in the health of forests and new threats. In order to prevent disease from spreading into healthy forests, early detection allows for timely treatments such as cutting out affected trees (Hicke et al., 2012).

11.1.9.4 Human Activities

Human activities are very destructive to forest ecosystems, mainly logging, agriculture, and pollution. Not only do these unsustainable practices reduce the forest cover but also cause pollution, fragmentation of habitats, and destruction of vital ecosystem functions. Logging is the main driver of deforestation, mostly illegal and legal logging in the tropics. While it is often necessary to produce timber and other forest products, destructive logging practices can seriously damage forest ecosystems. Targeting high-value trees through selective logging can disrupt habitats, making it difficult for species to migrate and procreate. Fragmentation of forests makes forests more vulnerable to secondary threats such as insect outbreaks and wildfires. This results in large patches of forests that are being harvested for timber or illegal logging (Asner et al., 2005).

11.1.9.5 Agricultural Encroachment

Forest ecosystems are also seriously threatened by the spread of agriculture into wooded regions. Agricultural encroachment has emerged as a major contributor to deforestation as populations rise and food demands rise. This is especially noticeable in tropical areas when forests are cut down to make way for extensive farming operations like soy plantations and palm oil. In addition to destroying forest cover, agricultural growth upsets the natural equilibrium, causing soil erosion, water scarcity, and biodiversity loss (Gibbs et al., 2020).

11.1.9.6 Pollution

Pollution from industrial processes, mining, and agricultural runoff is seriously threatening forest ecosystems. Agricultural runoff containing fertilizers and pesticides can contaminate the water and soil, endangering wildlife and forest vegetation. Mining also contaminates soil and water due to the release of heavy metals and other hazardous compounds into the environment (Singh et al., 2024). Pollutants can damage leaves, reduce photosynthesis, and weaken trees, making them more susceptible to pests and diseases. Industrial pollution, which includes air pollution from factories, affects the health of forests (Malmer et al., 2010).

11.1.9.7 Sustainable Practices

Sustainability practice helps significantly in minimizing the consequences of illegal human activity towards the forest. One of the key concepts used within SFM is the process by which forests are managed for protecting their ecological services, productivity, and biodiversity. Forest Stewardship Council certification programmes encourage responsible forest management by assuring that timber products have been harvested with minimal adverse impact on the environment (Forest Stewardship Council, 2015). Using agroforestry by integrating trees into agricultural systems is a promising strategy for achieving a balance between ecosystem health and human needs. Agroforestry enhances biodiversity, improves soil health, sequesters carbon by conserving tree cover in agricultural landscapes, enhances crop productivity, and economically benefits farmers (Jose, 2009).

11.1.10 Forest Adaptation Potential to Maintain the Flow of Ecosystem Services

Forest adaptation strategies mean to adapt to a shifting climate and explore variables that have potential to affect public choices (Hajjar & Kozak, 2015). The role of ESs in adaptation to climate change is shown in Figure 11.1. According to FAO (2013), adaptation strategies fall into two broad categories: (1) adaptation of forests, to increase forests' resilience to climate change, and (2) adaptation using forests, which involves measures to decline the susceptibility of people who rely on forest to climate change (Figure 11.2). As per Smithers and Smit (1997), adaptation measures are distinguished into two broad kinds: measures that aim to buffer a system from distresses by enhancing its resistance to change, and measures that accelerate a shift or an evolution of the system to a different state that encounters altered conditions. Buffering measures may only be effective over a short term, while measures that aim to facilitate a shift or evolution of a system do not resist changes but rather ease and achieve natural adaptation processes (Locatelli et al., 2010).

11.1.10.1 Natural Adaptation Processes

Forest tree species typically exhibit significant level of genetic variation which results from factors such as local adaptation, population size, and neutral evolutionary processes in a heterogeneous environment (Hamrick, 1989). Genetic changes endorsed to climate change include hybridization or interbreeding of species' as their habitats change (IUCN, 2019). Several tree species have survived past climate variability because they possess both significant phenotypic plasticity and high levels of genetic variation (Isabel et al., 2020). Fundamental responses such as range shifts, adaptability, and evolutionary responses are among the evolutionary responses that show up in response to climate change (Edelsparre et al., 2024). Moreover, when

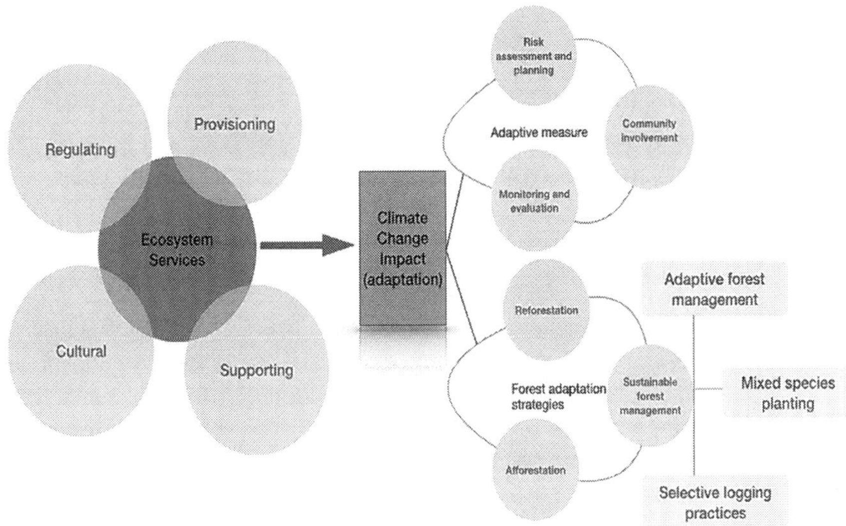

FIGURE 11.2 Climate change adaptation strategies and their role in ecosystem services.

a shifting species meets new habitat and population, it may have an impact on the evolutionary trajectory of entire systems, making the repercussions crucial from an evolutionary perspective (Suárez et al., 2022). Assisted migration aims to enhance the resilience of forest tree species to climate change, using species-specific climatic constraints and local modifications by seed provenance transfers (Chakraborty et al., 2024). According to Aitken et al. (2008), the three possible outcomes for forest tree populations are (1) migration to follow the geographical shift of favourable climatic conditions, (2) evolution-driven adaptation to a changing environment, or (3) extirpation. However, the restricted migration abilities of various species to reach appropriate areas and the increasing population maladaptation will eventually limit natural species reorganizing and endanger forest ES if increasingly mismatched local adaptations are overlooked (Chakraborty et al., 2024).

11.1.10.2 Human-Assisted Adaptation

Human-assisted adaptation in forest management refers to the intentional change in action in a way that the ecosystems of the forests cope and adapt with climate change. Assisted practices may comprise species migration, selective breeding for enhancing tree species' robustness, and increased conservation of genetic diversity. By introducing species or genotypes, which would be well adapted to the climate that is expected, forest managers can enhance the resilience of the ecosystem against stressors caused by increased temperature and drought with outbreaks of pests. There are two types of human-assisted adaptations, i.e., active and passive. Active adaptation involves the use of silvicultural methods (e.g., stand conversion, tending, thinning, tree species enrichment) to alter stand structures and tree species configuration in ways that make the resulting forest well adapted to climate that will occur during the life of the stand (Bolte et al., 2009). It also involves a revision of planting strategies to reflect upcoming site conditions rather than past. Passive adaptation of forest management is using the natural, inherent resilience, and succession processes of forest ecosystems and indicates a reduction or even termination of silvicultural intake (Jandl et al., 2019). Due to lack of promising measures of active adaptation and current high resilience of beech forest systems, mainly passive adaptation appears to be feasible using natural forest dynamics and possible succession for forest adaptation to climate change. Passive adaptation focuses on the utilization of natural forest dynamics for forest adaptation which can be thus regarded as congruent element of non-management (Bolte et al., 2009).

11.1.10.3 Reforestation Approach

This approach helps to mitigate the effects of climate change which involves re-establishment of former forest areas that have experienced deforestation or degradation (Osman et al., 2023). Reforestation in tropical regions is most likely to have cooling net impact, while in boreal forests, it may boost regional temperatures through albedo effects (Anderson et al., 2011). It also contributes to global biodiversity conservation by enhancing species' resilience to climate variability, which will otherwise amplify species' declines that are previously occurring because of ongoing loss of forest habitat (Travis, 2003).

11.1.10.4 Afforestation

It is one of the most effective ways to combat climate change by the establishment of forest on land that was not previously forested (Hudson, 2023). Commercial afforestation can carry out effective greenhouse gas emission (GHG) mitigation that is resilient to upcoming decarbonization pathways. National afforestation plans should encourage a range of commercial, conservation, and agroforestry systems suited to local circumstances and adaptable to climate change (Dymond et al., 2014), in order to provide the range of ESs needed by society, in addition to GHG mitigation (Li et al., 2015).

11.1.10.5 Sustainable Forest Management (SFM)

SFM is an evolving and dynamic concept that aims to enhance and maintain the economic, social, and environmental values of all types of forests, for the benefit of present and future generations (FAO, 2013). It can help to reduce the adverse effects of climate variability on forests and forest-dependent people, and it can also help to ensure that forests play their role in mitigating climate change. It constitutes primary approach to forest management, and its enactment requires, at the national or subnational levels, enabling laws and policies on the ground.

11.1.11 Policy and Management Implications

Forests play an important role in minimizing the effects of climate change. As carbon sinks, they balance the global carbon cycle. However, on the global scale, it is estimated that deforestation and forest degradation account for about 10% of total GHGs (IPCC, 2014). Global policy and legislation, such as the United Nations Framework Convention on Climate Change (UNFCCC) and its associated programme, REDD+, are implemented. These frameworks are global efforts towards sustainable management and are helping to manage the ill effects of climate change.

REDD+ is one of the international mechanisms designed to focus on reducing emissions from deforestation and forest degradation. The global initiative like REDD+ framework offers financial rewards to developing countries for reducing emission by deforestation and degradation, and conserving, managing forests sustainably, and enhancing forest carbon stocks (UNFCCC, 2007). REDD+ finds its premise in that developing nations can earn payments from industrialized countries and other international organizations if they effectively reduce deforestation and linked emissions. Another critical aspect of REDD+ is its focus on multiple benefits from forest conservation. While the mechanism strictly reduces carbon emissions, it focuses more on biodiversity protection, enhancement of ESs, and support for local livelihoods. Angelsen (2009) assert that there are very likely significant contributions that the REDD+ mechanism can make to climate change mitigation while promoting co-benefits from social and environmental lenses.

There are three stages in REDD+: national strategies and capacity building; policy and measures implementation; and results-based payments rewarding performance. A core principle under the mechanism of REDD+ will be to ensure that the rights of indigenous peoples and local communities are respected in forest management practices, making them party to decisions regarding their forests (UNFCCC, 2010). This approach is necessary to have sustainable success since local communities are the custodians of forest lands and mainly impacted by forest policies.

11.1.11.1 UNFCCC and Global Climate Policies

The UNFCCC, adopted in 1992 at the Rio Earth Summit, is the international treaty that mainly addresses global climate change. The ultimate objective of the UNFCCC is to stabilize greenhouse gas concentrations in the atmosphere at a level that prevents dangerous anthropogenic interference with the climate system. Its annual conferences, called Conferences of the Parties, or COPs, are the yearly meetings where governments discuss and enforce policies surrounding climate change – including those involving forestry and land-use.

Because of their ability to absorb carbon, forests have been recognized by the UNFCCC as part of the approach to addressing global climate change. In 2007, at COP13 in Bali, REDD+ was launched as one of the key mechanisms. In 2015, at COP21, where the Paris Agreement was signed, forests were further underlined as being important for attaining the objectives of global climate change (Kumar et al., 2021). Article 5 of the agreement encourages countries to strive further in REDD+ and SFM areas suggested by Agreement, (2015). This makes the UNFCCC play a pivotal role in the integration of efforts in climate adaptation and mitigation. For the case of forests, this means that policies not only focus on the storage of carbon but also enhance the resilience of forest ecosystems to impacts from climate change. Strengthening the link between mitigation and adaptation, forest conservation is through the framework of the UNFCCC set to contribute to environmental and social sustainability in a broader sense.

11.1.11.2 Forest Adaptation and Policy Frameworks

The Agreement (2015) and the *Convention on Biological Diversity* (CBD) highlight the importance of reducing deforestation and forest degradation to achieve climate targets through increased carbon sequestration (Agreement, 2015). Forest management is crucial in maintaining forest adaptation potential. SFM practices, especially ones keeping healthy, diverse, and productive conditions for forests, are critical aspects of adaptation measures. SFM practices have been endorsed in various reports, including those by Millennium Ecosystem Assessment (2005), highlighting the need for selective harvesting, agroforestry, and restoration efforts to enhance resilience.

11.1.11.3 Policy Gaps and Challenges

There are mainly policy gaps in forest management due to a lack of cohering adaptive strategies over the complex, interconnected issues facing climate change, deforestation, and biodiversity loss. The issues are related to insufficient support towards sustainable financing, an insufficient level of integration of adaptation and mitigation strategies about climate change, and insufficient legal frameworks to protect indigenous rights and local stewardship over the forest land. Community-based forest management in improving adaptation outcomes through local involvement of people is necessary in decision-making processes. A consensus on policy implications and joint forest management needs to be brought on a single platform where everybody benefits and thinks towards a sustainable future by the eye of forest management.

11.1.12 Challenges and Opportunities

Even though frameworks such as REDD+ and the UNFCCC provide wonderful platforms through which to address the issue of deforestation, several challenges accompany the implementation process. To begin with, sufficient funding for carrying out large-scale REDD+ projects is a challenge in its implementation. Even though countries can arrange funds through schemes such as the Green Climate Fund (GCF), there are still inadequate financial resources to meet the increasing demand for forest conservation projects. Good governance is necessary both at the local and national levels. Corruption, weak implementation of forest policies, and tenancy disputes undermine efforts aimed at reducing deforestation and promoting sustainable forestry. The long-term impacts of international forestry policy depend on strengthening accountability, transparency, and local participation in forest management. International frameworks and programmes such as the UNFCCC and REDD+ can significantly contribute to combating climate change by reducing forest degradation and improving SFM. These frameworks further underscore the many benefits of forests, including carbon sequestration, biodiversity conservation, and support for local livelihoods. Yet, such tactics could be effective only if some issues were corrected – like governance and funding guarantee. Mechanisms such as REDD+ and UNFCCC will play significant roles in the protection of forests and enhancing their role in climate mitigation because the dynamics of climate change pose a serious threat to global ecosystems.

11.2 CONCLUSION AND WAY FORWARD

Forests are one of the key components of international efforts in slowing down and adapting to climate change. They are important for lowering atmospheric GHGs and providing a buffer against climate extremes because of their capacity to absorb carbon, control temperatures, and sustain biodiversity. Protecting and restoring forest ecosystems are therefore critical actions in any comprehensive climate change response. Forests represent ecosystems that provide fundamental services, including carbon sequestration, water filtration, biodiversity conservation, and climate regulation. However, such highly valued ecosystems are under threat from climate change, land-use changes, and deforestation. In order for everyone to benefit and look towards a sustainable future through the lens of forest management, a consensus on policy implications and cooperative forest management must be brought to a unified platform. There is still a lack of knowledge about the variety of ESs that are still overlooked when developing forest management plans, aside from research on the ESs that our pristine forests offer. Although a lot has been written about the ESs that forests provide, there aren't many instances where this idea has been successfully applied to the management, protection, and planning of temperate forest ecosystems worldwide. Due to the numerous links and interactions between land-uses, forest types, and the substances that are exchanged between forest ecosystems and the atmosphere, broad-scale planning for ESs is required. Therefore, additional research is required to have a deeper understanding of these interactions.

REFERENCES

Aitken, S. N., Yeaman, S., Holliday, J. A., Wang, T., & Curtis-McLane, S. (2008). Adaptation, migration or extirpation: Climate change outcomes for tree populations. *Evolutionary Applications*, *1*(1), 95–111.

Allen, C. D., Macalady, A. K., Chenchouni, H., Bachelet, D., McDowell, N., Vennetier, M., Kitzberger, T., Rigling, A., Breshears, D. D., Hogg, E. H., Gonzalez, P., Fensham, R., Zhang, Z., Castro, J., Demidova, N., Lim, J.-H., Allard, G., Running, S. W., Semerci, A., & Cobb, N. (2010). A global overview of drought and heat-induced tree mortality reveals emerging climate change risks for forests. *Forest Ecology and Management*, *259*(4), 660–684

Anderegg, W. R., Kane, J. M., & Anderegg, L. D. (2013). Consequences of widespread tree mortality triggered by drought and temperature stress. *Nature Climate Change*, *3*(1), 30–36.

Anderson, R. G., Canadell, J. G., Randerson, J. T., Jackson, R. B., Hungate, B. A., Baldocchi, D. D., Ban-Weiss, G. A., Bonan, G. B., Caldeira, K., & Cao, L. (2011). Biophysical considerations in forestry for climate protection. *Frontiers in Ecology and the Environment*, *9*, 174–182.

Angelsen, A. (Ed.). (2009). *Realising REDD+: National strategy and policy options*. Cifor.

Asner, G. P., Knapp, D. E., Broadbent, E. N., Oliveira, P. J., Keller, M., & Silva, J. N. (2005). Selective logging in the Brazilian Amazon. *Science*, *310*(5747), 480–482.

Assessment, M. E. (2005). *Ecosystems and human well-being: wetlands and water*. World resources institute.

Bolte, A., Ammer, C., Löf, M., Madsen, P., Nabuurs, G. J., Schall, P., Spathelf, P., & Rock, J. (2009). Adaptive forest management in central Europe: Climate change impacts, strategies and integrative concept. *Scandinavian Journal of Forest Research*, *24*(6), 473–482.

Bonan, G. B. (2008). Forests and climate change: Forcings, feedbacks, and the climate benefits of forests. *Science*, *320*(5882), 1444–1449.

Borma, L. S., Costa, M. H., da Rocha, H. R., Arieira, J., Nascimento, N. C. C., Jaramillo-Giraldo, C., ... & Nobre, C. A. (2022). Beyond carbon: The contributions of South American tropical humid and subhumid forests to ecosystem services. *Reviews of Geophysics*, *60*(4), e2021RG000766.

Bowman, D. M. J. S., Balch, J. K., Artaxo, P., Bond, W. J., Cochrane, M. A., D'Antonio, C. M., Defries, R. S., Johnston, F. H., Keeley, J. E., Krawchuk, M. A., Kull, C. A., Mack, M., Moritz, M. A., Pyne, S. J., Roos, C. I., Scott, A. C., Sodhi, N. S., & Swetnam, T. W. (2009). Fire in the Earth system. *Science, 324*(5926), 481–484.

Boyd, I. L., Freer-Smith, P. H., Gilligan, C. A., & Godfray, H. C. J. (2013). The consequence of tree pests and diseases for ecosystem services. *Science, 342*(6160), 1235773.

Brando, P. M., Balch, J. K., Nepstad, D. C., Morton, D. C., Putz, F. E., Coe, M. T., Silvério, D., Massad, T. J., Davidson, E. A., Nobrega, C. C., Alencar, A., & Soares-Filho, B. S. (2014). Abrupt increases in Amazonian tree mortality due to drought-fire interactions. *Proceedings of the National Academy of Sciences, 111*(17), 6347–6352.

Brauch, H. G., Spring, Ú. O., Mesjasz, C., Grin, J., Kameri-Mbote, P., Chourou, B., Dunay, P., & Birkmann, J. (Eds.). (2011). *Coping with global environmental change, disasters and security: Threats, challenges, vulnerabilities and risks* (Vol. 5). Springer Science & Business Media.

Butler, C. D., & Oluoch-Kosura, W. (2006). Linking future ecosystem services and future human well-being. *Ecology and Society, 11*(1).

Chakraborty, D., Ciceu, A., Ballian, D., Benito Garzón, M., Bolte, A., Bozic, G., Brang, P., Corrigan, H., Dounavi, A., Garcia-Gil, M. R., Hubert, J., Knutzen, F., Kramer, K., Maaten, T., Marini, L., Mattioni, C., Muñoz-Reinoso, J. C., Nenadić, M., Patsiou, T. S., ... Schueler, S. (2024). Assisted tree migration can preserve the European forest carbon sink under climate change. *Nature Climate Change, 14*, 1–8.

Chand, H. B., Singh, S., Kumar, A., Kumar Kewat, A., Bhatt, R., & Bohara, R. (2021). Comprehensive overview of REDD+ in India: Status, opportunities and challenges. *Grassroots Journal of Natural Resources, 4*(3), 185–200.

Clark, D. A., Piper, S. C., Keeling, C. D., & Clark, D. B. (2010). Tropical rain forest tree growth and atmospheric carbon dynamics linked to interannual temperature variation during 1984–2000. *Proceedings of the National Academy of Sciences, 107*(15), 6859–6864.

Costanza, R. (1996). Ecological economics: Reintegrating the study of humans and nature. *Ecological Applications, 6*(4), 978–990.

Costanza, R., de Groot, R., Braat, L., Kubiszewski, I., Fioramonti, L., Sutton, P., Farber, S., & Grasso, M. (2017). Twenty years of ecosystem services: How far have we come and how far do we still need to go? *Ecosystem Services, 28*, 1–16.

Costello, A., Abbas, M., Allen, A., Ball, S., Bell, S., Bellamy, R., Friel, S., Groce, N., Johnson, A., Kett, M., Lee, M., Levy, C., Maslin, M., McCoy, D., McGuire, B., Montgomery, H., Napier, D., Pagel, C., Patel, J., … Patterson, C. (2009). Managing the health effects of climate change: Lancet and University College London Institute for Global Health Commission. *The Lancet, 373*(9676), 1693–1733.

Cudmore, T. J., Björklund, N., Carroll, A. L., & Staffan Lindgren, B. S. (2010). Climate change and range expansion of an aggressive bark beetle: Evidence of higher beetle reproduction in naïve host tree populations. *Journal of Applied Ecology, 47*(5), 1036–1043.

Dymond, C. C., Tedder, S., Spittlehouse, D. L., Raymer, B., Hopkins, K., McCallion, K., & Sandland, J. (2014). Diversifying managed forests to increase resilience. *Canadian Journal of Forest Research, 44*(10), 1196–1205.

Edelsparre, A. H., Fitzpatrick, M. J., Saastamoinen, M., & Teplitsky, C. (2024). Evolutionary adaptation to climate change. *Evolution Letters, 8*(1), 1–7.

Ellison, D., Morris, C. E., Locatelli, B., Sheil, D., Cohen, J., Murdiyarso, D., Gutierrez, V., van Noordwijk, M., Creed, I. F., Pokorny, J., Gaveau, D. L. A., Spracklen, D. V., Tobella, A. B., Ilstedt, U., Teuling, A. J., Gebrehiwot, S. G., Sands, D. C., Muys, B., Verbist, B., … Sullivan, C. A. (2017). Trees, forests and water: Cool insights for a hot world. *Global Environmental Change*, 43, 51–61

FAO. (2013). *Climate change guidelines for forest managers* [FAO Forestry Paper No. 172]. Food and Agriculture Organization of the United Nations.

Fernandez-Gimenez, M. E., Ballard, H. L., & Sturtevant, V. E. (2008). Adaptive management and social learning in collaborative and community-based monitoring: A study of five community-based forestry organizations in the western USA. *Ecology and Society, 13*(2).

Forest Stewardship Council. (2015). *The FSC principles and criteria for forest stewardship.* FSC.

Funk, J. M., Aguilar-Amuchastegui, N., Baldwin-Cantello, W., Busch, J., Chuvasov, E., Evans, T., Griffin, B., Harris, N., Ferreira, M. N., Petersen, K., Phillips, O., Soares, M. G., & van der Hoff, R. J. A. (2019). Securing the climate benefits of stable forests. *Climate Policy, 19*(7), 845–860.

Geist, H. J., & Lambin, E. F. (2002). Proximate causes and underlying driving forces of tropical deforestation. *BioScience, 52*(2), 143–150.

Ghazoul, J., & Sheil, D. (2010). *Tropical rain forest ecology, diversity, and conservation.* Oxford University Press.

Gibbs, H.K., Ruesch, A.S., Achard, F., Clayton, M.K., Holmgren, P., Ramankutty, N., and Foley, J.A. 2010. Tropical forests were the primary sources of new agricultural land in the 1980s and 1990s. *Proceedings of the National Academy of Sciences of the United States of America*, 107(38): 16732–16737.

Grimm, N. B., Chapin III, F. S., Bierwagen, B., Gonzalez, P., Groffman, P. M., Luo, Y., Melton, F., Nadelhoffer, K., Pairis, A., Raymond, P. A., Schimel, J., & Williamson, C. E. (2013). The impacts of climate change on ecosystem structure and function. *Frontiers in Ecology and the Environment, 11*(9), 474–482

Hajjar, R., & Kozak, R. A. (2015). Exploring public perceptions of forest adaptation strategies in Western Canada: Implications for policy-makers. *Forest Policy and Economics, 61*, 59–69.

Hall, C., & Knuth, M. (2019). An update of the literature supporting the well-being benefits of plants: A review of the emotional and mental health benefits of plants. *Journal of Environmental Horticulture, 37*(1), 30–38.

Hamilton, L. S. (2019). Forestry and watershed management. In *Deforestation* (pp. 99–131). Routledge.

Hamrick, J. L. (1989). Isozymes and the analysis of genetic structure in plant populations. In *Isozymes in plant biology* (pp. 87–105). Springer Netherlands.

Hegde, R., Suryaprakash, S., Achoth, L., & Bawa, K. S. (1996). Extraction of non-timber forest products in the forests of Biligiri Rangan Hills, India. 1. Contribution to rural income. *Economic Botany, 50*(3), 243–251.

Hicke, J. A., Allen, C. D., Desai, A. R., Dietze, M. C., Hall, R. J., Hogg, E. H., Kashian, D. M., Moore, D., Raffa, K. F., Sturrock, R. N., & Vogelmann, J. E. (2012). Effects of biotic disturbances on forest carbon cycling in the United States and Canada. *Global Change Biology, 18*(1), 7–34.

Hosonuma, N., Herold, M., De Sy, V., De Fries, R. S., Brockhaus, M., Verchot, L., & Romijn, E. (2012). An assessment of deforestation and forest degradation drivers in developing countries. *Environmental Research Letters, 7*(4), 044009.

Hudson, B. (2023). Resilient Forest Management and Climate Change. *Georgia Law Review, 58*, 1775.

IPCC. (2014). *Fifth assessment report (AR5) of the intergovernmental panel on climate change.* IPCC.

IPCC. (2021). *Climate change 2021: The physical science basis* [Contribution of Working Group I to the Sixth Assessment Report of the Intergovernmental Panel on Climate Change]. Cambridge University Press. https://doi.org/10.1017/9781009157896

Isabel, N., Holliday, J. A., & Aitken, S. N. (2020). Forest genomics: Advancing climate adaptation, forest health, productivity, and conservation. *Evolutionary Applications, 13*(1), 3–10.

IUCN. (2019). *Species and climate change.* Retrieved from: https://iucn.org/resources/issues-brief/species-and-climate-change

Jandl, R., Spathelf, P., Bolte, A., & Prescott, C. E. (2019). Forest adaptation to climate change—Is non-management an option?. *Annals of Forest Science*, *76*(2), 48.

Jenkins, M., & Schaap, B. (2018). *Forest ecosystem services: Background Analytical Study 1. Background study prepared for the thirteenth session of the United Nations Forum on Forests*. United Nations Forum on Forests Secretariat.

Jones, M. W., Smith, A. J. P., Betts, R., Canadell, J. G., Prentice, I. C., & Le Quéré, C. (2020). Climate change increases the risk of wildfires: January 2020. *ScienceBrief*.

Jose, S. (2009). Agroforestry for ecosystem services and environmental benefits: An overview. *Agroforestry Systems*, *76*(1), 1–10.

Kanta, C., Kumar, A., Chauhan, A., Singh, H., & Sharma, I. P. (2024). The interplay between plant functional traits and climate change. In N. Kumar & H. Singh, *Plant functional traits for improving productivity* (pp. 41–58). Singapore: Springer Nature Singapore.

Keenan, R. J. (2015). Climate change impacts and adaptation in forest management: A review. *Annals of Forest Science*, *72*, 145–167.

Keleş, S. (2019). An assessment of hydrological functions of forest ecosystems to support sustainable forest management. *Journal of Sustainable Forestry*, *38*(4), 305–326.

Kumar, A., & Singh, S. (2019). Invasive Alien Lantana: A global threat in Indian perspective. *Indian Journal of Tropical Biodiversity*, *27*, 55–67.

Kumar, A., Singh, S., Chand, H. B., Kewat, A. K., & Singh, K. P. (2021). Road towards restoring land degradation in India: A UNCCD perspective. *RJCES*, *9*(4), 27–38.

Kumar, A., Singh, S., Kumar, D., Singh, R. K., Gupta, A. K., Premkumar, K., Chand, H. B., & Kewat, A. K. (2024). Investigating the phenology and interactions of competitive plant species co-occurring with invasive *Lantana camara* in Indian Himalayan Region. *Scientific Reports*, *14*(1), 400.

Kumar, A., Singh, S., Kumar, R., Kumar, M., Rani, K., & James, J. (2024). Balancing conservation and development for achieving land degradation neutrality. *Trends in Current Biology*, *2*(4), 17–24.

Lal, R. (2003). Soil erosion and the global carbon budget. *Environment International*, *29*(4), 437–450.

Lewis, S. L., Wheeler, C. E., Mitchard, E. T. A., & Koch, A. (2019). Restoring natural forests is the best way to remove atmospheric carbon. *Nature*, *568*(7750), 25–28

Li, Y., Zhao, M., Motesharrei, S., Mu, Q., Kalnay, E., & Li, S. (2015). Local cooling and warming effects of forests based on satellite observations. *Nature Communications*, *6*, 6603.

Liebhold, A. M., Brockerhoff, E. G., Kalisz, S., Nuñez, M. A., Wardle, D. A., & Wingfield, M. J. (2017). Biological invasions in forest ecosystems: A global problem requiring international and multidisciplinary collaboration. *Biological Invasions*, *19*(12), 3073–3090.

Locatelli, B. (2010). *Local, global: integrating mitigation and adaptation* (Forests and Climate Change Perspective; 4 pp.) [Brief]. CIRAD–Forests & Climate Change, Montpellier, France.

Locatelli, B., Brockhaus, M., Buck, A., & Thompson, I. (2010). Forests and adaptation to climate change: Challenges and opportunities. *Forests and society: Responding to global drivers of change* (pp. 21–42). International Union of Forest Research Organizations (IUFRO).

Lugo, A. E. (2000). Effects and outcomes of Caribbean hurricanes in a climate change scenario. *The Science of the Total Environment*, *262*(3), 243–251.

Malmer, A., Murdiyarso, D., Bruijnzeel, L. A., & Ilstedt, U. (2010). Carbon sequestration in tropical forests and water: A critical look at the basis for commonly used generalizations. *Global Change Biology*, *16*(2), 599–604.

Millar, C. I., Stephenson, N. L., & Stephens, S. L. (2007). Climate change and forests of the future: Managing in the face of uncertainty. *Ecological Applications*, *17*(8), 2145–2151.

Nguyen, T. T., Ngo, H. H., Guo, W., Wang, X. C., Ren, N., Li, G., Ding, J., & Liang, H. (2019). Implementation of a specific urban water management – Sponge City. *Science of the Total Environment*, *652*, 147–162.

Osman, A. I., Fawzy, S., Lichtfouse, E., & Rooney, D. W. (2023). Planting trees to combat global warming. *Environmental Chemistry Letters*, *21*(6), 3041–3044.

Pan, Y., Birdsey, R. A., Fang, J., Houghton, R., Kauppi, P. E., Kurz, W. A., Phillips, O.L., Shvidenko, A., Lewis, S. S., Canadell, J. G., Ciais, P, Jackson, R. B., Pacala, S. W., Mcguire, A. D., Piao, S., Rautiainen. A., Sitch. S., Hayes, D. (2011). A large and persistent carbon sink in the world's forests. *Science*, *333*(6045), 988–993.

Paris Agreement, (2015). United nations framework convention on climate change. *Paris, France*, *25*.

Pfund, J.-L., Watts, J. D., Boissière, M., Boucard, A., Bullock, R. M., Ekadinata, A., Dewi, S., Feintrenie, L., Levang, P., Rantala, S., Sheil, D., Sunderland, T., & Urech, Z. L. (2011). Understanding and integrating local perceptions of trees and forests into incentives for sustainable landscape management. *Environmental Management*, *48*(2), 334–349.

Sadhu, S. D., Garg, M., & Kumar, A. (2018). Major environmental issues and new materials. In *New polymer nanocomposites for environmental remediation* (pp. 77–97). Elsevier.

Singh, S., Khatri, P. K., & Kumar, A. (2022). Phytodiversity of Bandatola Medicinal Plant Conservation Area, Chhattisgarh. In *Biodiversity in the service of mankind* (pp. 151–170). Walnut Publications.

Singh, S., Kumar, A., & Kumar, A. (2024). Harmony in extraction: Navigating mining, land degradation and ecological restoration for sustainable futures. In *Forests and climate change: Biological perspectives on impact, adaptation, and mitigation strategies* (pp. 513–534). Springer Nature Singapore.

Smithers, J., & Smit, B. (1997). Human adaptation to climatic variability and change. *Global Environmental Change*, *7*(2), 129–146.

Sodhi, N. S., Koh, L. P., Brook, B. W., & Ng, P. K. L. (2010). Southeast Asian biodiversity: An impending disaster. *Trends in Ecology & Evolution*, *19*(12), 654–660.

Suárez, D., Arribas, P., Jiménez-García, E., & Emerson, B. C. (2022). Dispersal ability and its consequences for population genetic differentiation and diversification. *Proceedings of the Royal Society B*, *289*(1975), 20220489.

Swamy, L., Drazen, E., Johnson, W. R., & Bukoski, J. J. (2018). The future of tropical forests under the United Nations Sustainable Development Goals. *Journal of Sustainable Forestry*, *37*(2), 221–256.

Thuiller, W., Lavorel, S., Araújo, M. B., Sykes, M. T., & Prentice, I. C. (2005). Climate change threats to plant diversity in Europe. *Proceedings of the National Academy of Sciences*, *102*(23), 8245–8250.

Tockner, K., Lorang, M. S., & Stanford, J. A. (2010). River flood plains are model ecosystems to test general hydrogeomorphic and ecological concepts. *Wetlands*, *30*(4), 705–723.

Travis, J. M. J. (2003). Climate change and habitat destruction: A deadly anthropogenic cocktail. *Proceedings of the Royal Society B: Biological Sciences*, *270*, 467–473.

Trumbore, S., Brando, P., & Hartmann, H. (2015). Forest health and global change. *Science*, *349*(6250), 814–818.

12 Forests and Climate
Synergies for Sustainable Development

Manish Kumar, Sandhya Goswami, and Vikaspal Singh Rawat

12.1 INTRODUCTION

The functioning of global climate depends on forests due to their significant ability to absorb carbon and essential role in managing atmospheric carbon dioxide amounts. Forest-climate interactions feature carbon sequestration, eco-system services, and conservation of biodiversity. Employing these advantages to help forests play a key role in reducing the pace of climate change and promoting sustainable development is termed sustainable forest management (SFM).

Essential functions of India's forests help in sustainable growth and regulate climate. About 25.17% (Indian State of Forest Report (ISFR), 2023) of the land of this nation includes forests that are vital for conservation of biodiversity and absorbing carbon and delivering essential ecosystem services. Forest ecosystems play a crucial role in India due to their vulnerability to severe damage caused by growing industrialization and urban growth.

The chapter will explore the fascinating links between forests and the local as well as global climate system in the quest to attain sustainable development and reduce climate change. The chapter highlights the ecological, economical, and social gains provided by forests and the potential collaborations achieved through sustainable forest practices. This chapter will feature relevant case examples and advancements in methodologies accompanied by legislative provisions that steer SFM in a climate-fitted approach.

12.2 CONCEPT OF SUSTAINABLE DEVELOPMENT

An all-encompassing plan under the name sustainable development pursues a solution balancing social welfare and economic progress with ecological preservation. Meeting the Sustainable Development Goals (SDGs) calls for weaving sustainable forest practices into national and international growth policies. Appealing for funds for forest conservation is crucial with climate change in mind because it boosts the toughness of forest systems and communities that rely on them (Busch & Mukherjee, 2017). To merge financial benefits with environmental conservation goals of sustainable forestry, India's tax mechanism encourages state governments to maintain and restore forests (Busch & Mukherjee, 2017).

 DOI: 10.1201/9781003588580-12

Furthermore, sustainable growth offers a significant role for community input in protecting forests. Involving rural folks in the decision-making raises social equity and supports the aspirations of conservation initiatives. Studies show that stronger conservation results and improved community well-being arise from community-led forest management practices (Khatun, 2013). Rural populations are key to local ecosystems' health and possess significant traditional knowledge; therefore, their active contribution is necessary for achieving SFM goals.

In India the Forest Rights Act from 2006 grants rights to communities living in the forest and encourages their role in governance (Malovrh et al., 2019). Giving communities' authority over their forest assets leads to better livelihoods and superior conservation results. Research indicates that when communities take part in managing forests they generally lead to better forest health and lower deforestation rates (Hahn & Knoke, 2010). Adding local customs and techniques to forest management improves the survival abilities of forest ecosystems during climate change (McGinley & Cubbage, 2011). State governments will likely get a push to conserve and rejuvenate forests through new ecological fiscal transfers by the Indian government. By merging financial benefits with environmental care, this plan pushes countries to devote resources to preserve forests (Lee et al., 2017). In order to effectively manage its forests and address climate change India will need to improve its financial investment.

For sustainable progress in India, forests and climate must work together. By recognizing the climate-regulating influence of forests along with making social equity and economic well-being a priority, India stands a chance to realize its climate ambitions. To deepen our insights into these complicated links and to establish useful policies for sustainable forest resource management require additional exploration and initiatives. Realizing the impact of forests on climate and integrating SFM into development strategies has the potential to enable countries to meet their climate targets and foster just societal structures and financial development. Ongoing exploration and tasks are vital to boost our insight into these sophisticated interactions and to build efficient laws that advocate for forest resource sustainability.

12.3 IMPORTANCE OF FORESTS IN CLIMATE REGULATION

Since forests can enhance and contain carbon dioxide causing global warming, forests are crucial for managing the climate. For effective action against climate change, carbon sequestration and biodiversity preservation can be gained from growing forests. As forestry-based mitigation measures, like afforestation and natural regeneration, have an important role to mitigate climate change, which stress the need for restoring, managing, and protecting forests (Raihan & Tuspekova, 2022). By managing humidity and precipitation, forests influence climate locally and have an effect on water availability and agricultural production (Senapati et al., 2013). The severe climate crisis endangers forests and leads to shifts in where different types of forests are located. Evidence suggests that climate change could cause changes in biomes throughout India as moist broadleaf forests evolve into drier species (Rasquinha & Sankaran, 2016). Biodiversity faces risk from these changes as they also interfere with the ecological benefits offered by forests and increase climate change issues for marginalized communities.

Large parts of forests serve as an important source of carbon for India's total carbon capacity needed to fulfill climate goals outlined in global agreements like Paris Agreement (Kruk & Kornatowska, 2014). The government of India has put forth several measures to raise the forest cover and strengthen forest management strategies recognizing the impact forests have on climate change. By expanding forest land and repairing damaged forest places, the National Afforestation Programme looks to enhance carbon sequestration efforts (D'Amato et al., 2017). Controlling regional climates depends on the presence of forests. Both affect the levels of humidity and affect crucial water resources for agriculture. Research indicates that reducing forests and harming them can influence climate outcomes by boosting temperatures and modifying rainfall patterns allowing for more severe droughts and floods in areas of India that are susceptible (Gilani & Innes, 2020). To maintain climate stability and resilience forests must be in good condition.

12.4 THE ROLE OF FORESTS IN CLIMATE REGULATION

12.4.1 Carbon Sequestration Process in Forests

In forestry terms 'carbon sequestration' refers to the process of tree functions that draw down air-borne carbon dioxide. Forests maintain an annual amount of carbon close to 3 Pg that functions as important carbon sinks for absorbing anthropogenic emissions (Sandeep & Kavileveettil, 2023). About 45% of the land carbon's storage is found in forest standing biomass made up of trees and their branches. This biomass stands out as an important collection point for carbon (Jiao et al., 2021). A large amount of carbon is present in forest soils which significantly boosts their place in the global carbon cycle (Holmberg et al., 2019).

Several elements such as tree species variety and land use techniques determine the performance of carbon storage. Experiences with pests and fires show they can greatly impact how well forests retain carbon. To enhance carbon sequestration by forests necessary is protecting the vitality of their ecosystems.

The current assessment shows that India has reached to 30.43 billion tonnes of CO_2 equivalent, which indicates that as compared to the base year of 2005, we have already reached to 2.29 billion tons of additional carbon sink (ISFR, 2023). Soil carbon determinants are significantly influenced by deforestation and shifting land use patterns with the Western Ghats being vital to the process. Increasing carbon reserves via afforestation and reforestation is a vital focus for the Indian government as it develops its climate action plans. Sustainable land management methods can enhance Indian forests' ability to store carbon while helping to counter climate change and safeguard biodiversity.

12.4.2 Impact of Forests on Regional and Global Climate Patterns

Forests influence both local and worldwide climate dynamics by means of biogeochemical and physical interactions. According to the data from Jiao et al. (2021), forests function to manage greenhouse gasses such as CO_2 and significantly influence global warming. In 2017 Sun *et al.* noted that trees influence temperatures nearby through processes like evapotranspiration and water changes in energy fluxes. This may cause an improvement in humidity and specific local cooling. Research shows that forests are linked to climate dynamics and have a significant impact on weather patterns. For example, in Northeast China's forests act crucially in regulating temperature and rainfall. Studies reveal that forests have the ability to alter local temperatures and patterns of precipitation, especially in places where they occupy a sizable portion of the land (Jiao et al., 2021). The hydrological process gains from forests that control soil moisture and direct river flows crucial for supplying water locally.

India's rich forests significantly change local weather and carbon processes. Through controls over processes such as evapotranspiration that decrease temperatures and boost humidity levels, trees increase rainfall and preserve temperature stability. Managing rainfall distribution is crucial for the Western Ghats forests and shapes both water supply and crop production. By sequestering carbon forests globally help mitigate the consequences of climate change. Reducing trees to a fraction and harming them interferes with these processes causing more heat and changed precipitation that worsen the climate change situation (Figure 12.1).

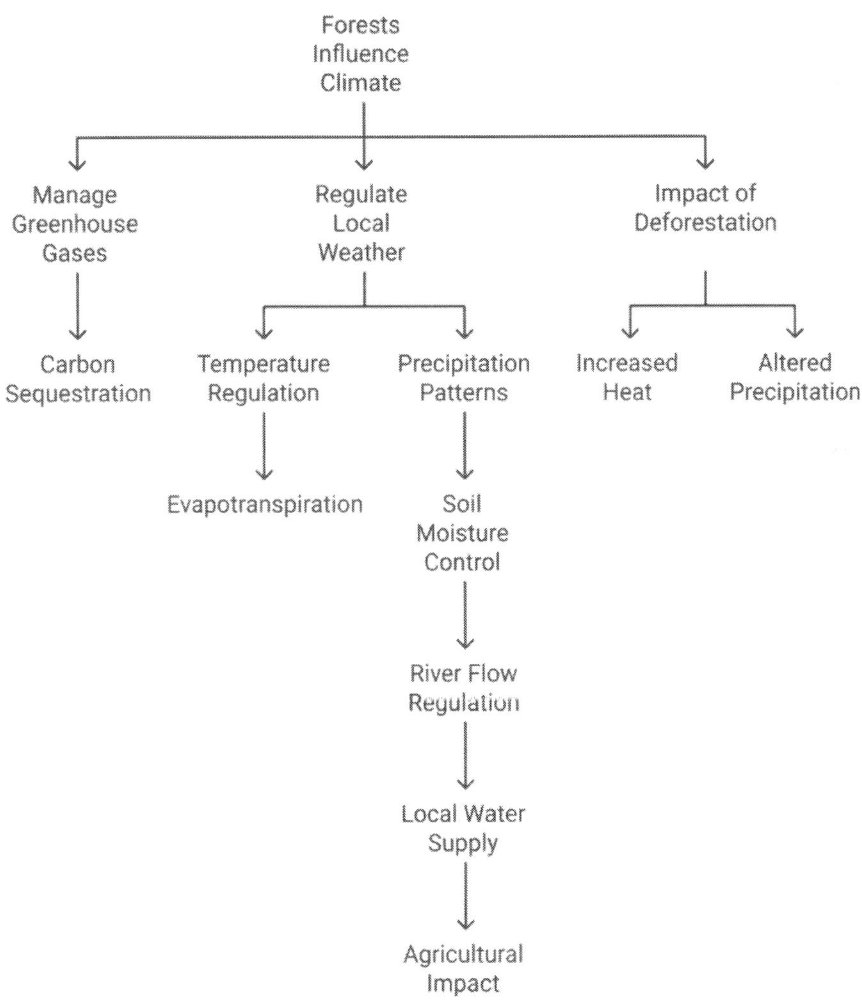

FIGURE 12.1 Impact of forests and deforestation on climate patterns.

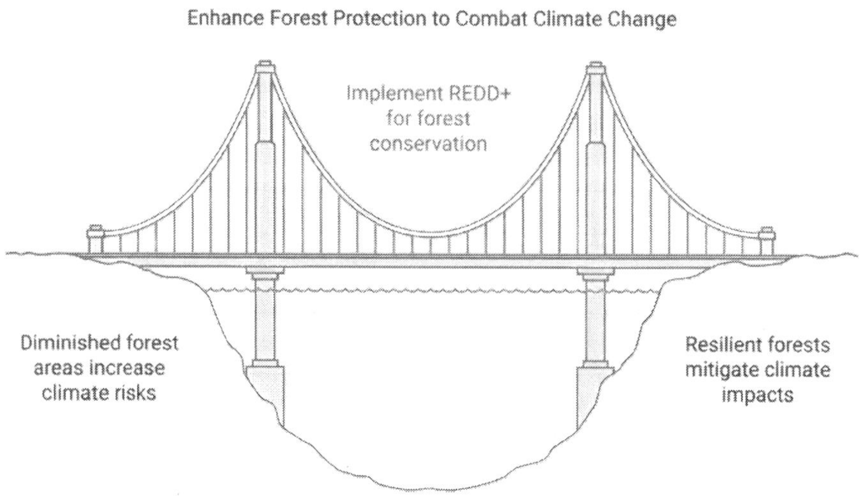

FIGURE 12.2 Thepillars of climate resilience: Forest conservation and REDD+.

12.4.3 Role of Forests in Mitigating Climate Change

Efforts to limit climate change rely heavily on forests for their carbon storage and ability to manage regional temperatures. To combat climate change effectively global initiatives should emphasize forest ecosystem protection. By serving as carbon sinks, forests neutralize the impacts of climate change by lowering emissions of greenhouse gases in the environment. Through the protection of biodiversity and the provision of key services that assist with climate adaptation, forests increase ecosystem resilience (Ahmad et al., 2023).

By offering vital ecosystem services that enhance resilience to climate impacts, forests are essential in reducing carbon emissions from climate change. Changes caused by climate change lessen flood causes and drought effects and aid in managing soil moisture levels (Figure 12.2). In India clear-cutting trees leads to higher risks of intense climate events because of the relationship between diminished forest areas and overall exposure to disasters. The REDD+ program seeks to attract rewards for effective management and conservation of forests to support climate change mitigation. Furthermore, rehabilitating destroyed forests may boost their potential for capturing carbon and assist with managing climate. As the occurrence of forest disturbances like fires and invasive species grows larger and more common, it becomes harder for them to fulfill their climate regulatory function. Devoid, practical forest management techniques aimed at restoration and preservation are necessary to keep climate regulation services active.

12.5 FOREST DEGRADATION AND CLIMATE CHANGE

12.5.1 Causes of Deforestation and Forest Degradation

Deforestation stems from humans' actions mainly in less developed areas where financial constraints push forests to become locations for urban growth and production

(Malík et al., 2023). The rise of commercial agriculture accounts for a large part of the 10–15% of global greenhouse gas emissions related to deforestation in tropical regions. In areas such as the Brazilian Amazon, cattle ranching is a key cause of deforestation. Logging and mining along with infrastructure build-up worsen the decline of forest ecosystems.

Expansion of agricultural activity urban and construction leads to forest clearance in India. An essential contributor is the demand for greater agricultural land; altering agricultural practices in the northeastern states has caused significant acreage depletion of forests (Figure 12.3). The practice leads to major losses of biomass and carbon in northeastern India by clearing about 0.05 million hectares of forest each year as estimated (Sandeep & Kavileveettil, 2023). Also, commercial farming has triggered the shifting of wooded regions into agricultural plots in provinces such as Punjab and Haryana. The deterioration of forests is intensified by road construction and mining as well as the urbanization process (Lundmark et al., 2014). In addition to economic hardship and limited job options, local communities often resort to unsustainable forest resource use (Marcos-Martínez et al., 2022) which ultimately causes deforestation (Hosonuma et al., 2012).

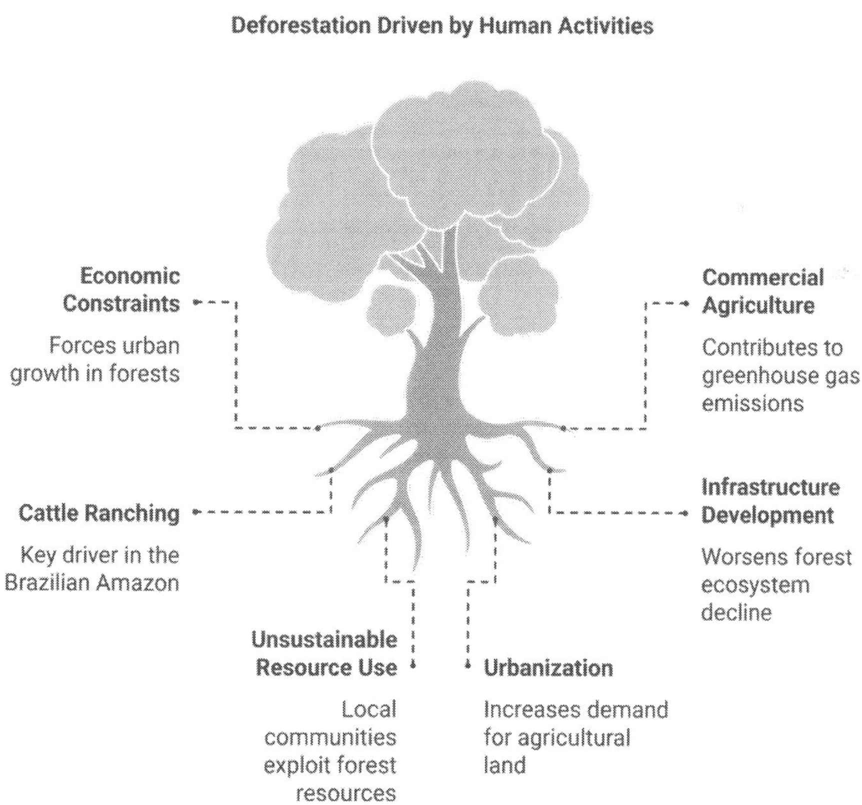

FIGURE 12.3 Deforestation – a web of human activities.

12.5.2 Analyze the Impact of Deforestation on Greenhouse Gas Emissions

Loss of forests plays an important role in carbon dioxide emissions. When forests are cleared or damaged, soils and trees release the carbon that affects global climate change. The transition of carbon sink status in the Amazon relates to land conversions and corresponding forest fires that emit CO_2. Existing forests cannot hold as much carbon because of warmer and dryer conditions related to climate change. Efficient forest conservation strategies are vital to reduce greenhouse gas emissions and avert climate change as shown by this loop.

Deforestation greatly increases the amount of greenhouse gases released by India. Realizing the critical role of forests in carbon holding prompted the Indian government to implement REDD+ efforts to increase forest carbon absorption and fight climate change. Continued clearance of forests jeopardizes efforts at sequestration and raises India's total levels of greenhouse gases set to rise severely if current trends remain (Sun et al., 2017). Deforestation and climate change interact in a way that lowers the effectiveness of remaining trees as carbon sinks.

12.5.3 Consequences of Forest Loss on Biodiversity and Ecosystem Services

Loss of trees considerably damages biodiversity and ecosystem services. Silva and Rodgers (2018) state that habitat fragmentation caused by deforestation threatens numerous species and upsets natural functioning. The survival of numerous creatures depends on forest ecosystems; hence deforestation is presumed to be a key factor in the high rate of extinctions. When forests are lost, they diminish essential ecosystem services needed for environmental wellness and human survival such as water balance and soil quality. Changes in habitat due to deforestation can alter flows and alter both flood risks and water purity. When forests face degradation their ability to deal with climate change can diminish and their ability to change in response to ecological shifts will be compromised.

Declining forests influence both ecosystem services and biodiversity more than just impacts from climate change. India hosts a great number of life forms including trees and wildlife; nevertheless, loss of these forests jeopardizes their continued existence. According to Ahmad et al. (2023), forest fragmentation caused by cutting trees affects ecological systems and reduces genetic richness. Also, the constant destruction of forests affects the productivity of important ecosystem services for secession of agricultural activities and well-being of people such as the functions of water and climate. The destruction of forests leads to disturbances in the hydrology of regions some of which experience increased flooding and poor water quality (Rajasugunasekar, 2023). Moreover, as a result of forest degradation, the ability to adapt to climate change and variability of forest ecosystems has been reduced (Devaraju et al., 2015).

To sum up, deforestation and forest degradation in India are attributed to several interlinked socioeconomic and environmental determinants and undermine ecosystem services and biodiversity and emanate high levels of greenhouse gases emissions. To overcome these challenges, more holistic polices which enhance communities' efforts toward forest protection, proper land use, and management practices as well as forest conservation are imperative.

12.6 SYNERGIES BETWEEN FORESTS AND CLIMATE CHANGE MITIGATION

12.6.1 Potential of Forests for Carbon Sequestration

Forests comprise one of the largest carbon sinks. Over a trillion tonnes of carbon dioxide (CO_2) are currently stored in biomass in the world's forests. Even without a carbon sequestration policy, forests appear to be sequestering an additional 4 billion tonnes of CO_2 per year (Mendelsohn et al., 2012). Forests via trees that photosynthesize capture CO_2 in their biomass and soil facilitating sequestration of carbon. However, potential carbon sequestration differs between forest types and treatment practices. Tropical forests, however, are more efficient in it due to high biomass density while well-managed mixed forests with mature trees are also more efficient with carbon compared to younger or monoculture plantations (Verma & Ghosh, 2022). According to the Intergovernmental Panel on Climate Change, forest carbon stores in the upsurge are a condition sine qua non for the Paris Agreement and other climate objectives. Many countries have availed themselves to such schemes as REDD+ to enhance carbon retention and promote forest conservation. However, the potential capacity and utility of forests to store carbon because of climate change, landuse change, and population pressure are imperiled and should be acted upon immediately.

12.6.2 Role of Forests in Renewable Energy Production

Forests are also important for the generation of renewable energy. Forest biomass can be harnessed as a sustainable energy resource which can help reduce greenhouse gas emissions and energy dependence on fossil fuels. The energy production from forest resources is being adopted in all corners of the world, with Finland and Sweden leading in biomass utilization for energy needs. This approach enhances the income levels of rural communities by creating new income sources against the backdrop of reduced carbon emissions. Most importantly, where access to conventional energy sources is limited, biomass from forest can be utilized for energy. Due to SFM practices, countries are able to enhance their energy security and also address climate change simultaneously.

Tree biomass is a renewable source of energy that can help reduce the reliance on fossil fuels and other sources of greenhouse gas emission (Figure 12.4). The

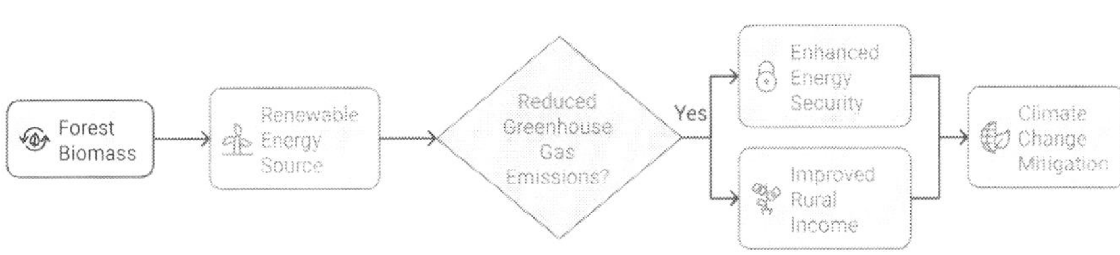

FIGURE 12.4 Climate change mitigation pathway through forest biomass as a renewable energy source.

promotion of wood and other products from forestry for bioenergy use policies is assisting the uptake of forest biomass energy use in India. In addition to reducing carbon emissions, this change provides an alternative source of income for rural people. Furthermore, integrating biomass from forests into energy networks can support SFM practices and improve energy security (Šafařík et al., 2021).

12.6.3 Benefits of Forests in Building Climate Resilience

Forests are thus recognized as beneficial by enhancing water intake and storage and moderating extreme weather events. The effects of droughts and floods, whose occurrences are increasing because of global warming, are effectively managed with forests. In particular, forested catchments preserve and augment the seasonable flow of rivers, which supports the people and agriculture in the region. In addition, forests provide important ecosystem services which enhance the resilience of both people and nature to climate change. Through the conservation and restoration of forest ecosystems, countries are able to enhance their capacity for resilience toward climate change impacts and maintain the viability of their climate-related resources and livelihood.

Because of climate change, floods and droughts have become more common in India, and forests are very beneficial to mitigate such impacts. For example, forested watersheds maintain river and selvage, helping re-charge aquifers. Additionally, they also enhance adaptive capacity of human, physical, and natural systems in natural resources under changing climates through relevant provisioning of view functions (Joshi & Bharucha, 2020).

12.6.4 Analyze the Contribution of Forests to Biodiversity Conservation

In conserving biological diversity, forests are environmentally important because they are the habitats of various species of plants and animals. This is the main reason forests help in upholding ecological equilibrium, as these areas support approximately 80% of the world's terrestrial species. The other threats to biodiversity are mainly caused by clearance of lands and logging which usually lead to the survival of only a few species. Such ecological processes that are fostered by forests are vital in maintaining genetic diversity and the health of ecosystems, which are key in ensuring adaptability of the environment. Even more so, healthy forest ecosystems aid in the conservation of biodiversity against climate change by providing critical resources and niches, especially for many organisms. Hence, it can be concluded that forest protection is both necessary for the maintenance of biological diversity and the habitat as well as the health of ecosystems that provide for human existence.

The Indian subcontinent is diverse and hosts a number of organisms, most of which depend on the existence of forest-based ecosystems. Deforestation and habitat alteration are some of the problems facing the biodiversity of the world today and several species are on the verge of extinction as a result of these activities. In order to survive environmental changes, forests contain such biological phenomena as genetic preservation and homeostasis of ecosystems structure (Verma & Ghosh, 2022).

In addition, it has been shown that healthy forest ecosystems also greatly augment climate change adaptation through biological and habitat resources (Ge et al., 2016). So, to protect biodiversity, forest conservation is fundamental for the protection of ecosystems that support human beings and for the well-being of all life on Earth.

The forests are vital in mitigating global warming due to their role in carbon storage, generation of clean energy, enhancing adaptation to climate changes and preserving biodiversity. There is a need for effective management and conservation if the ecosystems are to be safeguarded from adverse effects of climate change and as well perform optimally.

12.7 SUSTAINABLE FOREST MANAGEMENT (SFM) FOR CLIMATE AND DEVELOPMENT

As defined by FAO (2017), SFM refers to the management of forest to maintain its biodiversity, productivity, and ecological processes providing diversified benefits over long time scales while meeting the social, economic, and cultural needs of the local people and society at large. According to the report of the Brundtland Commission, SFM is to be defined as the type of management that maintains and enhances benefits over time and across ecosystems for realizing such needs while recognizing differences in biological characteristics among species combined with people's legitimate ecological multi-dimensional quality-of-life values (Kruk & Kornatowska, 2014). SFM is designed to ensure that forest ecosystems continue to provide essential benefits, such as carbon sequestration, water purification, and wildlife habitat, and improving local livelihoods at the same time.

SFM plays a significant role in the facilitation of both socio-ecological growth and environmental conversations. In case of India, forest ecosystems are highly diverse and provide various products and ecosystem services to the vast population of the country as well as other parts of the world so optimum SFM need to be inculcated to cope up with increasing urbanization, industrialization, and emerging threats from global climate change that will perturb Indian forests during this century (D'Amato et al., 2017; Kruk & Kornatowska, 2014). SFM ensures the management and use of forest resources that guarantees long-term capabilities for maintaining their socioeconomic, environmental, and social functions; the implementation of principles aimed at supplying elementary biological goods since forever while keeping up higher level diversity of trees. Indian forest cover is only 24.62% of their lands (Yadav, 2024) and the National Forest Policy has set a target of 33%. This emphasizes how critical it is to implement efficient SFM techniques that can increase forest cover and support ecosystem services and biodiversity.

12.7.1 Key Principles of Sustainable Forest Management

Inclusion of ecological, social, and economic aspects in forest management processes is a basis of SFM (Figure 12.5). In order to achieve it, it is necessary to

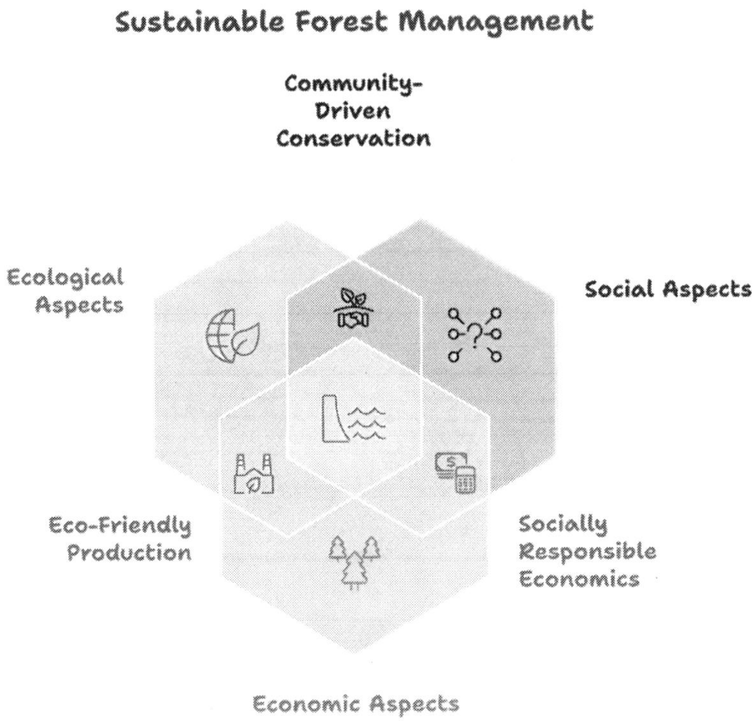

FIGURE 12.5 Integrated pillars of sustainable forest management.

maintain biodiversity and forest health, ensure sustainable production of forest products, and promote rights of local populations and their participation in decision-making (Gilani & Innes, 2020; Malovrh et al., 2019). Additionally, SFM strongly focuses on the use of adaptive management methods allowing for responses to changing societal needs and environmental conditions (Hahn & Knoke, 2010). The importance of certification systems such as those implemented by Forest Stewardship Council are indicated also as a tool for promoting ethical forest management practices that serve the objectives of sustainability (Malovrh et al., 2019).

Inclusion of ecological, social, and economic aspects in forest management processes is one of the core principles of SFM in India. It involves conserving biodiversity, ensuring sustainable yields of forest products, and recognizing rights of and fostering the involvement of local populations in the governance of forests (Gilani & Innes, 2020; Malovrh et al., 2019). One important policy reform to promote participatory governance in forest management is the Forest Rights Act enacted in 2006, which aims at empowering people living near forests by recognizing their rights over forest resources (Macura et al., 2011). Moreover, for sustainable use of forest resources, local communities partner with forest departments under Joint Forest Management (JFM) that was formalized in the 1970s (Bhattacharya et al., 2010).

12.7.2 Case Studies of Successful Sustainable Forest Management Practices

There are many successful global cases for SFM. In Costa Rica, through governmental regulation and non-governmental certification, the forest management has been greatly improved (McGinley & Cubbage, 2011). The forest policy development also makes substantial progress in British Columbia province in Canada with over 99% of total forest land governed under legislation in 2015 (Gilani & Innes, 2020). The community-based forest management (CBFM) practices in Indonesia are one of the successful approaches to grant local indigenous communities rights for SFM and promoting their livelihood simultaneously (Lee et al., 2017).

In India, there are quite a few case studies that can be cited to demonstrate the best practices of SFM. The Mahatma Gandhi National Rural Employment Guarantee Act (MGNREGA) has an influence on reforestation and forest conservation efforts. This leads to stronger environmental resilience and creates green jobs (Yadav, 2024). Also, communities involved in forest management have shown great results at the Similipal Biosphere Reserve. They have demonstrated expertise in managing wildfires and protecting natural habitats (Mishra et al., 2022). This shows how important local knowledge and participation are for conservation work (Yadav & Dugaya, 2012). Additionally, Yadav and Dugaya (2012) mention that providing certificates for non-timber forest products (NTFPs) appears to be a promising way to create sustainable jobs while also preserving forests.

12.8 CHALLENGES AND OPPORTUNITIES FOR SUSTAINABLE FOREST MANAGEMENT

Even with improvements in SFM a variety of issues still stand. Stakeholders often encounter disputes originating from their differing requirements for forest resources. Additionally financial constraints could hinder attempts to enhance sustainability as there is a focus on timber production instead of environmental issues. Given that past practices often sidelined at-risk groups they are now at the forefront of calls for better-inclusive policies that resonate with their needs. There is opportunity to progress SFM using pioneering techniques and technologies that strengthen forest productivity and resilience (Bringezu et al., 2012).

India remains to deal with multiple difficulties that impede the advancement of SFM. Resource and land disputes lead to frequent conflicts among stakeholders particularly those involving industrial and local community interests. Insufficient policy execution and the lack of understanding of rights and obligations in the community can weaken current policies. A significant danger exists in climate change that intensifies challenges like pest invasions and forest blazes and threatens the health of forests. There are ways to enhance SFM through innovative practices such as agroforestry and management approaches that meld traditional knowledge with modern conservation methods.

There is a large strategy in SFM focused on balancing ecological care and community and economic factors. By following its key ideas and gaining insights from effective models, stakeholders can transform sustainability opportunities and barriers linked to forest management into a brighter future for forests and the communities that rely on them. In managing sustainable forests with a balanced approach, India employs a complicated method to meet ecological requirements alongside social and financial goals. Stakeholders can utilize the challenges and opportunities presented by SFM to advance a sustainable future for forests and their reliant communities while following its essential principles and insights from successful practices.

12.9 POLICY AND GOVERNANCE FOR SUSTAINABLE FOREST MANAGEMENT

12.9.1 Role of International Agreements and Policies

International commitments greatly affect the creation of national strategies for SFM. UNFCCC and the Convention on Biological Diversity (CBD) define international systems that push nations to implement eco-friendly practices in forestry to reduce climate change and safeguard biodiversity as observed by Savaresi (2013). The project aims to grant monetary incentives for nations with low incomes decrease the emission of deforestation and enhance sustainable practices. Government policies usually influence the performance of these global pacts and the way local administrations can execute them. The REDD+ initiative intends to deliver economic rewards for developing nations to lower the rate of emissions from deforestation while advancing sustainable forest practices. National policies often determine how well these international agreements work and how effectively local governance can apply them.

Agreements with other countries are essential for helping India develop its forest management plans. To achieve global sustainability goals with its policies, India utilizes the United Nations Framework Convention on Climate Change (UNFCCC) and the CBD (Forsyth & Sikor, 2013; Sotirov et al., 2020). Financial incentives for sustainable forest practices become crucial for India, thanks to the focus of the REDD+ initiative. International agreements motivate India to implement rules that reduce climate change and boost conservation of biodiversity alongside fostering sustainable livelihoods for those dependent on forests.

12.9.2 Analyze the Importance of National Forest Policies and Regulations

Local strategies to implement international commitments depend heavily on the existence of national forest policies and regulations. The framework for forest management is determined by these policies and sustainable practices are defined through them alongside the allocation of roles among various participants including local groups and governmental institutions. Many countries transformed their national frameworks from authoritarian systems to participative strategies that allow local groups to control forest assets. This transformation plays a vital role in understanding how local expertise and actions contribute to sustainable achievement. The success of these policies tends to rely on their precise definition and the level of enforcement measured. Unrealized implementation and obliviousness in the community can reduce the success of national guidelines contributing to regular decline in forestry resources.

Forest strategies in India are key for converting worldwide commitments into real strategies. The National Forest Policy of 1988 emphasizes the need for sustainable management of forest resources while recognizing the rights of local communities (Gordon Kofi Sarfo-Adu, 2021). By conferring legal rights through the 2006 Forest Rights Act, the act significantly improves this system and promotes community participation in forest management. These policies frequently succeed based on how they are applied locally. Since the 1990s JFM has remained a significant policy. Nevertheless, its success falls differently across areas depending on local governance and community participation.

For rural communities to thrive economically, national policies must include NTFPs. The economy of forest-dependent families greatly benefits from NTFPs. Nevertheless, regulations historically favored timber extraction over NTFPs. For SFM in India to succeed a focused strategy that values both timber and non-timber resources is crucial.

12.9.3 Role of Community-Based Forest Management (CBFM)

Promoting SFM is essentially the function of community involvement. Connecting local populations with forest resource governance raises the likelihood of achieving environmentally sustainable outcomes by combining conventional wisdom with current conservation approaches. When local groups assume control and accountability over forest management practices, they tend to adopt sustainability methods that aid both nature and their lives. In Bhutan's forests, forest management is handled efficiently by community-based user groups highlighting the success of participatory governance (Buffum et al., 2010). For CBFM to succeed in effectively managing forests sustainably it requires ongoing development of local structures and adequate assistance from governments (Figure 12.6).

Increasingly accepted as an essential strategy for SFM in India is CBFM. Through CBFM local groups can engage in forest resource management thereby achieving superior conservation benefits and improved livelihoods (Massiri et al., 2020; Nzunda & Manyanda, 2023). When communities engage directly in decision-making about sustainability measures, they tend to follow practices that mirror their expertise and demands.

In West Bengal's forests areas are improving significantly due to community involvement in the JFM model (Jana et al., 2017). Research demonstrates that initiatives led by communities in forest management can result in enhanced ecological health and enhanced social fairness (Macura et al., 2011; Mutia et al., 2019). Furthermore, issues continue including the involvement of marginalized communities in management and the protection of their rights (Kumar & Kerr, 2012). For SFM in India to succeed seamlessly requires the cooperation of national policies and global commitments. Without proper implementation of national policies, local communities will not gain the benefits from international frameworks that promote change. Through cooperation among various governments and community

FIGURE 12.6 Key determinants of sustainable forest management in India.

representatives, India can strive for a prosperous sustainable forest practice that serves both their environment and populace.

12.10 CASE STUDIES

The association between climate change and forests is significant for sustainable development in India because if it's diverse and ecologically rich natural resources. The amalgamation of climate action with strategies of forest management not only backs carbon sequestration but also supports biodiversity and improves ecosystem services. Some examples are included in this association to demonstrate the connections between forests and climate, emphasizing different programs, laws, and scientific discoveries that support sustainable development.

12.10.1 Case Study 1: Integrating Forests and Biodiversity in Nepal's National Adaptation Plan

National Adaptation Plan of Nepal has been necessary for dealing with impacts of climate change by assimilation of biodiversity with forests. The country's forests, which cover about 45% of its area, are indispensable for regulating local and global climates and supporting the economy (Karki et al., 2018). This case highlights the importance of incorporating ecological considerations into national climate strategies, ensuring that forest management contributes to both adaptation and mitigation efforts.

12.10.2 Case Study 2: Ecosystem Services Assessment in Morocco

The Maamora forest in Morocco was framed as a case study aimed at mapping ecosystem services in order to facilitate SFM. The spatial identification of areas in their spatial structure, i.e., service provision hotspots and degraded areas is very much needed for the decision-makers to have better rational and equitable practices of forest resource management (Benabou et al., 2022). The approach underlines the enormous need for understanding ecosystems and services flow with various dimensions impacting sustainable development and equity in increasing climate impacts.

12.10.3 Case Study 3: Disturbances in European Forests and Climate Change

Research into forest disturbances under climate change for Europe states that in the face of changing disturbance regimes, management strategies should adapt. Such strategies can promote stability in forests and ensure continued delivery of ecosystem services, even as climate variability increases (Seidl et al., 2017). This case stresses the demand for adaptive management practices that should be adopted along with a working consideration of the changing dynamics of forest ecosystems in the context of climate change.

12.10.4 Case Study 4: Dynamics of the Carbon Cycling Flow Models in Forest Management

A review of carbon cycle flow models further crucially highlights the centrality forests occupy in mitigating climate change. Changes to forest management practices bear considerable implications on the carbon balance that have downstream consequences for atmospheric CO_2 levels and adaptable management practices that incorporate social learning into the adaptation of forest ecosystem management (Nunes et al., 2019). This case elucidates the need for integration of carbon management into forest policies aimed at sustainable development goals.

12.10.5 Case Study 5: Climate Change Adaptation in the Australian Forest Management Practice

Forest managers in Australia are realizing more and more the need to change their ways of working in connection with climate change. The response to problems created by the impacts of climate includes the formulation of 'climate-smart' forest management frameworks (Keenan, 2015). This case study demonstrates the role of proactive adaptation strategies in sustaining the health and resilience of ecosystems.

12.10.6 Case Study 6: Sustainability Impact Assessment in European Forest Management

The assessment of sustainability impacts of alternative forest management in Europe shows promise with respect to a framework incorporating both ecological and economic indicators of forest sustainability (Carnus et al., 2012). The framework aims to assist policymakers in making informed decisions that will roll a balance between the production of wood and ecological well-being, therefore, encouraging the sustainable management of forests.

12.10.7 Case Study 7: Restoration of Old-Forest Ecosystems in North America

Efforts to restore the old-forest ecosystems of North America rely on efforts to improve the forest in terms of structure and function to increase resilience against disturbances like fire and disease. Such restoration not only preserves biodiversity but also enhances the adaptation capacity of forests to climate change (Jones et al., 2017). This case illustrates how ecological restoration serves to promote forest resilience.

12.10.8 Case Study 8: Trade-Offs in Timber Production and Carbon Sequestration in the USA

Some studies analyzing trade-offs between timber production, carbon sequestration, and water yield in *Pinus elliottii* forests in the southeastern USA provide insights in the management of subtropical ecosystems for optimal service provision (Cademus et al., 2014). The case demonstrates that guidelines are necessary for valuing trade-offs among multiple ecosystem services that need sustainable management of the forest.

12.10.9 Case Study 9: Collaborative Forest Management in Indonesia

Collaborative management of forest ecosystems in Petungkriyono forest, Central Java forest, Indonesia, provides further suggestions to bolster climate change mitigation and adaptation efforts. Decisions made together with the local communities through partnerships are successfully ensuring SFM and incorporating good sustainable landuse practices (Damayanti et al., 2020). This case highlights a practical instance where participatory governance has played its part in presenting the case for sustainable development.

12.11 CONCLUSION

In short, forests-climate change link presents a set of real opportunities for a leap forward in sustainable development. The case studies discussed show that in many cases, sound forest management strategies enhance carbon sequestration, protect biodiversity, and enhance resilience to climate impacts. But such synergies could only be achieved with an integrated approach that incorporates adaptive management, the water-energy-food nexus, and biodiversity conservation as a priority.

Growing challenges posed by climate change to forest ecosystems necessitate a requirement from policymakers to embrace a more holistic framework whereby forest management, climate actions, and socioeconomic development are interlinked. In this way, it becomes possible to develop innovative solutions by working together with relevant stakeholders, including local communities, governments, and conservation organizations. The sustainable management of forests can significantly aid in confronting climate change in a manner that directly and positively affects present and future generations.

ACKNOWLEDGMENT

All Images are generated using Text to Image Generator AI. Authors are thankful to Dolphin (PG) Institute of Biomedical & Natural Sciences (An Autonomous Institute), Dehradun, Uttarakhand, India, for providing administrative and academic support.

REFERENCES

Ahmad, M. M., Imtiaz, M. A., Rasheed, F., Jameel, M. A., Asif, M., & Ahmad, H. (2023). The crucial role of forests in combating climate change. *Trends in Biotechnology and Plant Sciences*, *1*(1), 1–6. https://doi.org/10.62460/TBPS/2023.009

Benabou, A., Moukrim, S., Laaribya, S., Aafi, A., Chkhichekh, A., Maadidi, T., ... &, Aboudi, A. (2022). Mapping ecosystem services of forest stands: Case study of Maamora, Morocco. *Geography Environment Sustainability*, *15*(1), 141–149. https://doi.org/10.24057/2071-9388-2021-047

Bhattacharya, P., Pradhan, L., & Yadav, G. (2010). Joint forest management in India: Experiences of two decades. *Resources Conservation and Recycling*, *54*(8), 469–480. https://doi.org/10.1016/j.resconrec.2009.10.003

Bringezu, S., O'Brien, M., & Schütz, H. (2012). Beyond biofuels: Assessing global land use for domestic consumption of biomass. *Land Use Policy*, *29*(1), 224–232. https://doi.org/10.1016/j.landusepol.2011.06.010

Buffum, B., Lawrence, A., & Temphel, K. (2010). Equity in community forests in Bhutan. *The International Forestry Review*, *12*(3), 187–199. https://doi.org/10.1505/ifor.12.3.187

Busch, J., & Mukherjee, A. (2017). Encouraging state governments to protect and restore forests using ecological fiscal transfers: India's tax revenue distribution reform. *Conservation Letters*, *11*(2). https://doi.org/10.1111/conl.12416

Cademus, R., Escobedo, F., McLaughlin, D., & Abd-Elrahman, A. (2014). Analyzing trade-offs, synergies, and drivers among timber production, carbon sequestration, and water yield in *Pinuselliottii* forests in southeastern USA. *Forests*, *5*(6), 1409–1431. https://doi.org/10.3390/f5061409

Carnus, J., Hengeveld, G., & Mason, B. (2012). Sustainability impact assessment of forest management alternatives in Europe: An introductory background and framework. *Ecology and Society*, *17*(4). https://doi.org/10.5751/es-04838-170449

D'Amato, A., Jokela, E., O'Hara, K., & Long, J. (2017). Silviculture in the United States: An amazing period of change over the past 30 years. *Journal of Forestry*. https://doi.org/10.5849/jof-2016-035

Damayanti, I., Bambang, A., & Soeprobowati, T. (2020). The analyze of collaborative management perspective of Petungkriyono forest. *E3s Web of Conferences*, *202*, 05014. https://doi.org/10.1051/e3sconf/202020205014

Devaraju, N., Bala, G., & Nemani, R. (2015). Modelling the influence of land-use changes on biophysical and biochemical interactions at regional and global scales. *Plant Cell & Environment*, *38*(9), 1931–1946. https://doi.org/10.1111/pce.12488

FAO (2017). World Aquaculture 2015: A Brief Overview. FAO Fisheries and Aquaculture Circular No. 1140. FAO, Rome

Forsyth, T., & Sikor, T. (2013). Forests, development and the globalisation of justice. *Geographical Journal*, *179*(2), 114–121. https://doi.org/10.1111/geoj.12006

Ge, X., Xu, F., & Li, Y. (2016). Solid-state anaerobic digestion of lignocellulosic biomass: Recent progress and perspectives. *Bioresource Technology, 205*, 239–249. https://doi.org/10.1016/j.biortech.2016.01.050

Gilani, H., & Innes, J. (2020). The state of British Columbia's forests: A global comparison. *Forests, 11*(3), 316. https://doi.org/10.3390/f11030316

Gordon Kofi Sarfo-Adu.(2021). Role of forest related policies and laws on sustainable forest management practice: A critical overview. *JRDM*. https://doi.org/10.7176/jrdm/73-06

Hahn, W., & Knoke, T. (2010). Sustainable development and sustainable forestry: Analogies, differences, and the role of flexibility. *European Journal of Forest Research, 129*(5), 787–801. https://doi.org/10.1007/s10342-010-0385-0

Holmberg, M., Aalto, T., Akujärvi, A., Arslan, A., Bergström, I., Böttcher, K., & Forsius, M. (2019). Ecosystem services related to carbon cycling – Modeling present and future impacts in boreal forests. *Frontiers in Plant Science, 10*. https://doi.org/10.3389/fpls.2019.00343

Hosonuma, N., Herold, M., Sy, V., Fries, R., Brockhaus, M., Verchot, L., & Romijn, E. (2012). An assessment of deforestation and forest degradation drivers in developing countries. *Environmental Research Letters, 7*(4), 044009. https://doi.org/10.1088/1748-9326/7/4/044009

Indian State of Forest Report (2023). ISFR- Volume-I. Forest survey of India. Ministry of Environment, Forests & Climate Change, New Delhi. www.fsi.nic.in.

Jana, S., Ahmed, M., & Heubach, K. (2017). Dependency of rural households on non-timber forest products (NTFPs). *International Journal of Sustainable Economies Management, 6*(2), 37–50. https://doi.org/10.4018/ijsem.2017040104

Jiao, Y., Bu, K., Yang, J., Li, G., Shen, L., Liu, T., & Heng-qing, Z. (2021). Biophysical effects of temperate forests in regulating regional temperature and precipitation pattern across Northeast China. *Remote Sensing, 13*(23), 4767. https://doi.org/10.3390/rs13234767

Jones, G., Keane, J., Gutiérrez, R., & Peery, M. (2017). Declining old-forest species as a legacy of large trees lost. *Diversity and Distributions, 24*(3), 341–351. https://doi.org/10.1111/ddi.12682

Joshi, G. S., & Bharucha, E. (2020). Coherence in cross-sectoral policies is imperative for sustainable management of ecosystems. *International Journal of Innovative Technology and Exploring Engineering, 9*(4S), 33–41. https://doi.org/10.35940/ijitee.d1005.094s20

Karki, G., Paudel, B., & Uprety, B. (2018). Integrating forests and biodiversity in Nepal's national adaptation plan: A review and synthesis of knowledge stock on opportunities and way forward. *Banko Janakari, 27*(2), 21–31. https://doi.org/10.3126/banko.v27i2.21220

Keenan, R. (2015). Climate change impacts and adaptation in forest management: A review. *Annals of Forest Science, 72*(2), 145–167. https://doi.org/10.1007/s13595-014-0446-5

Khatun, K. (2013). Integrating national forestry initiatives in India with international climate change policy. *Climate Policy, 13*(3), 384–402. https://doi.org/10.1080/14693062.2013.768924

Kruk, H., & Kornatowska, B. (2014). Sustainable forest management in Poland – Theory and practice. *Folia Forestalia Polonica, 56*(1), 45–55. https://doi.org/10.2478/ffp-2014-0005

Kumar, K., & Kerr, J. M. (2012). *Territorialisation and marginalisation in the forested landscapes of Orissa, India*. http://dx.doi.org/10.1016/j.landusepol.2012.06.015

Lee, Y., Rianti, I., & Park, M. (2017). Measuring social capital in Indonesian community forest management. *Forest Science and Technology, 13*(3), 133–141. https://doi.org/10.1080/21580103.2017.1355335

Lundmark, T., Bergh, J., Höfer, P., Lundström, A., Nordin, A., Poudel, B., & Werner, F. (2014). Potential roles of Swedish forestry in the context of climate change mitigation. *Forests, 5*(4), 557–578. https://doi.org/10.3390/f5040557

Macura, B., Zorondo-Rodríguez, F., Grau-Satorras, M., Demps, K., Laval, M., García, C., & Reyes-García, V. (2011). Local community attitudes toward forests outside protected areas in India. Impact of legal awareness, trust, and participation. *Ecology and Society, 16*(3). https://doi.org/10.5751/es-04242-160310

Malík, A., Santi, N., & Hamka, H. (2023). Spatial-temporal analysis of forest deforestation in Lore-Lindu national park. *Agroland the Agricultural Sciences Journal (E-Journal)*, *10*(1), 58–69. https://doi.org/10.22487/agroland.v10i1.1750

Malovrh, Š, Bećirović, D., Marić, B., Nedeljković, J., Posavec, S., Petrović, N., … & Avdibegović, M. (2019). Contribution of forest stewardship council certification to sustainable forest management of state forests in selected southeast European countries. *Forests*, *10*(8), 648. https://doi.org/10.3390/f10080648

Marcos-Martínez, R., Sánchez, J., Srivastava, L., Soonsawad, N., & Bachelet, D. (2022). Valuing the impact of forest disturbances on the climate regulation service of western U.S. Forests. *Sustainability*, *14*(2), 903. https://doi.org/10.3390/su14020903

Massiri, S., Malík, A., Golar, G., & Nugroho, B. (2020). Institutional capacity of forest management unit in promoting sustainable community-based forest management. Case study of forest management unit in CentralSulawesi Province, Indonesia. *Jurnal Manajemen Hutan Tropika (Journal of Tropical Forest Management)*, *26*(2), 169–177. https://doi.org/10.7226/jtfm.26.2.169

McGinley, K., & Cubbage, F. (2011). Governmental regulation and nongovernmental certification of forests in the tropics: Policy, execution, uptake, and overlap in Costa Rica, Guatemala, and Nicaragua. *Forest Policy and Economics*, *13*(3), 206–220. https://doi.org/10.1016/j.forpol.2010.10.002

Mendelsohn, R., Sedjo, R., & Sohngen, B. (2012). Forest carbon sequestration. In *Fiscal policy to mitigate climate change: A guide for policymakers* (pp. 89–102). International Monetary Fund. www.elibrary.imf.org/downloadpdf/display/book/9781616353933/9781616353933.pdf

Mishra, S., Mishra, A., Kumar, C., Sahu, D., & Mishra, S. (2022). Distressed lives and livelihood in biosphere reserves during Anthropocene; Similipal forest blaze—2021. *Current Journal of Applied Science and Technology*, 17–27. https://doi.org/10.9734/cjast/2022/v41i2531772

Mutia, T., Sumarmi, S., Bachri, S., & Budidjanto, B. (2019). *A study on Bayan community perception towards awiq-awiq local wisdom based forest management*. https://doi.org/10.2991/icskse-18.2019.8

Nunes, L., Meireles, C., Gomes, C., & Ribeiro, N. (2019). Forest management and climate change mitigation: A review on carbon cycle flow models for the sustainability of resources. *Sustainability*, *11*(19), 5276. https://doi.org/10.3390/su11195276

Nzunda, E., & Manyanda, B. (2023). A SWOT analysis of community-based forest management policy as a basis for REDD+ in Tanzania. *GSC Advanced Research and Reviews*, *15*(3), 022–037. https://doi.org/10.30574/gscarr.2023.15.3.0157

Raihan, A., & Tuspekova, A. (2022). Nexus between emission reduction factors and anthropogenic carbon emissions in India. *Anthropocene Science*, *1*(2), 295–310. https://doi.org/10.1007/s44177-022-00028-y

Rajasugunasekar, D. (2023). An integrative review for the role of forests in combating climate change and promoting sustainable development. *International Journal of Environment and Climate Change*, *13*(11), 4331–4341. https://doi.org/10.9734/ijecc/2023/v13i113614

Rasquinha, D., & Sankaran, M. (2016). Modelling biome shifts in the Indian subcontinent under scenarios of future climate change. *Current Science*, *111*(1), 147. https://doi.org/10.18520/cs/v111/i1/147-156

Šafařík, D., Hlaváčková, P., & Michal, J. (2021). Potential of forest biomass resources for renewable energy production in the Czech Republic. *Energies*, *15*(1), 47. https://doi.org/10.3390/en15010047

Sandeep, S., & Kavileveettil, S. (2023). *Perspective chapter: Forest degradation under global climate change*. https://doi.org/10.5772/intechopen.106992

Savaresi, A. (2013). REDD+ and human rights: Addressing synergies between international regimes. *Ecology and Society*, *18*(3). https://doi.org/10.5751/es-05549-180305

Seidl, R., Thom, D., Kautz, M., Martín-Benito, D., Peltoniemi, M., Vacchiano, G., & Reyer, C. (2017). Forest disturbances under climate change. *Nature Climate Change*, *7*(6), 395–402. https://doi.org/10.1038/nclimate3303

Senapati, M., Behera, B., & Mishra, S. (2013). Impact of climate change on Indian agriculture & its mitigating priorities. *American Journal of Environmental Protection*, *1*(4), 109–111. https://doi.org/10.12691/env-1-4-6

Silva, A., & Rodgers, J. (2018). Deforestation across the world: Causes and alternatives for mitigating. *International Journal of Environmental Science and Development*, *9*(3), 67–73. https://doi.org/10.18178/ijesd.2018.9.3.1075

Sotirov, M., Pokorny, B., Kleinschmit, D., & Kanowski, P. (2020). International forest governance and policy: Institutional architecture and pathways of influence in global sustainability. *Sustainability*, *12*(17), 7010. https://doi.org/10.3390/su12177010

Sun, X., Li, X., Guan, Z., Liu, J., & Zhang, S. (2017). The use of meteorological data to assess the cooling service of forests. *Ecosystem Services*, *25*, 28–34. https://doi.org/10.1016/j.ecoser.2017.03.016

Verma, P., & Ghosh, P. (2022). REDD+ strategy for forest carbon sequestration in India. *The Holistic Approach to Environment*, *12*(3), 117–130. https://doi.org/10.33765/thate.12.3.4

Yadav, M., & Dugaya, D. (2012). Non-timber forest products certification in India: Opportunities and challenges. *Environment Development and Sustainability*, *15*(3), 567–586. https://doi.org/10.1007/s10668-012-9393-1

Yadav, P. (2024). MGNREGA-assisted afforestation for climate moderation in India: An overview. *International Journal of Environment and Climate Change*, *14*(2), 310–321. https://doi.org/10.9734/ijecc/2024/v14i23946

13 Challenges of the Forestry Sector in Adaptation and Mitigation to Climate Change

Suraj Goswami, Suruchi Devi, Akrita Gupta, Bhaskar Sinha, and Jigyasa Bisaria

13.1 INTRODUCTION

Forests are critical in stabilizing climate, regulating ecosystem services, preserving biodiversity, and managing the carbon cycle. Sustainable forest management (SFM) is essential to optimize climate benefits from the forest landscape (Funk et al., 2019; IUCN, 2021; Thompson et al., 2009; Watson et al., 2018). It is estimated that forests absorb 638 Gt of carbon, which is nearly 20% of worldwide anthropogenic-based greenhouse gas (GHG) emissions like deforestation (Tebeje, 2020). Apart from sequestrating atmospheric carbon, the forests also provide many intangible services that improve the community's resilience to cope with climate change. Thus, it is essential to combine forest restoration as a national strategy with adaptation and mitigation initiatives, such as Reducing Emissions from Deforestation and Forest Degradation (REDD+) and other carbon projects for enhancing people's participation and effective implementation (Rizvi et al., 2015). Globally, 350 million ha of degraded land under the Bonn Challenge can sequester 1.7 Gt of CO_2 equivalents yearly (IUCN, 2021).

Global forests can be increased by approximately 1 billion ha through afforestation and reforestation, which is considered a golden opportunity to create additional carbon sequestration and to reduce the impact of climate change (Bastin et al., 2019). However, the additional forest may conflict with other land uses, especially agriculture production. This conflict can be avoided by SFM, which will not only contribute to enhancing carbon sequestration without changing land use but also reduce conflict with farmers. While new forests can enrich biodiversity, they may disrupt existing ecosystems if established in naturally treeless areas (like grasslands), underscoring the need for careful planning to optimize mitigation efforts through effective forest management (Kauppi et al., 2022). Moreover, forests are being recognized for addressing gaps in existing adaptation programs to mitigate climate-associated risks (Seddon et al., 2019; UNEP, 2022) and in fulfilling multiple sustainable development goals (SDGs) and nurturing interactions across several sectors (Friends of EbA, 2022).

DOI: 10.1201/9781003588580-13

Forest ecosystems are severely impacted worldwide by increasing temperatures and carbon dioxide concentration, fluctuating precipitation patterns, and high frequency of extreme weather events (Graham et al., 1990; Seidl et al., 2017). These variations adversely affect forest ecosystem health, diversity and carbon storage capacity, and climate regulation of the planet (Omann et al., 2009; Trumbore et al., 2015). Loss of biodiversity is a major consequence of climate change as flora struggle to adapt to these intensified challenges, resulting in species distribution changes and accelerated wildfire risks due to long droughts (Muluneh, 2021). Amazon forests are highly vulnerable to climate change due to severe droughts, resulting in the dieback of trees and low carbon storage capacity (Bennett et al., 2021; Brando et al., 2019). Temperate forests face problems like heat stress and shifting precipitation patterns, which affect pollination, species habitat, and forest configuration (Morellato et al., 2016; Teshome et al., 2020). Boreal forests, constituting 30% of global forests, are at extreme risk from increased frequency of forest fires and pest infestation while reducing their carbon sequestration capacity (Gauthier et al., 2015; Psistaki et al., 2024; Stocks & Ward, 2011). These changes have directly impacted the socio-economics of forest-dependent communities (Ofoegbu et al., 2017) by changing the distributions and compositions of species and worsening the food security and economic vulnerability of these people (Chakraborty et al., 2018). To address these problems, suitable adaptation and mitigation plans that mix traditional knowledge and local people's participation are vital (Alessa et al., 2016; Bhatia & Shukla, 2024).

Monitoring of climate variation and its associated impact is important for reducing the vulnerabilities of dependent communities (Ladi et al., 2022). Remote sensing (RS) and ground observations are the major monitoring techniques used to observe the changes in flora distribution, soil health, and land use/cover. Additionally, new techniques such as deep learning (DL) and machine learning (ML) are helping in decision-making and predicting the impacts earlier (Abduljabbar et al., 2019; Koc & Acar, 2021). The major global policy frameworks like United Nations Framework Convention on Climate Change (UNFCCC), REDD+, Paris Agreements and United Nations Strategic Plan for Forests (UNSPF) recognize SFM as one of the effective means to tackle climate crises (Sharma et al., 2020; Wertz-Kanounnikoff, 2007). However, political and government issues, socio-economic conflicts, and financial constraints preclude the effective implementation of these policies and, therefore, a need for an integrated approach that involves communities and fosters international partnership is emphasized (Klein et al., 2005). These international frameworks will further support timely interventions of forestry activities and strengthen the resilience of forest and dependent communities.

13.1.1 Mitigation and Adaptation Potential of Forest

Well-managed forests act as effective carbon sinks and absorb two-fold carbon as they emit, which is equivalent to 7.6 billion Mt of CO_2 each year. Thus, forests offer an important capacity to store atmospheric GHG through the photosynthesis process and reduce the adverse impacts of climate change (Psistaki et al., 2024; UNDP Climate Promise, 2024). In addition to climate change mitigation, forests also provide livelihood to one-fourth of the global population (1.6 billion people) and also provide 15–100 billion US$ valued goods and services like clean water and fertile soils (IUCN, 2022). Adopting low-emission technologies is vital to achieving the target of the Paris Agreement to limit global warming to 1.5°C with a particular focus on enhancing forest carbon sinks. Afforestation on barren land is a key strategy with significant potential for GHG sequestration (Forster et al., 2021). The mitigation potential of afforestation/reforestation ranges from 0.0 to 10.1 $GtCO_2$ per year for the 2020–2050 period (Katila et al., 2024). However, various factors such as climate, choice of species, forest management, and product substitution are some of the challenges in designing forest-based suitable mitigation strategies (Forster et al., 2021). In developing countries, the forest is a lifeline for local communities and plays a dual role in mitigating climate change and supporting adaptation efforts through multiple ecosystem services. However, overexploitation and degradation of forest land are the major threats to forest-dependent communities and the forests' ability to sequester carbon effectively (Hosonuma et al., 2012; Pandey et al., 2016). Many countries' existing adaptation and mitigation plans already include forests as an adaptation and mitigation strategy for climate change.

Tropical forests have the highest carbon sequestration potential and cover more land area than forests in any other biome. Therefore, they have a crucial role in this effort, capturing between 2.2 and 2.7 Gt of carbon annually (Goodman & Herold, 2014). The Indian forest sector has a massive mitigation target for climate change by creating an additional 2.5–3 billion tons of carbon sequestration by 2030. However, bold efforts are required to achieve these goals, such as prioritizing suitable planning and implementing SFM practices (Roy, 2023). It is estimated that higher carbon prices lead to early carbon gain and vice versa and 51–78% carbon gain with avoided deforestation by 2100 (Sathaye et al., 2006). In the Indian scenario, a 100 US$ carbon price (compared to 50 US$) will increase the carbon stock area by 85–100% (Ravindranath et al., 2007). Additionally, adaptation programs such as watershed conservation, shelter belts, coastal mangrove plantations, and agroforestry should be promoted as mitigation activities (Murthy et al., 2011).

13.1.2 Impact of Climate Change on Forest

Forests throughout the world are affected by climate change. Approximately 420 million ha of forest land were converted into other land use between 1990 and 2020 (FAO, 2024). With increasing climate change, the impacts are manifested in the form of alterations in forest structure and distribution, impacting health and decreasing the forest's ability to store carbon, replacing the native species with exotic species, biodiversity loss, impacting water yields, increasing intensity of fires and pest infestation (Boyd et al., 2013; Jones et al., 2022; Kour et al., 2024; Moore & Allard, 2008; Nunes et al., 2021; Price et al., 2013; Singh et al., 2022). However, the effects and challenges of climate change vary with different forest types.

Tropical forests cover 3.6% of the Earth's land surface and store about 250 billion tons of carbon (Butler, 2020). Rainforests like the Amazon and Congo Basin are prominently affected by climate change (Papastefanou et al., 2022). In Amazon, the drought conditions have led to tree dieback and decreasing carbon storage (Cardeli et al., 2024; Doughty et al., 2015), whereas in the Congo Basin, shifts in precipitation patterns are disrupting the hydrological balance and resulting in the lower availability of water causing extreme weather conditions (Karam et al., 2023). Temperate forests cover around 10 million sq km area and are known for carbon sequestration, preserving biodiversity, and maintaining hydrological cycles (Bonan, 2008; Myers, 1990). These temperate forests are experiencing alterations in growth, delays in soil water balance, phenological changes (Gilliam, 2016), shifts in flowering patterns, and disruption in the pollination cycles (Morellato et al., 2016). The boreal forest is the northernmost forest, covering 11% of the Earth's surface (Frelich, 2020; Safford & Vallejo, 2019). The impacts of climate change on the Boreal forests are mixed, i.e., positive and negative. Conditions for the northern boreal forests are becoming adaptable and stable for species to survive. In contrast, the southern part is experiencing slow growth of trees, deaths, and species migration, causing biodiversity loss (Cooper et al., 2023). According to Corning et al. (2024) and Islam et al. (2024), the number of forest fires in burned areas might triple by 2100, and the amount of sequestered carbon will decrease by 56–76%.

India has a forest cover of 713,789 sq km, and it is projected that 315,667 sq km and 449,919 sq km of forest cover will become climate change hotspots by 2030, under RCP 4.5 and RCP 8.5 scenarios, respectively. More than 36% of forest area is fire-prone, and 4% is extremely prone to fire (FSI, 2021), at the same time, the frequency and intensity of fire incidences have been accelerated by climate change. The prolonged droughts in the Western Ghats have led to the depletion of biodiversity (Gopalakrishnan et al., 2011). Additionally, the rise in temperature is causing heat stress in forest flora, making them susceptible to disease and pests (Ravindranath et al., 2005). According to Chaturvedi et al. (2011), it is projected that 77% of the forest area of India will observe shifts under high emission scenarios by 2085, which will lead to the migration of species. In central India, the teak forest is likely to increase, and the sal forest will decrease due to changes in suitable habitat range due to changing climate variables (Patasaraiya et al., 2023). Furthermore, teak and sal forests are more sensitive to changes in past temperature as compared with rainfall, with sal forests more sensitive to minimum temperature and teak forests to maximum temperature (Patasaraiya et al., 2021). Such findings implicate that suitable silvicultural strategies for forest management should be designed based on the future distribution of species in relation to the climate variables based on focused research to understand the changing patterns of species with climate change.

13.1.2.1 Socio-Economic Consequences for Forest-dependent Communities

Globally, around 1.6 billion people are dependent on forests for income, food, energy, medicine, shelter, fresh air, and clean water (Li et al., 2019; World Bank, 2002). Forests help minimize erosion, regulate the climate, absorb GHGs, assist species movement and pollination, and are places to live for 80% of terrestrial species (Brockerhoff et al., 2017; UN DESA, 2021). Forest ecosystem plays a major role in maintaining the water cycle, protecting biodiversity, making forest-dependent communities or rural communities economically stable, and providing them with direct and indirect benefits (Obeng et al., 2011; Singh et al., 2022; Wale et al., 2022).

The impacts of climate change are affecting ecosystem services and degrading forest regions, which directly impacts the majority of forest-dependent communities

(Batar et al., 2017; Saalu et al., 2020). Due to rising climate change, forests are facing problems such as a decrease in the resource, disruption in the composition, changes in agricultural productivity timings, causing problems of food security, financial risks, and vulnerability to forest-dependent populations (Gandharba, 2023; Garekae et al., 2017; US EPA, 2016). About 300 million people in India depend on forests for their basic needs, such as fuelwood, timber, fodder, bamboo, and NTFPs (Non Timber Forest Produces) (FSI, 2019). The forests serve as the major CO_2 sink and remove around 11.25% of total GHG emissions (Kishwan et al., 2009). The majority of forest-dependent communities are dependent on NTFPs as income sources. About 275 million rural people depend on NTFPs for subsistence livelihoods, whereas 70% of NTFPs collection is done in the tribal belt (Mitchell et al., 2003; Pandey et al., 2016). The impact of climate change is adversely affecting the forests causing damage to the ecosystem services and loss of forest goods. These losses and damage make it very challenging for India to provide livelihood and shelter to forest-dependent communities (FAO, 2023), and therefore, comprehensive monitoring of social, economic, and ecological aspects of forest resources assumes significance (Chaki et al., 2022). A study done in the Khunti district of Jharkhand observed the change in lac yield and degrading quality of NTFPs is occurring due to rising temperature, showing that the forest-dependent communities are highly at risk (Magry et al., 2022). Therefore, minimizing the socio-economic changes of climate change on forest-dependent communities requires integrating effective adaptation strategies with traditional knowledge and community involvement (Devi et al., 2023; Keenan, 2015; Wale et al., 2022).

13.1.3 Techniques for Monitoring the Impacts of Climate Change on Forest

Monitoring climate variation and its associated impact is important for making suitable strategies for vulnerable communities (Ladi et al., 2022) and climate hotspot areas. There are many monitoring techniques to detect the impact of climate change on forests, such as RS techniques (Ferchichi et al., 2022), ML, and DL (Reichstein et al., 2019). RS technique includes normalized difference vegetation index (NDVI) and enhanced vegetation index (EVI) to monitor the forest ecosystem change by assessing the biophysical and biochemical properties of vegetation (Huete, 2012; Zeng et al., 2022). NDVI uses multispectral satellites like NOAA-AVHRR and MODIS that provide high-frequency data at the global scale (Anyamba & Tucker, 2012; Fan & Liu, 2016). Furthermore, most advanced RS technology have developed hyperspectral imaging satellites such as the compact airborne spectrographic imager (CASI) and the airborne prism experiment (APEX) that provide fine-scale data for forest range edges studies (Jump et al., 2010). Additionally, there are different methods like change detection methods and time series methods that are helpful to assess the impact. It is vital to integrate climate models and RS, which could be more useful for understanding present and future impacts (Devi et al., 2018).

Emerging technologies, particularly ML and DL, have evolved in monitoring the impacts of climate change (Koc & Acar, 2021; Ladi et al., 2022). As a subset of artificial intelligence, ML utilizes large datasets to detect patterns and predict future events

(Abduljabbar et al., 2019; Koc & Acar, 2021). Recent advancements in technology, ML and DL methodology in climate change, such as boosted regression trees, fuzzy inference systems (Jain et al., 2020), maximum entrophy, topic modeling (Singha et al., 2024), and Classification and Regression Tree (CART) (Ladi et al., 2022) are being utilized for various applications, like analyzing land surface temperatures, planning forest ecosystems, predicting natural hazards, and modeling species distribution. They also play a vital role in land cover monitoring, assessing soil health, integrating climate policies, studying biogeoclimatic ecosystems, monitoring agricultural productivity, and evaluating vulnerability and resilience. Furthermore, for climate change mitigation, ML and DL are applied as geoengineering solutions for assessing soil organic carbon, analyzing land cover changes and managing forest ecosystems by using various techniques such as convolutional neural networks (CNNs), random forest (RF), long short-term memory networks, decision trees, and logistic regression (Ladi et al., 2022; Padarian Campusano, 2019). Recently, forest fire susceptibility has been assessed and mapped by integrating Geographic Information System (GIS), RS, and ML methods in the Similipal Tiger Reserve (STR) considering climate change impacts. Additionally, future climate projections suggested persistent fire risk along STR's edges through 2100 (Singha et al., 2024). Integrating GIS, RS, and ML methods can provide us with more significant results and help in creating robust strategies for adaptation and mitigation through forests.

13.1.4 Global Policy Frameworks

The significant threats to forests due to climate change have led to the creation of several international policy frameworks aimed at mitigating climate change impacts and promoting SFM activities. Different frameworks, such as the UNFCCC, REDD+, Paris Agreement, and UNSPF, emphasize the significant role of forests in addressing climate change (Figure 13.1).

13.1.4.1 United Nations Framework Convention on Climate Change (UNFCCC)

The UNFCCC is a major climate treaty ratified by 197 countries for climate agreements in 1992. The UNFCCC aimed to stabilize GHG concentrations without imposing enforceable emissions reductions and cooperative action, which provided the foundation for more specific agreements like the Kyoto Protocol and Paris Agreements (United Nations Framework Convention on Climate Change. Secretariat, 1992). The Kyoto Protocol introduced the afforestation/reforestation (A/R CDM) under clean development mechanism, which allows rich countries to earn carbon credits for investing in reforestation initiatives in developing countries, thereby boosting both carbon sequestration and sustainable development (Thomas et al., 2010).

13.1.4.2 REDD+ (Reducing Emissions from Deforestation and Forest Degradation)

REDD+ emerged under the UNFCCC to combat climate change by addressing emissions from deforestation and promoting SFM. The framework was first proposed in 2005 during COP 11 in Montreal, with countries like Papua New Guinea and Costa Rica advocating for global cooperation to reduce forest-related emissions

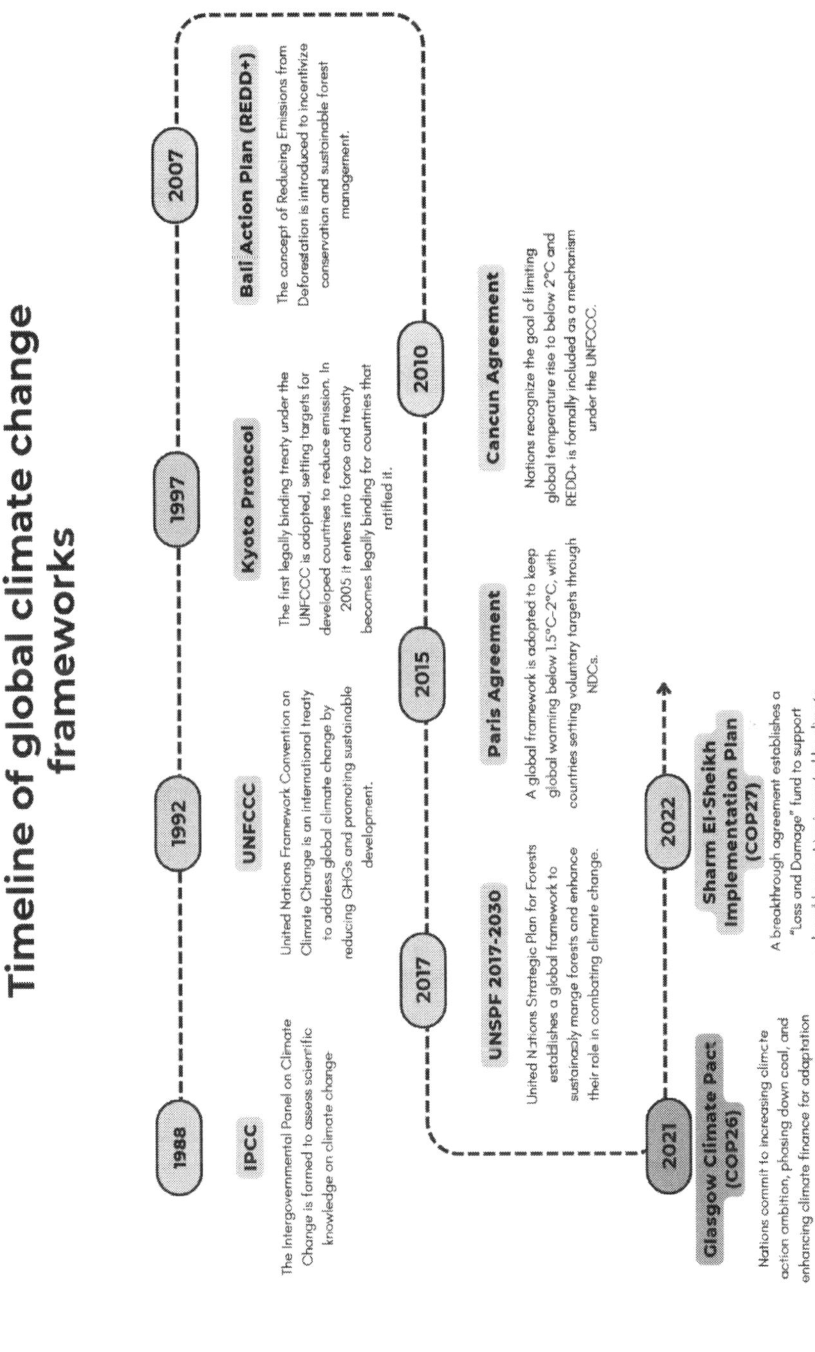

FIGURE 13.1 Timeline of major global climate change framework.

in developing countries. REDD+ was launched in COP 13 in Bali under the Bali Action Plan as a key strategy for climate mitigation (Corbera et al., 2010; Fay & Denduangrudee, 2018; Martin, 2020; Matsson, 2008; Wertz-Kanounnikoff, 2007). It included numerous approaches for SFM and enhancing carbon stock. The Cancun Agreement at COP 16 in 2010 was important as finance, national forest monitoring systems, and safeguards for shielding biodiversity and tribal communities were included in REDD+ (Maniatis et al., 2019; Sanz & Penman, 2016). REDD+ finance is vital to its successful implementation and is dependent on public and private finance, as well as results-based approaches, where nations receive monetary benefits for reducing deforestation/degradation and promoting SFM.

The Warsaw Framework for REDD+ was established at COP 19 in 2013, which led to preparing guidelines for result-based finance and MRV (measurement, reporting, and verification) of emissions reductions. The Warsaw Framework also defined the role of forest reference emission levels in benchmarking countries' activities under REDD+ (Arts et al., 2019; Voigt & Ferreira, 2016). Furthermore, many challenges are faced in the distribution of incentives among the stakeholders who contributed to the net increase of carbon stock (Jha & Sinha, 2017; Jha et al., 2017; Sinha, 2017). In addition, aligning REDD+ with more inclusive targets like carbon neutrality, a deforestation-free supply chain, and nature-based solutions (NbS) can expedite implementation (Cosslett, 2013; Horstmann & Hein, 2017).

13.1.4.3 Paris Agreement

In COP 21 (December 12, 2015), the UNFCC achieved an important landmark by introducing the Paris Agreement. This agreement seeks to unite all countries in fighting against climate change issues and promoting a low-carbon economy. The key aim of the Paris Agreement is to keep the average global temperature rise below 2°C and try to limit it to 1.5°C above pre-industrial levels (Chien et al., 2021; Sharif et al., 2020; Sharma et al., 2020). The agreement also aims to support countries' resilience to climate impacts and align financial flows with low greenhouse emissions and climate-resilient pathways (Anwar et al., 2022; Shahbaz et al., 2019). However, developing nations need funds, advanced technical support, and capacity-building opportunities to achieve these targets.

The main reason for climate change is human-induced activities, including fuel-based mechanization in farming, residue burning, deforestation, and fossil fuel consumption, all of which contribute to high GHG levels, cause significant environmental damage, affecting infrastructure, health, and productivity (Balsalobre-Lorente et al., 2022; Huang et al., 2016; Ishikawa-Ishiwata & Furuya, 2022). In this regard, nationally determined contributions (NDCs) are the target set by all countries according to their capacity to reduce emissions and mandate the submission of NDC synthesis reports every five years (Balsalobre-Lorente et al., 2020). Main mitigation approaches include reducing global temperature and emission levels and enhancing adaptation, especially in developing nations. Adaptation approaches focus on community resilience and wider international collaboration for vulnerable nations.

Article 5 of the Paris Agreement emphasized promoting forest conservation, the role of SFM, and the implementation of REDD+ activities. By including forest-related strategies in their NDCs, countries can contribute to global carbon sequestration goals while promoting sustainable development and poverty alleviation. The Paris Agreement's framework provides flexibility for countries to pursue voluntary cooperation, including carbon market mechanisms like Internationally Transferred Mitigation Outcomes (ITMOs), which allow forest carbon credits to be traded between nations (Michaelowa et al., 2019; Voigt et al., 2024). This framework boosts international cooperation on forests while confirming that mitigation efforts contribute to wider ecological and socio-economic targets.

13.1.4.4 United Nations Strategic Plan for Forests (2017–2030)

The UN General Assembly adopted the UNSPF in 2017 for SFM. There are six global forest goals and 26 associated targets aligned with SDGs and the Paris Agreement, which include increasing forest cover by 3% (120 million ha) globally by 2030. The UNSPF inspires nations to adopt SFM-based policy at the national level and boost cross-sectoral collaboration for sustainably managing the world forest (Economic & Counci, 2017; Picard & Garavaglia, 2021; Sotirov et al., 2020). The strengthening of forest governance is the main element of UNSPF to ensure transparency, accountability, and inclusivity in suitable decision-making. It also emphasizes the role of financing SFM through public and private investments (Arce, 2019). The UNSPF provides a guideline for nations to manage forests sustainably, encourage the integration of forests into national development policies, and address key problems like illegal felling, forest degradation, and habitat loss (Arroyo, 2019; Tegegne et al., 2018).

The forestry sector is a major part of climate change adaptation and mitigation strategies under the UNFCCC and the Paris Agreement. REDD+ also promote SFM, carbon sequestration activities, improving forest cover and incentivizing countries for better management (Venter & Koh, 2012). The LULUCF (Land Use, Land Use Change, and Forestry) framework allows countries to report the carbon emissions and removals related to land-use changes, including afforestation and reforestation, thereby allowing for an inclusive assessment of their contributions to climate goals (Schlamadinger et al., 2007) and incorporate these efforts into NDCs, leveraging carbon market mechanisms and improve the SFM (Girsang et al., 2024; Ozor et al., 2020). However, forestry also faces many challenges, like deforestation and forest degradation, thus necessitating careful accounting as a deduction in national carbon balances and impacting total emission reduction goals. The forestry sector can significantly contribute to global climate resilience and mitigation efforts, aligning with the overarching objectives of the Paris Agreement. However, the forestry sector needs rigorous management strategies to balance effectively by SFM and robust accounting frameworks (Fry, 2008; Briner & Konrad, 2014; Thuy et al., 2018).

13.1.5 Opportunities for Enhancing the Adaptation and Mitigation of Climate Change

13.1.5.1 Sustainable Forest Management (SFM)

SFM aims to balance forests' economic, environmental, social, and cultural values tailored to each country's goals and needs. The key strategies for climate resilience include maintaining healthy ecosystems, restoring degraded forests, and conserving biodiversity. These strategies support both natural ecosystem adaptation and planned human-guided adaptation (Braatz, 2012). By addressing current and future societal needs for forest products, SFM also protects the whole ecosystem from climate change impacts. Thus, it is crucial to support a forest-based economy, alleviate poverty, and conserve natural resources and biodiversity in adaptation strategies. SFM remains one of the best ways to combat climate change, along with agroforestry, especially in developing nations where people's reliance on forests for their livelihood is significant. Agroforestry aids SFM through tree-based conservation and production functions, addressing sustainability issues like resource availability, biodiversity, wood and non-timber products, ecosystem services, and environmental benefits, including soil and water conservation, carbon sequestration, and socio-economic gains (El Tahir, 2015).

SFM in tropical regions faces substantial challenges, including weak governance, overexploitation of forest resources, and deforestation, which compromise conservation efforts and species protection. The loss of intact forests and conventional logging practices further hinder SFM and become a challenge to preserving biodiversity and essential ecosystem services for long-term forest health (Patel et al., 2024). Thus, SFM can overcome these challenges by using six concepts: best management practices/reduced impact logging, biodiversity conservation, forest protection, multi-scale planning, participatory forestry, and sustained forest production. These concepts form the basis for an iterative decision-making framework that supports the planning, implementation, and assessment of SFM practices (Foster et al., 2010).

13.1.5.2 Nature-Based Solutions (NbS) for Forest Resilience and Climate Mitigation

NbS are actions where humans use natural processes to conserve nature and are considered the best way to cope with the climate change problem and resolve other societal challenges (Seddon, 2022; Turner et al., 2022). The solutions include practices like restoring forests, plantation, and agroforestry; preserving and re-establishing wetlands, woods, and other ecosystems; enhancing soil health via sustainable agricultural methods; and developing green infrastructure, such as parks and urban gardens (Turner et al., 2022). The importance of NbS in adaptation has become popular recently (Masson-Delmotte et al., 2021; Seddon et al., 2020; Turner et al., 2022) due to their cost-effectiveness. NbS works as an umbrella concept that includes various ecosystem-based approaches, such as ecosystem-based adaptation and mitigation, natural water retention measures, sustainable management, blue and green infrastructure, and disaster risk reduction (Calliari et al., 2022; Seddon et al., 2020).

The European Union (EU) has become a global leader in implementing NbS to enhance climate resilience and has set the target to transform into a sustainable, climate-neutral, climate-resilient EU by 2050. Furthermore, it has recently released a biodiversity strategy for 2030 along with a strategy on adaptation to climate change and forest management, which are fundamental components of the European Green Deal (Calliari et al., 2022). Currently, various NBS are increasingly used to design resilient landscapes and cities that achieve economic development while benefiting the environment and society across Africa. There are some recommended NbS actions, which include (1) urban regeneration, (2) improving well-being in urban areas, (3) coastal resilience, (4) watershed management and ecosystem restoration,

(5) sustainable use of matter and energy, (6) enhancing the insurance value of ecosystems, and (7) increasing carbon sequestration (Lafortezza et al., 2018).

13.1.6 Challenges in Adaptation and Mitigation of Climate Change

At the global level, various policies and agreements aim to address the impacts of climate change on forests. However, implementing these frameworks faces several challenges in adaptation and mitigation, which can be categorized into ecological and scientific, financial, policy and governance, technological, and socio-economic challenges (Figure 13.2). The unpredictability of climate change and the intricate response of ecosystems make it challenging to develop suitable policies for ecological and scientific challenges (Schneider & Kuntz-Duriseti, 2002). The dynamic nature of forests makes it challenging to forecast how they will be impacted by variations in temperature, precipitation patterns, and other climatic conditions (Lindner et al., 2014). Therefore, it is necessary to have a dynamic adaptive management technique that can sync with changing climate. Conflicting land-use priorities is one of the main socio-economic challenges. Deforestation and other degradation frequently result from forests competing with industrial development, urbanization, and agriculture. In addition, when indigenous groups depend on forest resources for their subsistence, social and cultural conflict occurs. Conservation efforts could encounter opposition if local needs and expertise are not considered while developing policies (Froese & Schilling, 2019; Riahi et al., 2017).

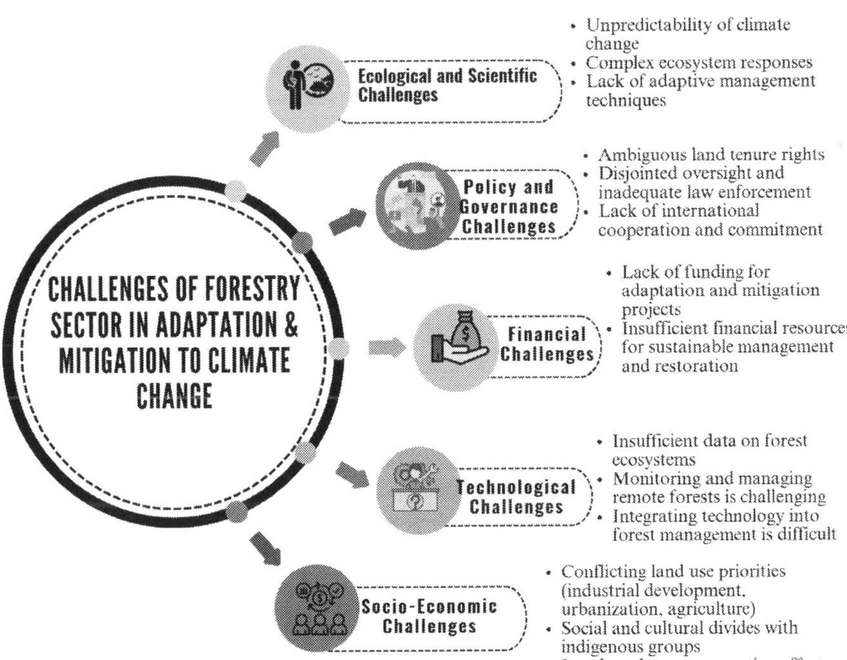

FIGURE 13.2 Challenges of the forestry sector in adaptation and mitigation to climate change.

Ambiguous land tenure rights, disjointed oversight, and inadequate law enforcement are the major governance challenges for efficient policy implementation (Klein et al., 2005) and require strong national commitment and cooperation. However, it can be challenging to keep states working together, particularly when there are conflicting national interests or tax implementation of international agreements (Laukkonen et al., 2009). Updated technological support and insufficient data related to forest resources and their interaction with climate change (Verburg et al., 2011) result in the creation of weak policies. Additionally, there is a need to monitor forests, particularly in isolated or degraded areas (Adenle et al., 2015; Weaver et al., 2013). Financial support is another obstacle to effectively implementing adaptation and mitigation techniques worldwide (Adenle, 2015). The availability of financial resources is critical to the success of adaptation and mitigation programs, whether focused on funding for technical developments, SFM, or forest restoration projects. Even well-designed policies can fall short without sufficient finance (Clark et al., 2018; Melo et al., 2013). Addressing these challenges requires an inclusive approach that considers the ecological, socio-economic, governance, technological, and financial factors with a strong emphasis on international collaboration and the engagement of local communities.

13.2 CONCLUSION

Leveraging climate finance and incentive-based mechanisms is crucial to improving forests' role in climate adaptation and mitigation. Strategies such as carbon finance and trading and REDD+ can provide crucial funding for SFM, while the Green Climate Fund (GCF) can support projects that promote resilience. Emphasizing NbS can restore degraded landscapes and improve ecosystem services. By integrating these approaches, stakeholders can strengthen the adaptive capacity of forests and their dependent communities, ensuring a sustainable future in the face of climate change.

REFERENCES

Abduljabbar, R., Dia, H., Liyanage, S., & Bagloee, S. A. (2019). Applications of artificial intelligence in transport: An overview. *Sustainability*, *11*(1), 189.

Adenle, A. A., Azadi, H., & Arbiol, J. (2015). Global assessment of technological innovation for climate change adaptation and mitigation in developing world. *Journal of Environmental Management*, *161*, 261–275.

Alessa, L., Kliskey, A., Gamble, J., Fidel, M., Beaujean, G., & Gosz, J. (2016). The role of Indigenous science and local knowledge in integrated observing systems: Moving toward adaptive capacity indices and early warning systems. *Sustainability Science*, *11*, 91–102.

Anwar, A., Sinha, A., Sharif, A., Siddique, M., Irshad, S., Anwar, W., & Malik, S. (2022). The nexus between urbanization, renewable energy consumption, financial development, and CO_2 emissions: Evidence from selected Asian countries. *Environment, Development and Sustainability*, *24*, 1–21.

Anyamba, A., & Tucker, C. J. (2012). Historical perspective of AVHRR NDVI and vegetation drought monitoring. *Remote Sensing of Drought: Innovative Monitoring Approaches*, *23*, 20.

Arce, J. J. C. (2019). Forests, inclusive and sustainable economic growth and employment. In *Background study prepared for the fourteenth session of the United Nations Forum on Forest (UNFF), Forests and SDG8*. United Nations Department of Economic and Social Affairs, UNFF Secretariat.

Arroyo, M. M. *Expert workshop on strengthening the global core set of forest indicators to support the implementation of the 2030 Agenda and the UN Strategic Plan for Forests 2030*, 22–24 October 2019 FAO, Rome.

Arts, B., Ingram, V., & Brockhaus, M. (2019). The performance of REDD+: From global governance to local practices. *Forests*, *10*(10), 837.

Balsalobre-Lorente, D., Driha, O. M., Bekun, F. V., Sinha, A., & Adedoyin, F. F. (2020). Consequences of COVID-19 on the social isolation of the Chinese economy: Accounting for the role of reduction in carbon emissions. *Air Quality, Atmosphere & Health*, *13*, 1439–1451.

Balsalobre-Lorente, D., Ibáñez-Luzón, L., Usman, M., & Shahbaz, M. (2022). The environmental Kuznets curve, based on the economic complexity, and the pollution haven hypothesis in PIIGS countries. *Renewable Energy*, *185*, 1441–1455.

Bastin, J. F., Finegold, Y., Garcia, C., Mollicone, D., Rezende, M., Routh, D., Zohner, C. M., & Crowther, T. W. (2019). The global tree restoration potential. *Science*, *365*(6448), 76–79.

Batar, A., Watanabe, T., & Kumar, A. (2017). Assessment of land-use/land-cover change and forest fragmentation in the Garhwal Himalayan region of India. *Environments*, *4*(2), 34. https://doi.org/10.3390/environments4020034

Bennett, A. C., Dargie, G. C., Cuni-Sanchez, A., Tshibamba Mukendi, J., Hubau, W., Mukinzi, J. M., Phillips, O. L., Malhi, Y., Sullivan, M. J. P., Cooper, D. L. M., Adu-Bredu, S., Affum-Baffoe, K., Amani, C. A., Banin, L. F., Beeckman, H., Begne, S. K., Bocko, Y. E., Boeckx, P., Bogaert, J., … Lewis, S. L. (2021). Resistance of African tropical forests to an extreme climate anomaly. *Proceedings of the National Academy of Sciences*, *118*(21), e2003169118.

Bhatia, S., & Shukla, N. (2024). Integrating traditional knowledge and cultural heritage with climate adaptation and disaster risk reduction: The role of training and tools. *Journal of Cultural Heritage Management and Sustainable Development*. *14*(5), 773–776.

Bonan, G. B. (2008). Forests and climate change: Forcings, feedbacks, and the climate benefits of forests. *Science*, *320*(5882), 1444–1449.

Boyd, I. L., Freer-Smith, P. H., Gilligan, C. A., & Godfray, H. C. J. (2013). The consequence of tree pests and diseases for ecosystem services. *Science*, *342*(6160). https://doi.org/10.1126/science.1235773

Braatz, S. (2012). Building resilience for adaptation to climate change through sustainable forest management. *Building Resilience for Adaptation to Climate Change in the Agriculture Sector*, *23*, 117.

Brando, P. M., Paolucci, L., Ummenhofer, C. C., Ordway, E. M., Hartmann, H., Cattau, M. E., Yin, Y., Higgins, M., Kueppers, L. M., Macedo, M. N., Schwartz, N. B., Tabor, K., Verchot, L. V., & Balch, J. (2019). Droughts, wildfires, and forest carbon cycling: A pantropical synthesis. *Annual Review of Earth and Planetary Sciences*, *47*(1), 555–581.

Briner, G., & Konrad, S. (2014). *Planting the foundations of a post-2020 land sector reporting and accounting framework*. Organisation for Economic Co-operation and Development (OECD). https://www.oecd.org/content/dam/oecd/en/publications/reports/2014/11/planting-the-foundations-of-a-post-2020-land-sector-reporting-and-accounting-framework_g17a2646/5js1qf2p7pzw-en.pdf

Brockerhoff, E. G., Barbaro, L., Castagneyrol, B., Forrester, D. I., Gardiner, B., González-Olabarria, J. R., Lyver, P. O. B., Meurisse, N., Oxbrough, A., Taki, H., Thompson, I. D., van der Plas, F., & Jactel, H. (2017). Forest biodiversity, ecosystem functioning and the provision of ecosystem services. *Biodiversity and Conservation*, *26*(13), 3005–3035. https://doi.org/10.1007/s10531-017-1453-2

Butler, R. (2020). *Rainforests – A place out of time: Tropical rainforests and the perils they face – Information on tropical forests, deforestation, and biodiversity*. Mongabay. https://rainforests.mongabay.com/

Calliari, E., Castellari, S., Davis, M., Linnerooth-Bayer, J., Martin, J., Mysiak, J., Nieto, A., Väänänen, E., Rincón, L., Soini, K., Ghosh, A., De Blasio, S., Lienhoop, N., Hernández-Morcillo, M., Kivimaa, P., Ramazanova, M., Ševčík, M., Smith, A., Tinch, R., & Zandersen, M. (2022). Building climate resilience through nature-based solutions in Europe: A review of enabling knowledge, finance and governance frameworks. *Climate Risk Management*, *37*, 100450.

Cardeli, B. R., Lapola, D. M., Hickler, T., & Paula, M. D. de. (2024). *A new modeling approach to evaluate the effects of climate change on plant functional diversity in the Amazon rainforest*. https://doi.org/10.5194/egusphere-egu24-16900

Chaki, A., Sinha, B., Bisaria, J., & Srivastav, P. (2022). Need for a comprehensive monitoring framework of Indian forests. *Economic & Political Weekly*, *57*(2), 25–28.

Chakraborty, A., Saha, S., Sachdeva, K., & Joshi, P. K. (2018). Vulnerability of forests in the Himalayan region to climate change impacts and anthropogenic disturbances: A systematic review. *Regional Environmental Change*, *18*, 1783–1799.

Chaturvedi, R. K., Gopalakrishnan, R., Jayaraman, M., Bala, G., Joshi, N. V., Sukumar, R., & Ravindranath, N. H. (2011). Impact of climate change on Indian forests: A dynamic vegetation modeling approach. *Mitigation and Adaptation Strategies for Global Change*, *16*(2), 119–142. https://doi.org/10.1007/s11027-010-9257-7

Chien, F., Anwar, A., Hsu, C. C., Sharif, A., Razzaq, A., & Sinha, A. (2021). The role of information and communication technology in encountering environmental degradation: Proposing an SDG framework for the BRICS countries. *Technology in Society*, *65*, 101587.

Clark, R., Reed, J., & Sunderland, T. (2018). Bridging funding gaps for climate and sustainable development: Pitfalls, progress and potential of private finance. *Land Use Policy*, *71*, 335–346.

Cooper, D. J., Berner, L., Dial, R., & Rotbarth, R. (2023). *The world's boreal forests may be shrinking as climate change pushes them northward*. The Conversation.

Corbera, E., Estrada, M., & Brown, K. (2010). Reducing greenhouse gas emissions from deforestation and forest degradation in developing countries: Revisiting the assumptions. *Climatic Change*, *100*(3), 355–388.

Corning, S., Krasovskiy, A., Kiparisov, P., San Pedro, J., Viana, C. M., & Kraxner, F. (2024). Anticipating future risks of climate-driven wildfires in boreal forests. *Fire*, *7*(4), 144–144. https://doi.org/10.3390/fire7040144

Cosslett, C. E. (2013, March). Understanding the potential impacts of REDD+ on the financing and achievement of sustainable forest management. In *A report prepared for the Secretariat of the United Nations Forum on Forests (UNFFS)*.https://www.un.org/esa/forests/wp-content/uploads/2015/03/Understanding_Impacts_REDD+.pdf

Devi, R. M., Patasaraiya, M. K., Sinha, B., & Bisaria, J. (2023). Major drivers for reducing vulnerability of forest-fringe communities in Kanha Tiger Reserve, Madhya Pradesh and designing suitable adaptation strategies. *Asia-Pacific Journal of Regional Science*, *7*(2), 569–590.

Devi, R. M., Patasaraiya, M. K., Sinha, B., Saran, S., Dimri, A. P., & Jaiswal, R. (2018). Understanding the linkages between climate change and forest. *Current Science*, *114*(5), 987–996.

Doughty, C. E., Metcalfe, D. B., Girardin, C. A. J., Amézquita, F. F., Cabrera, D. G., Huasco, W. H., Silva-Espejo, J. E., Araujo-Murakami, A., da Costa, M. C., Rocha, W., Feldpausch, T. R., Mendoza, A. L. M., da Costa, A. C. L., Meir, P., Phillips, O. L., & Malhi, Y. (2015). Drought impact on forest carbon dynamics and fluxes in Amazonia. *Nature*, *519*(7541), 78–82. https://doi.org/10.1038/nature14213

Economic, U. N., & Counci, S. (2017). *United Nations strategic plan for forests 2017–2030 and quadrennial programme of work of the United Nations forum on forests for the period 2017–2020*. United Nations.

El Tahir, B. A. (2015). Climate change adaptation through sustainable forest management in Sudan: Needs to qualify agroforestry application. *Sudan Academy of Sciences Journal, 11*, 162–185.

Fan, X., & Liu, Y. (2016). A global study of NDVI difference among moderate-resolution satellite sensors. *ISPRS Journal of Photogrammetry and Remote Sensing, 121*, 177–191.

FAO. (2023). *Forests and climate change*. Climate Smart Agriculture Sourcebook, Food and Agriculture Organization of the United Nations. https://www.fao.org/3/i3325e/i3325e.pdf

FAO. (2024). *The state of the world's forests 2024 – Forest-sector innovations towards a more sustainable future*. FAO. FAO eBooks. https://doi.org/10.4060/cd1211en

Fay, C., & Denduangrudee, H. M. S. (2018). *An uneven path toward rights and REDD+ in Indonesia* [Background Paper]. Washington, DC: Center for Global Development.

Ferchichi, A., Abbes, A. B., Barra, V., & Farah, I. R. (2022). Forecasting vegetation indices from spatio-temporal remotely sensed data using deep learning-based approaches: A systematic literature review. *Ecological Informatics, 68*, 101552.

Forster, E. J., Healey, J. R., Dymond, C., & Styles, D. (2021). Commercial afforestation can deliver effective climate change mitigation under multiple decarbonisation pathways. *Nature Communications, 12*(1), 3831.

Foster, B. C., Wang, D., Keeton, W. S., & Ashton, M. S. (2010). Implementing sustainable forest management using six concepts in an adaptive management framework. *Journal of Sustainable Forestry, 29*(1), 79–108.

Frelich, L. E. (2020). Boreal and taiga biome. *Forests – Trees of Life*, 103–115. https://doi.org/10.1016/B978-0-12-409548-9.11926-8

Friends of EbA. (2022). *Ecosystem-based Adaptation and the successful implementation and achievement of the sustainable development goals*. https://iucn.org/resources/file/ecosystem-based-adaptation-and-successful-implementation-and-achievement-sustainable

Froese, R., & Schilling, J. (2019). The nexus of climate change, land use, and conflicts. *Current Climate Change Reports, 5*, 24–35.

Fry, I. (2008). Reducing emissions from deforestation and forest degradation: Opportunities and pitfalls in developing a new legal regime. *Review of European Community & International Environmental Law, 17*(2), 166–182.

FSI. (2019). *India state of forest report 2019*. Forest Survey of India, Ministry of Environment, Forest & Climate Change Government of India.

FSI. (2021). *ISFR report, Chapter 2 Forest cover, Forest survey of India* (pp. 18–53). https://fsi.nic.in/forest-report-2021-details

Funk, J. M., Aguilar-Amuchastegui, N., Baldwin-Cantello, W., Busch, J., Chuvasov, E., Evans, T., Goodman, L., Hansen, M. C., Hoff, R. J. V. D., Leonard, S., Minnemeyer, S., Pearson, T. R. H., Pistorius, T., Radeloff, V. C., Seymour, F., Shrestha, A., Sizer, N., Steil, L., Suh, J., & van der Hoff, R. J. (2019). Securing the climate benefits of stable forests. *Climate Policy, 19*(7), 845–860.

Gandharba, K. (2023). *Climate change, forests, and communities: Examining the complex interactions and implications for livelihoods*. https://doi.org/10.31235/osf.io/zag7x

Garekae, H., Thakadu, O. T., & Lepetu, J. (2017). Socio-economic factors influencing household forest dependency in Chobe enclave, Botswana. *Ecological Processes, 6*(1). https://doi.org/10.1186/s13717-017-0107-3

Gauthier, S., Bernier, P., Kuuluvainen, T., Shvidenko, A. Z., & Schepaschenko, D. G. (2015). Boreal forest health and global change. *Science, 349*(6250), 819–822.

Gilliam, F. S. (2016). Forest ecosystems of temperate climatic regions: From ancient use to climate change. *New Phytologist, 212*(4), 871–887. https://doi.org/10.1111/nph.14255

Girsang, L. W. P., Simbolon, N., Saputri, R. N., & Lubis, R. K. (2024). Optimizing sustainability: Exploring the intersection of carbon trading and social forestry initiatives. *Mahadi: Indonesia Journal of Law*, *3*(01), 23–30.

Goodman, R., & Herold, M. (2014, November 4). *Why maintaining tropical forests is essential and urgent for a stable climate* (Center for Global Development Working Paper No. 385). Center for Global Development. https://doi.org/10.2139/ssrn.2622758

Gopalakrishnan, R., Jayaraman, M., Govindasamy, B., & Ravindranath, N. (2011). Climate change and Indian forests. *Current Science*, *101*(3), 354–361.

Graham, R. L., Turner, M. G., & Dale, V. H. (1990). How increasing CO_2 and climate change affect forests. *BioScience*, *40*(8), 575–587.

Horstmann, B., & Hein, J. (2017). *Aligning climate change mitigation and sustainable development under the UNFCCC: A critical assessment of the clean development mechanism, the green climate fund and REDD+ (No. 96).* Studies. German Development Institute/Deutsches Institut für Entwicklungspolitik (DIE). https://hdl.handle.net/10419/199221

Hosonuma, N., Herold, M., De Sy, V., De Fries, R. S., Brockhaus, M., Verchot, L., Angelsen, A., & Romijn, E. (2012). An assessment of deforestation and forest degradation drivers in developing countries. *Environmental Research Letters*, *7*, 044009.

Huang, W., Gao, Q. X., Cao, G. L., Ma, Z. Y., Zhang, W. D., & Chao, Q. C. (2016). Effect of urban symbiosis development in China on GHG emissions reduction. *Advances in Climate Change Research*, *7*(4), 247–252.

Huete, A. R. (2012). Vegetation indices, remote sensing and forest monitoring. *Geography Compass*, *6*(9), 513–532.

Murthy, I. K., Tiwari, R., & Ravindranath, N. H. (2011). Climate change and forests in India: Adaptation opportunities and challenges. *Mitigation and Adaptation Strategies for Global Change*, *16*, 161–175.

Ishikawa-Ishiwata, Y., & Furuya, J. (2022). Economic evaluation and climate change adaptation measures for rice production in Vietnam using a supply and demand model: Special emphasis on the Mekong River Delta region in Vietnam. In *Interlocal adaptations to climate change in east and southeast Asia: Sharing lessons of agriculture, disaster risk reduction, and resource management* (pp. 45–53). Springer International Publishing.

Islam, M. R., Jönsson, A. M., Bergkvist, J., Lagergren, F., Lindeskog, M., Mölder, M., Scholze, M., & Kljun, N. (2024). *Integrated assessment of climate change and forest management impacts on carbon fluxes and biomass in a southern boreal forest.* https://doi.org/10.5194/egusphere-egu24-341

IUCN. (2021). *Forests and climate change.* https://iucn.org/sites/default/files/2022-04/forests_and_climate_change_issues_brief_2021.pdf

IUCN (2022). *Forests and climate change.* Available at: https://iucn.org/resources/issues-brief/forests-and-climate-change (Accessed: 10 December 2024).

Jain, P., Coogan, S. C., Subramanian, S. G., Crowley, M., Taylor, S., & Flannigan, M. D. (2020). A review of machine learning applications in wildfire science and management. *Environmental Reviews*, *28*(4), 478–505.

Jha, S. K., Mishra, S., Sinha, B., Alatalo, J. M., & Pandey, R. (2017). Rural development program in tribal region: A protocol for adaptation and addressing climate change vulnerability. *Journal of Rural Studies*, *51*, 151–157.

Jha, S. K., & Sinha, B. (2017). Benefit sharing mechanism for operationalization of REDD+ in India. *Indian Forester*, *143*(5), 421–430.

Jones, M. W., Abatzoglou, J. T., Veraverbeke, S., Andela, N., Lasslop, G., Forkel, M., Smith, A. J. P., Burton, C., Betts, R. A., Werf, G. R., Sitch, S., Canadell, J. G., Santín, C., Kolden, C., Doerr, S. H., & Le Quéré, C. (2022). Global and regional trends and drivers of fire under climate change. *Reviews of Geophysics*, *60*(3). https://doi.org/10.1029/2020rg000726

Jump, A. S., Cavin, L., & Hunter, P. D. (2010). Monitoring and managing responses to climate change at the retreating range edge of forest trees. *Journal of Environmental Monitoring*, *12*(10), 1791–1798.

Karam, S., Zango, B.-S., Seidou, O., Perera, D., Nagabhatla, N., & Tshimanga, R. (2023). Impacts of climate change on hydrological regimes in the Congo River basin. *Sustainability, 15*(7), 6066–6066. https://doi.org/10.3390/su15076066

Katila, P., Colfer, C. J. P., de Jong, W., Galloway, G., Pacheco, P., & Winkel, G. (Eds.). (2024). *Restoring forests and trees for sustainable development: Policies, practices, impacts, and ways forward.* Oxford University Press.

Kauppi, P. E., Stål, G., Arnesson-Ceder, L., Sramek, I. H., Hoen, H. F., Svensson, A., Bergh, J., Fridman, J., Lundmark, T., Nordin, A., & Nordin, A. (2022). Managing existing forests can mitigate climate change. *Forest Ecology and Management, 513*, 120186.

Keenan, R. J. (2015). Climate change impacts and adaptation in forest management: A review. *Annals of Forest Science, 72*(2), 145–167. https://doi.org/10.1007/s13595-014-0446-5

Kishwan, J., Pandey, R., & Dadhwal, V. K. (2009). *India's forest and tree cover: Contribution as a carbon sink* [Technical paper]. Indian Council of Forestry Research and Education. Dehradun.

Klein, R. J., Schipper, E. L. F., & Dessai, S. (2005). Integrating mitigation and adaptation into climate and development policy: Three research questions. *Environmental Science & Policy, 8*(6), 579–588.

Koc, M., & Acar, A. (2021). Investigation of urban climates and built environment relations by using machine learning. *Urban Climate, 37*, 100820.

Kour, D., Ahluwalia, K. K., Ramniwas, S., Kumar, S., Rustagi, S., Singh, S., Rai, A. K., Yadav, A. N., & Ahluwalia, A. S. (2024). Environment and climate change: Influence on biodiversity, present scenario, and future prospect. *Journal of Applied Biology and Biotechnology*. https://doi.org/10.7324/jabb.2024.148189

Ladi, T., Jabalameli, S., & Sharifi, A. (2022). Applications of machine learning and deep learning methods for climate change mitigation and adaptation. *Environment and Planning B: Urban Analytics and City Science, 49*(4), 1314–1330.

Lafortezza, R., Chen, J., Van Den Bosch, C. K., & Randrup, T. B. (2018). Nature-based solutions for resilient landscapes and cities. *Environmental Research, 165*, 431–441.

Laukkonen, J., Blanco, P. K., Lenhart, J., Keiner, M., Cavric, B., & Kinuthia-Njenga, C. (2009). Combining climate change adaptation and mitigation measures at the local level. *Habitat International, 33*(3), 287–292.

Li, Y., Mei, B., & Linhares-Juvenal, T. (2019). The economic contribution of the world's forest sector. *Forest Policy and Economics, 100*, 236–253. https://doi.org/10.1016/j.forpol.2019.01.004

Lindner, M., Fitzgerald, J. B., Zimmermann, N. E., Reyer, C., Delzon, S., van Der Maaten, E., Schelhaas, M.-J., Lasch, P., Eggers, J., van der Maaten-Theunissen, M., Suckow, F., Psomas, A., Poulter, B., & Hanewinkel, M. (2014). Climate change and European forests: What do we know, what are the uncertainties, and what are the implications for forest management?. *Journal of Environmental Management, 146*, 69–83.

Magry, M. A., Cahill, D., Rookes, J., & Narula, S. A. (2022). An integrated value chain analysis of non-timber forest products: A case of Jharkhand state of India. *Small-Scale Forestry, 21*(4), 621–645.

Maniatis, D., Scriven, J., Jonckheere, I., Laughlin, J., & Todd, K. (2019). Toward REDD+ implementation. *Annual Review of Environment and Resources, 44*(1), 373–398.

Martin, P. (2020). *The changing climate of development: REDD+ in Papua New Guinea* [Doctoral dissertation, UNSW Sydney].

Masson-Delmotte, V., Zhai, P., Pirani, A., Connors, S. L., Péan, C., Berger, S., N. Caud, Y. Chen, L. Goldfarb, M. I. Gomis, M. Huang, K. Leitzell, É. Lonnoy, J. B. R. Matthews, T. K. Maycock, T. Waterfield, O. Yelekçi, R. Yu, & Zhou, B. (2021). Climate change 2021: The physical science basis. *Contribution of Working Group I to the Sixth Assessment Report of the Intergovernmental Panel on Climate Change, 2*(1), 2391.

Melo, F. P., Pinto, S. R., Brancalion, P. H., Castro, P. S., Rodrigues, R. R., Aronson, J., & Tabarelli, M. (2013). Priority setting for scaling-up tropical forest restoration projects: Early lessons from the Atlantic Forest Restoration Pact. *Environmental Science & Policy, 33*, 395–404.

Michaelowa, A., Shishlov, I., & Brescia, D. (2019). Evolution of international carbon markets: Lessons for the Paris Agreement. *Wiley Interdisciplinary Reviews: Climate Change, 10*(6), e613.

Mitchell, C. P., Corbridge, S. E., Jewitt, S. L., Mahapatra, A. K., & Kumar, S. (2003). *Non timber forest products: Availability, production, consumption, management and marketing in eastern India*. https://assets.publishing.service.gov.uk/media/57a08ce8ed915d3cfd0016d4/R6916FTR.pdf

Moore, B., & Allard, G. (2008). *Climate change impacts on forest health*. Food and Agriculture Organization of the United Nations (FAO). https://www.fao.org/forestry-fao/15905-0dc804ee7d97e656f06507bdcecddc721.pdf

Morellato, L. P. C., Alberton, B., Alvarado, S. T., Borges, B., Buisson, E., Camargo, M. G. G., Cancian, L. F., Carstensen, D. W., Escobar, D. F. E., Leite, P. T. P., Mendoza, I., Rocha, N. M. W. B., Soares, N. C., Silva, T. S. F., Staggemeier, V. G., Streher, A. S., Vargas, B. C., & Peres, C. A. (2016). Linking plant phenology to conservation biology. *Biological Conservation, 195*, 60–72. https://doi.org/10.1016/j.biocon.2015.12.033

Muluneh, M. G. (2021). Impact of climate change on biodiversity and food security: A global perspective—A review article. *Agriculture & Food Security, 10*(1), 1–25.

Myers, N. (1990). The world's forests and human populations: The environmental interconnections. *Population and Development Review, 16*, 237. https://doi.org/10.2307/2808073

Nunes, L. J. R., Meireles, C. I. R., Gomes, C. J. P., & Ribeiro, N. M. C. A. (2021). The impact of climate change on forest development: A sustainable approach to management models applied to Mediterranean-type climate regions. *Plants, 11*(1), 69. https://doi.org/10.3390/plants11010069

Obeng, E. A., Owusu-Sekyere, E., Dumenu, W. K., Nutakor, E. E., Samar, S. B., & Opoku, E. A. (2011). *Climate change and forest dependent livelihoods; vulnerabilities and impacts* [Technical report, Forest Livelihood and Sustainable Development Division, Government of Ghana Project]. https://www.researchgate.net/publication/261438526_CLIMATE_CHANGE_AND_FOREST_DEPENDENT_LIVELIHOODS_VULNERABILITIES_AND_IMPACTS

Ofoegbu, C., Chirwa, P. W., Francis, J., & Babalola, F. D. (2017). Socio-economic factors influencing household dependence on forests and its implication for forest-based climate change interventions. *Southern Forests: A Journal of Forest Science, 79*(2), 109–116.

Omann, I., Stocker, A., & Jäger, J. (2009). Climate change as a threat to biodiversity: An application of the DPSIR approach. *Ecological Economics, 69*(1), 24–31.

Ozor, N., Nyambane, A., Onuoha, C. M., Makokha, M. O., & M'mboyi, F. (2020). Nationally determined contributions (NDCs) implementation index, monitoring and tracking tools for selected countries in Africa. Available at: chrome-extension://efaidnbmnnnibpcajpcglclefindmkaj/https://atpsnet.org/wp-content/uploads/2019/07/NDCs-Implementation-Report-final.pdf (Accessed: 7 November 2023).

Padarian Campusano, J. S. (2019). *Machine learning to generate soil information* [Doctoral dissertation].

Pandey, A. K., Tripathi, Y. C., & Kumar, A. (2016). Non timber forest products (NTFPs) for sustained livelihood: Challenges and strategies. *Research Journal of Forestry, 10*(1), 1–7. https://doi.org/10.3923/rjf.2016.1.7

Pandey, S. S., Cockfield, G., & Maraseni, T. N. (2016). Assessing the roles of community forestry in climate change mitigation and adaptation: A case study from Nepal. *Forest Ecology and Management, 360*, 400–407.

Papastefanou, P., Zang, C. S., Angelov, Z., de Castro, A. A., Jimenez, J. C., De Rezende, L. F. C., Ruscica, R. C., Sakschewski, B., Sörensson, A. A., Thonicke, K., Vera, C., Viovy, N., Von Randow, C., & Rammig, A. (2022). Recent extreme drought events in the Amazon rainforest: Assessment of different precipitation and evapotranspiration datasets and drought indicators. *Biogeosciences*, *19*(16), 3843–3861. https://doi.org/10.5194/bg-19-3843-2022

Patasaraiya, M. K., Devi, R. M., Sinha, B., & Bisaria, J. (2023). Predicting impacts of climate change on teak and Sal forests in central India using maximum entropy modeling: An approach for future conservation and silvicultural strategies. *Forest Science*, *69*(4), 463–472.

Patasaraiya, M. K., Devi, R. M., Sinha, B., Bisaria, J., Saran, S., & Jaiswal, R. (2021). Understanding the resilience of Sal and Teak forests to climate variability using NDVI and EVI time series. *Forest Science*, *67*(2), 192–204. https://doi.org/10.1093/forsci/fxaa051

Patel, R., Mukherjee, S., Sahu, B., Dash, B., Jaison, M., Avinash, K., & Singh, P. (2024). Sustainable forest management (SFM) for C footprint and climate change mitigation. In *Agroforestry to combat global challenges: Current prospects and future challenges* (pp. 203–217). Springer Nature Singapore.

Picard, N., & Garavaglia, V. (2021). Mediterranean forests and the United Nations sustainable development goals. In G. Ne'eman, Y. Osem, (Eds.), *Pines and their mixed forest ecosystems in the Mediterranean basin* (pp. 603–616). Springer. https://doi.org/10.1007/978-3-030-63625-8_28.

Price, D. T., Alfaro, R. I., Brown, K. J., Flannigan, M. D., Fleming, R. A., Hogg, E. H., Girardin, M. P., Lakusta, T., Johnston, M., McKenney, D. W., Pedlar, J. H., Stratton, T., Sturrock, R. N., Thompson, I. D., Trofymow, J. A., & Venier, L. A. (2013). Anticipating the consequences of climate change for Canada's boreal forest ecosystems. *Environmental Reviews*, *21*(4), 322–365. https://doi.org/10.1139/er-2013-0042

Psistaki, K., Tsantopoulos, G., & Paschalidou, A. K. (2024). An overview of the role of forests in climate change mitigation. *Sustainability*, *16*(14), 6089.

Ravindranath, N. H., Joshi, N. V., Sukumar, R., & Saxena, A. (2005). Impact of climate change on forests in India. *Current Science Association*, *90*(3), 354–361.

Ravindranath, N. H., Murthy, I. K., Chaturvedi, R. K., Andrasko, K., & Sathaye, J. A. (2007). Carbon forestry economic mitigation potential in India, by land classification. *Mitigation and Adaptation Strategies for Global Change*, *12*, 1027–1050.

Reichstein, M., Camps-Valls, G., Stevens, B., Jung, M., Denzler, J., Carvalhais, N., & Prabhat, F. (2019). Deep learning and process understanding for data-driven earth system science. *Nature*, *566*(7743), 195–204.

Riahi, K., Van Vuuren, D. P., Kriegler, E., Edmonds, J., O'neill, B. C., Fujimori, S., Bauer, N., Calvin, K., Dellink, R., Fricko, O., Lutz, W., Popp, A., Cuaresma, J. C., Samir KC, Leimbach, M., Jiang, L., Kram, T., Rao, S., Emmerling, J., … Tavoni, M. (2017). The shared socioeconomic pathways and their energy, land use, and greenhouse gas emissions implications: An overview. *Global Environmental Change*, *42*, 153–168.

Rizvi, A. R., Baig, S., Barrow, E., & Kumar, C. (2015). *Synergies between climate mitigation and adaptation in forest landscape restoration*. IUCN.

Roy, A. (2023) *Harnessing the power of India's forests for climate change mitigation*. orfonline.org. Available at: https://www.orfonline.org/research/harnessing-the-power-of-indias-forests-for-climate-change-mitigation#:~:text=The%20forest%20sector%20in%20India,regeneration%20of%20the%20country's%20forests (Accessed: 10 December 2024).

Saalu, F. N., Oriaso, S., & Gyampoh, B. (2020). Effects of a changing climate on livelihoods of forest dependent communities. *International Journal of Climate Change Strategies and Management*, *12*(1), 1–21. https://doi.org/10.1108/ijccsm-01-2018-0002

Safford, H. D., & Vallejo, V. R. (2019). Ecosystem management and ecological restoration in the Anthropocene: Integrating global change, soils, and disturbance in boreal and Mediterranean forests. *Global Change and Forest Soils*, 259–308. https://doi.org/10.1016/b978-0-444-63998-1.00012-4

Sanz, M. J., & Penman, J. (2016). An overview of REDD+. *Unasylva*, *67*(246), 21.

Sathaye, J., Makundi, W., Dale, L., Chan, P., & Andrasko, K. (2006). GHG mitigation potential, costs and benefits in global forests: A dynamic partial equilibrium approach. *The Energy Journal*, *27*(3_suppl), 127–162.

Schlamadinger, B., Bird, N., Johns, T., Brown, S., Canadell, J., Ciccarese, L., Dutschke, M., Fiedler, J., Fischlin, A., Fearnside, P., Forner, C., Freibauer, A., Frumhoff, P., Höhne, N., Kirschbaum, M., Labat, A., Marland, G., Michaelowa, A., Montanarella, L., & Yamagata, Y. (2007). A synopsis of land use, land-use change and forestry (LULUCF) under the Kyoto Protocol and Marrakech Accords. *Environmental Science & Policy*, *10*(4), 271–282.

Schneider, S. H., & Kuntz-Duriseti, K. (2002). Uncertainty and climate change policy. In S. H. Schneider, A. Rosencranz, & J. O. Niles (Eds.), *Climate change policy: A survey* (pp. 53–87). Island Press.

Seddon, N. (2022). Harnessing the potential of nature-based solutions for mitigating and adapting to climate change. *Science*, *376*(6600), 1410–1416.

Seddon, N., Chausson, A., Berry, P., Girardin, C. A., Smith, A., & Turner, B. (2020). Understanding the value and limits of nature-based solutions to climate change and other global challenges. *Philosophical Transactions of the Royal Society B*, *375*(1794), 20190120.

Seddon, N., Turner, B., Berry, P., Chausson, A., & Girardin, C. A. (2019). Grounding nature-based climate solutions in sound biodiversity science. *Nature Climate Change*, *9*(2), 84–87.

Seidl, R., Thom, D., Kautz, M., Martin-Benito, D., Peltoniemi, M., Vacchiano, G., Wild, J., Ascoli, D., Petr, M., Honkaniemi, J., Lexer, M. J., Trotsiuk, V., Mairota, P., Svoboda, M., Fabrika, M., Nagel, T. A., Reyer, C. P. O., Jandl, R., Morrissey, R. C., & Reyer, C. P. (2017). Forest disturbances under climate change. *Nature Climate Change*, *7*(6), 395–402.

Shahbaz, M., Balsalobre-Lorente, D., & Sinha, A. (2019). Foreign direct Investment–CO_2 emissions nexus in Middle East and North African countries: Importance of biomass energy consumption. *Journal of Cleaner Production*, *217*, 603–614.

Sharif, A., Mishra, S., Sinha, A., Jiao, Z., Shahbaz, M., & Afshan, S. (2020). The renewable energy consumption-environmental degradation nexus in top-10 polluted countries: Fresh insights from quantile-on-quantile regression approach. *Renewable Energy*, *150*, 670–690.

Sharma, R., Sinha, A., & Kautish, P. (2020). Examining the impacts of economic and demographic aspects on the ecological footprint in south and southeast Asian countries. *Environmental Science and Pollution Research*, *27*(29), 36970–36982.

Singh, M., Sinha, B., Bisaria, J., Thomas, T., & Srivastava, P. (2022). Understanding synergies and tradeoffs between forests, water, and climate change. *Wiley Interdisciplinary Reviews: Water*, *9*(6), e1614.

Singha, C., Swain, K. C., Moghimi, A., Foroughnia, F., & Swain, S. K. (2024). Integrating geospatial, remote sensing, and machine learning for climate-induced forest fire susceptibility mapping in Similipal Tiger Reserve, India. *Forest Ecology and Management*, *555*, 121729.

Sinha, B. (2017). Benefit sharing mechanisms under REDD+: Learning from the joint forest management program. In P. P. Bhojvaid, R. K. Rai, & M. K. Jha (Eds.), *Managing India's forests in a changing climate* (pp. 233–247). TERI Press.

Sotirov, M., Pokorny, B., Kleinschmit, D., & Kanowski, P. (2020). International forest governance and policy: Institutional architecture and pathways of influence in global sustainability. *Sustainability*, *12*(17), 7010.

Stocks, B. J., & Ward, P. C. (2011). *Climate change, carbon sequestration, and forest fire protection in the Canadian boreal zone* (Canadian Council of Forest Ministers Climate Change Technical Report Series, No. CCRR-20, pp. v–26). Canadian Forest Service.

Tebeje, Y. (2020). A review paper on the role of terrestrial carbon stocks for climate change mitigation mechanisms. *Journal of Environment and Earth Science*, *10*(8), 32–50.

Tegegne, Y. T., Cramm, M., & Van Brusselen, J. (2018). Sustainable forest management, FLEGT, and REDD+: Exploring interlinkages to strengthen forest policy coherence. *Sustainability*, *10*(12), 4841.

Teshome, D. T., Zharare, G. E., & Naidoo, S. (2020). The threat of the combined effect of biotic and abiotic stress factors in forestry under a changing climate. *Frontiers in Plant Science*, *11*. https://doi.org/10.3389/fpls.2020.601009

Thomas, S., Dargusch, P., Harrison, S., & Herbohn, J. (2010). Why are there so few afforestation and reforestation clean development mechanism projects?. *Land Use Policy*, *27*(3), 880–887.

Thompson, I., Mackey, B., McNulty, S., & Mosseler, A. (2009). Forest resilience, biodiversity, and climate change. In *A synthesis of the biodiversity/resilience/stability relationship in forest ecosystems* (Vol. 43, No. 1, pp. 1–67). Secretariat of the Convention on Biological Diversity. Technical Series.

Pham, T.T., Moeliono, M., Angelsen, A., Brockhaus, M., Gallo, P., Hoang, T.L., Đào Thi, L.C., Ochoa, C., Bocanegra, K. 2018. Strategic alignment: Integrating REDD+ in NDCs and national climate policies. In A. Angelsen, C. Martius, V. de Sy, A.E. Duchelle, A.M. Larson, Pham T.T. (Eds.), *Transforming REDD+: Lessons and new directions* (pp. 69–80). Center for International Forestry Research (CIFOR).

Trumbore, S., Brando, P., & Hartmann, H. (2015). Forest health and global change. *Science*, *349*(6250), 814–818.

Turner, B., Devisscher, T., Chabaneix, N., Woroniecki, S., Messier, C., & Seddon, N. (2022). The role of nature-based solutions in supporting social-ecological resilience for climate change adaptation. *Annual Review of Environment and Resources*, *47*(1), 123–148.

UN DESA. (2021). *The global forest goals report 2021.* United Nations Department of Economic and Social Affairs, United Nations Forum on Forests Secretariat.

UNDP Climate Promise. (2023). *Forests can help us limit climate change – Here is how UNDP climate promise.* Available at: https://climatepromise.undp.org/news-and-stories/forests-can-help-us-limit-climate-change-here-how (Accessed: 10 December 2024).

UNEP. (2022). *Harnessing nature to build climate resilience: Scaling up the use of ecosystem-based adaptation.* https://wedocs.unep.org/20.500.11822/40415

United Nations Framework Convention on Climate Change. Secretariat. (1992). *United Nations framework convention on climate change.* UNFCCC.

US EPA. (2016, December 22). *Climate impacts on ecosystems, Climate change impacts.* US EPA. https://19january2017snapshot.epa.gov/climate-impacts/climate-impacts-ecosystems_.html

Venter, O., & Koh, L. P. (2012). Reducing emissions from deforestation and forest degradation (REDD+): Game changer or just another quick fix?. *Annals of the New York Academy of Sciences*, *1249*(1), 137–150.

Verburg, P. H., Neumann, K., & Nol, L. (2011). Challenges in using land use and land cover data for global change studies. *Global Change Biology*, *17*(2), 974–989.

Voigt, C., & Ferreira, F. (2016). The Warsaw Framework for REDD+: Implications for national implementation and results-based finance. In *Research handbook on REDD-plus and international law* (pp. 30–60). Edward Elgar Publishing.

Voigt, C., Nemitz, D., Brana-Varela, J., Sanz-Sanchez, M. J., & Ferreira, F. (2024). *The Paris Agreement and the relevance of the Warsaw Framework for REDD+ (WFR). Review of European, comparative and international environmental law.*

Wale, E., Nkoana, M. A., & Mkuna, E. (2022). Climate change-induced livelihood adaptive strategies and perceptions of forest-dependent communities: The case of Inanda, KwaZulu-Natal, South Africa. *Trees, Forests and People*, *8*, 100250. https://doi.org/10.1016/j.tfp.2022.100250

Watson, J. E., Evans, T., Venter, O., Williams, B., Tulloch, A., Stewart, C., Thompson, I., Ray, J. C., Murray, K., Salazar, A., McAlpine, C., Potapov, P., Walston, J., Robinson, J. G., Painter, M., Wilkie, D., Filardi, C., Laurance, W. F., Houghton, R. A., Maxwell, S., Grantham, H., Samper, C., Wang, S., Laestadius, L., Runting, R. K., Silva-Chávez, G. A., Ervin, J., & Lindenmayer, D. (2018). The exceptional value of intact forest ecosystems. *Nature Ecology & Evolution*, *2*(4), 599–610.

Weaver, C. P., Lempert, R. J., Brown, C., Hall, J. A., Revell, D., & Sarewitz, D. (2013). Improving the contribution of climate model information to decision making: The value and demands of robust decision frameworks. *Wiley Interdisciplinary Reviews: Climate Change*, *4*(1), 39–60.

Wertz-Kanounnikoff, S. (2007). *Reducing emissions from deforestation and degradation in developing countries (REDD): Insights from the UNFCCC COP-13 in Bali* (Working Paper). Institut du Développement Durable et des Relations Internationales (IDDRI).

World Bank. (2004). *Sustaining forests: A World Bank strategy.* World Bank. https://documents.worldbank.org/curated/en/657831468781809431

Zeng, Y., Hao, D., Huete, A., Dechant, B., Berry, J., Chen, J. M., Xiao, X., Liu, L., Carter, S., Ban, Y., Yin, G., Wang, S., Yin, Y., Zhang, Y., Wu, D., Zhang, Y., Zhang, L., Jiang, C., Ma, X & Chen, M. (2022). Optical vegetation indices for monitoring terrestrial ecosystems globally. *Nature Reviews Earth & Environment*, *3*(7), 477–493.

Zhang, X. Y., Goldberg, M., Tarpley, D., Friedl, M. A., Morisette, J., Kogan, F., & Yu, Y. Y. (2010). Drought-induced vegetation stress in southwestern North America. *Environmental Research Letters*, *5*, 024008. https://doi.org/10.1088/1748-9326/5/2/024008

14 Exploring the Prospects of Climate-Forest Nexus for Sustainable Landscapes

Rishabh Srikar, Manisha Singh, Bhaskar Sinha, and Jigyasa Bisaria

14.1 INTRODUCTION

The pervasive influence of changing climatic patterns on different sectors such as water, health, energy, and its social, ecological, and economic implications are well-established and evidenced phenomena within the scientific community (Intergovernmental Panel on Climate Change (IPCC), 2023). Global industrialisation and relentless capitalistic economic growth have led to a tremendous increase in greenhouse gas (GHG) emissions, accelerating climate change (De Matteis, 2019; Rosenzweig et al., 2008; Stern & Kaufmann, 2014). Changing atmospheric compositions and rising temperatures have severely impacted natural climate systems, shifted regional weather patterns, and increased climate-induced natural hazards and disasters (Wheeler & Von Braun, 2013). These changes in climate patterns have had several cascading implications on biodiversity, water security, agricultural productivity, and the economy at local, regional, and continental scales (Tol, 2018).

Forests are vital ecosystems supporting the life and livelihood of 1.6 billion people globally, providing habitats to 80% of the world's terrestrial biodiversity, and are a natural storehouse for resources and services, such as clean water and healthy soils (worth USD 75–100 billion), and act as natural barriers against floods, landslides, soil erosion, avalanches, etc. (Mansourian et al., 2009). While these are some broad categorisations of the services received from forests, an actual inventory of the multitude of regulatory, provisionary, supporting, and cultural ecosystem services provided by them is vast and is being dynamically revised with new findings (Aznar-Sánchez et al., 2018; Krieger, 2001; Ninan & Inoue, 2014). Forest distribution and ecological functions depend on climate regimes, with future forest dynamics projected to have important feedback to global climate systems (Thom et al., 2017). Subsequently, global warming-induced climatic changes have had direct and indirect connotations for forest ecosystems, influencing their distribution and productivity, and deteriorating their ecosystem services and functions (Albrich et al., 2018).

The global terrestrial land sink is the largest carbon storehouse after oceans, with a cumulative sequestration of 210±45 GtC since 1850, removing 31% of total

DOI: 10.1201/9781003588580-14

anthropogenic GHG emissions until 2021 (Friedlingstein et al., 2022). Recent assessments report that forests comprise a significant portion of this sink, sequestering a net carbon sink of 39.83±3.45 GtC between 1990 and 2019 (Pan et al., 2024). However, human-led deforestation, degradation, and land use conversion have significantly weakened the carbon-capturing capacity of forests, making them a source of emissions in some cases (Gatti et al., 2021). Since 1850, about 30% of total anthropogenic GHG emissions have been attributed to land use conversion of forests, with the contribution of deforestation and degradation towards total emissions still ranging about 12–20% annually (Friedlingstein et al., 2022). Despite this, forests sequester a net sink of approximately 1.33±0.63 GtC annually, signifying the role of healthy forests as important carbon sinks (Pan et al., 2024). Therefore, stable and mature forests store a considerable stock of terrestrial carbon and are amassing popularity as a nature-based solution for reducing atmospheric carbon concentrations (Canadell & Raupach, 2008).

Forests and their ecosystem functions have complex and layered interlinkages with climate systems (Singh et al., 2022). However, with estimates indicating an alarming decline in global forest cover by 178 million ha from 1990 to 2020 (FAO, 2020), forest management and conservation have become pivotal for sustaining the derived ecosystem services as well as advancing progress towards attaining multiple Sustainable Development Goals (Baumgartner, 2019; United Nations, 2015). Utilising forests and forestry to combat climate change is particularly lucrative because the sector can remove significant quantities of atmospheric carbon cost-effectively whilst enabling mitigation within other sectors, such as energy and manufacturing, through the provision of raw materials in addition to sustaining vital ecosystem services (IPCC, 2023). Therefore, understanding the relationship between forests and climate change is imperative in designing mitigation and adaptation policies and programmes. In light of this, the chapter attempts to summarise the trends and advancements in understanding the climate-forest nexus along with opportunities and challenges in employing forests for securing a climate-resilient future (Figure 14.1).

14.2 UNDERSTANDING THE DYNAMICS BETWEEN CLIMATE AND FORESTS

The influence forests have on regulating climate and vice versa has been widely acknowledged and accepted in the scientific community (Bonan et al., 1992; De Frenne et al., 2021; Jackson et al., 2008; Thom et al., 2017). However, these feedback and response mechanisms are fairly complex and usually a collective result of plant physiological processes such as photosynthesis, respiration, and evapotranspiration (Lorenz & Lal, 2010). In order to engage forest ecosystems in climate change mitigation and adaptation, it is essential to understand the forest-climate interface. Several approaches have been used to explore and better understand these interactions, from experimental measurements using eddy covariance fluxes to satellite remote sensing-derived mathematical models and ensemble physical process-based modelling such as dynamical global vegetation models (Bonan, 2008). The following sub-sections briefly summarise how the methods

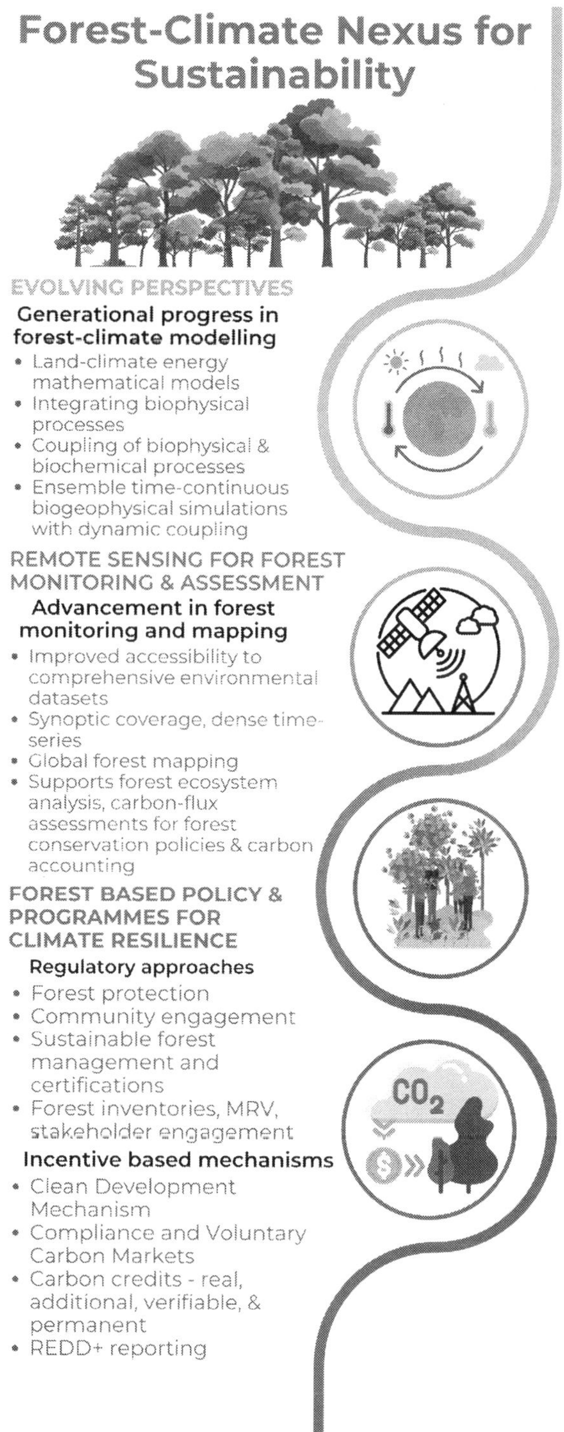

FIGURE 14.1 Prospects of forest-climate nexus for sustainable landscapes.

of understanding climate-forest relationships and corresponding inferences have evolved over the years.

14.2.1 Remote Sensing in Forest Monitoring and Assessment

Developments in Earth observation technologies have radically transformed the mapping and monitoring of forest ecosystems and resources over the years. There are several advantages that remote sensing and geographical information techniques have over traditional methods in assessing natural resources. Satellite-based remote sensing provides synoptic coverage of the Earth's surface, offering a comprehensive and precisely mapped perspective of the environment with highly accurate geolocations and wall-to-wall coverage, an advantage in forest assessments as compared to field surveys that face logistical and financial constraints (Boyd & Danson, 2005). With advancements in sensors aboard satellites and unmanned aerial systems, remote sensing data is now available at multiple spatial and temporal resolutions, allowing flexible analysis of diverse environmental phenomena (Lechner et al., 2020). Furthermore, remote sensing technologies reduce discrepancies in environmental measurements that arise due to variations in sampling methodologies, which often stem from differences in scientific, technical, and financial capacities across individuals, organisations, and nations (Pandey & Arellano, 2022).

The integration of remote sensing datasets with ground-based measurements has proven particularly advantageous in forestry and ecological studies, facilitating data collation and validation, which have historically posed significant challenges in these fields. Land use/land cover change (LULCC) is a key application of remote sensing for forest assessment and has undergone rapid democratisation with the advent of optical sensors such as Landsat and Sentinel missions, enabling global time-series analysis (Aziz et al., 2024). These sensors are widely used to map deforestation, land degradation, and land use conversion, playing a crucial role in forest conservation policies, carbon accounting frameworks like Reducing Emissions from Deforestation and Degradation plus the sustainable management of forests, and the conservation and enhancement of forest carbon stocks (REDD+), national forest inventories (NFI), and legal or economic jurisdictions related to forest rights (Calders et al., 2020; Chen et al., 2021; Poortinga et al., 2019; Reiche et al., 2018). By exploiting the physical response of different materials and landscapes to different wavelengths of the electromagnetic spectrum, remote sensing has enabled the development of indicators for forest health, productivity, and their ecosystem interactions with land, climate, and hydrology (Huete, 2012; Lausch et al., 2016, 2017). For instance, numerous vegetation indices to monitor vegetation extent, health, and dynamics have been developed using optical sensors that are sensitive to the visible and near-infrared spectrum. Examples include ratio-based indices such as NDVI, EVI, SAVI, OSAVI, and MSAVI, which serve as proxies for vegetation condition, as they are highly sensitive to chlorophyll content and photosynthetic activity (Huete, 1988; Huete et al., 2002; Qi et al., 1994; Rondeaux et al., 1996; Rouse et al., 1974).

Biomass assessments are of immense value in evaluating forest resources, particularly in climate science, as they provide crucial insights into existing carbon stocks and their fluxes. Remote sensing datasets have significantly improved our ability to scale up biomass measurements over regional scales using optical datasets and machine learning/artificial intelligence techniques (Xiao et al., 2019). Even in regions where optical remote sensing faces limitations, such as tropical forests with persistent cloud cover, microwave-based active remote sensing techniques like synthetic aperture radar (SAR) allow penetration through clouds to provide detailed information on canopy structure, understory vegetation, and biomass distribution (Ramachandran et al., 2023; Sinha et al., 2015). These advancements have significantly improved biomass and carbon mapping capabilities across diverse weather and topographies. Additionally, LiDAR (Light Detection and Ranging) has enabled the generation of high-resolution, three-dimensional forest structure models with unprecedented accuracy (Zolkos et al., 2013). Biomass assessments using optical sensors are typically conducted at spatial resolutions of 5–30 m (e.g. LISS series, Landsat series, Sentinel-2), whereas LiDAR can support sub-metre-level assessments, offering far greater precision (Zolkos et al., 2013). Even the densest tropical forests, such as those in the Amazon, Congo, and Asia-Pacific, have been successfully modelled using unmanned aerial LiDAR systems, revealing novel insights into canopy architecture, vertical stratification, and forest structure at local and regional scales (Dubaya et al., 2010; May et al., 2024; Migolet et al., 2022).

Beyond biomass, remote sensing has also revolutionised ecosystem productivity assessments by combining thermal remote sensing products with field measurements modeling energy fluxes and water cycling to estimate ecosystem productivity in forested landscapes (Coops, 2015; Maselli et al., 2009). For example, the Moderate Resolution Imaging Spectroradiometer (MODIS) provides critical insights into vegetation productivity and phenology through freely available data products, including gross/net primary productivity (GPP/NPP), leaf senescence patterns, land surface temperature, water use efficiency, evapotranspiration rates, and forest disturbances caused by nutrient stress, insect outbreaks, or fire regimes (Frazier et al., 2013; Ishtiaque et al., 2016). The ability to characterise energy fluxes and hydrological responses (evapotranspiration) at moderate spatial resolutions (typically hundreds of metres) makes MODIS particularly valuable for large-scale ecosystem monitoring (Turner et al., 2006).

Along with improved accessibility to datasets, hyperspectral remote sensing is another recent frontier in remote sensing of forests. It capitalises on the unique spectral reflectance of materials across hundreds to thousands of narrow contiguous spectral bands (as opposed to a few dozen spectral bands in most multispectral optical sensors) to discern materials with very similar spectral signatures (e.g. different forest types and species). The differentiation in spectral response enabled by several hundreds of bands allows detailed insights into plant phenology, nutrient stress, species distribution modelling, plant pathology, and invasive species (Upadhyay & Kumar, 2018). It has revealed a fresh understanding of plant physiology through biochemical changes in chlorophyll pigments, nitrogen content, and water use with a level of detail that was previously impossible (Adão et al., 2017).

Current and upcoming 'state-of-the-art' missions such as NASA's ECOSTRESS and the NASA-ISRO SAR (NISAR) missions continue to push the boundaries of forest monitoring by providing new datasets on evapotranspiration, heat fluxes, ecosystem productivity, water use efficiency, vegetation undergrowth, and leaf/stem structure at much finer spatial resolutions and with greater repeat coverage (Fisher et al., 2020; Kellogg et al., 2020). With ongoing advancements in sensor technology, data analytics, and artificial intelligence, remote sensing products are poised to greatly improve biogeophysical models and further elevate our understanding of forest-climate interactions.

14.2.2 The Evolution of Climate-Forest Models

Studies have largely inferred ecological influence on the climate through paired climate simulations using control and experimental setups wherein land use dynamics are represented using numerical parameterisations (Bonan, 2015). Early approaches to understanding the coupling between land use and climate focused on simulating energy fluxes and their influence on atmospheric composition (Edwards, 2011). However, these studies simplified hydrological components within their models (e.g. latent heat derived from soil water holding capacities) and did not explicitly represent vegetative functions (such as primary productivity, evapotranspiration, and net ecosystem exchange) of land use/land cover influencing climate (Sellers et al., 1997). They were succeeded by models that began incorporating vegetation-influenced radiative fluxes, biophysical processes, and interactions within and above canopies by factoring in crucial ecohydrological processes such as evapotranspiration (Sellers et al., 1997). The next generation of vegetation-climate models integrated biochemical and biophysical processes such as photosynthesis and vascular conduction in plants and modelled their influence on evapotranspiration (Bonan, 2008). With the evolution of synoptic coverage of the Earth through various satellite missions, models were developed to simulate vegetation growth and productivity by combining and scaling photosynthetic processes using remotely sensed nutrient stress, moisture availability, and photosynthetically active radiation (Bonan, 2015; Levis, 2010). These models demonstrated vegetation-influenced changes in local and regional climate through atmospheric fluxes, albedo, and evapotranspiration with a reliable degree of confidence; however, they still employed a static representation of the vegetation (Bonan, 2015). The latest models are now an ensemble of biogeophysical simulations, specifically designed to represent atmospheric and biosphere processes with dynamic coupling, wherein climatic processes influence ecological systems, while simulated ecosystem functions give feedback to the climate (Blanco & Lo, 2023).

Modern advancements in remote sensing datasets, analytical techniques and computation power have led to the rise of Dynamic Vegetation Growth Models (DGVMs), which are the most comprehensive models coupling vegetation growth dynamics and ecosystem functions with global climate models (Quillet et al., 2010). The vegetation dynamics comprise continuous processes such as photosynthesis, respiration, surface-energy fluxes, and cycling of major nutrients such as carbon, nitrogen, and phosphorus. Their strength lies in their simulation of various plant functional types with differentiated growth patterns through external forcings from climate and soil parameters (Foley et al., 2000). These transient changes in plant growth are then used to inform global climate models, thus dynamically representing land use changes in climate simulations (Peng, 2000). DGVMs incorporate seasonality and cyclicity of vegetation growth and have specifically designed soil-vegetation-atmosphere interaction modules that allow for a realistic transfer of vegetation parameters such as LAI (leaf area index), rooting depth, water, momentum, and energy fluxes to the climate module. At the same time, variables such as soil moisture, vegetation albedo, and canopy temperature are transferred to the vegetation modules to inform growth and productivity. The differing time scales of these processes (sub-daily and daily physiological fluxes versus seasonal productivity fluxes) are then accommodated and appropriately parameterised within climate modules of DGVMs (Scheiter et al., 2013).

14.2.3 Current Limitations of DGVMs

Despite the generational advancements in the theoretical basis of DGVMs and computational power, they still have had certain limitations over the years. Older generation DGVMs modelled surface-energy and mass fluxes (e.g. water balance) asynchronously. This meant that several iterations of simulations were required to reach model equilibrium, and therefore, the calculation of these fluxes could sometimes violate the laws of conservation of mass and energy (Betts et al., 2000). DGVMs also employ certain bioclimatic thresholds as triggers/stops to simulate vegetation growth for different plant functional types. However, they do not account for ecological processes that dynamically evolve with changing climate scenarios, such as early onset of leaf senescence in dry sub-tropics due to advancing summers (Arora & Boer, 2005). Furthermore, models simplify individual and community-based interactions such as competition, species-specific responses, mortality, regeneration, and recruitment, by aggregating physiological responses at regional and global scales (Kucharik et al., 2006; Moorcroft, 2006). This problem has been tackled recently by replacing plant functional types with individual-centric ecosystem demography models. Weaknesses in modelling competition between individuals and communities extend to the fact that heterogeneity among individuals of a species is ignored, and subtle interactions such as community co-dominance are not represented accurately. This is of particular importance as empirical observations report greater community resilience to climate change with increasing ecosystem heterogeneity (Purves & Pacala, 2008).

Another present limitation in climate-forest modelling is nutrient cycling across nutrient pools and transfer between sub-pools (e.g. carbon sub-pools), often being greatly simplified (Schröter et al., 2004). Competing nutrient processes, such as nitrogen becoming a limiting factor for carbon uptake in climate change scenarios, are also difficult to model realistically (Luo et al., 2006). Other simplifications, such as the representation of respiration as a fixed ratio of photosynthesis or linearly scaled estimates from productivity, must also be deliberated carefully when inferring results from climate-vegetation models (DeLucia et al., 2007). Landscape changes are often limited to land use and land cover change scenarios, ignoring important disturbances such as grazing, fire, insect pest disease, and topographic influences (Gruber & Galloway, 2008; Kloster et al., 2010; Kurz et al., 2008). Finally, there are also difficulties in including processes that are not fully understood yet, such as carbon fluxes in wetlands (Lafleur, 2009).

14.2.4 Inferences from Climate-Vegetation Models

Despite these limitations, DGVMs and other coupled vegetation-climate models have revealed several insights and broadened our understanding of the interplay between ecological processes and the climate. Most modelling exercises concur over a positive relationship between tropical forest cover/health and precipitation due to higher rates of evapotranspiration and reduced surface air temperature compared to agricultural and grazing lands (Leite-Filho et al., 2021). In undisturbed states, not only do tropical forests serve as carbon sinks, but they also have a net cooling effect through evapotranspiration, especially during the dry seasons (Bala et al., 2007). Under changing climate scenarios, however, both models and empirical studies depict a net release of carbon from the tropics during drier and warmer El Nino years (Hashimoto et al., 2004). Simulations of energy-flux-based models also indicate positive feedback between drier/warmer years and carbon release from the tropics, suggesting changing climate would impair the cooling effect provided by tropical forests and, in some scenarios, could also initiate forest dieback (Betts et al., 2004).

Boreal forests, too, significantly influence climate, particularly global carbon balance and mean temperatures, with models indicating increased albedo-based warming under decreasing forest cover with reasonable confidence (Snyder et al., 2004). While degradation and fire disturbances in boreal forests may lead to reduced annual net radiation and sensible heat fluxes in the short to midterm, in the long term, this is projected to result in increased albedo (Randerson et al., 2006). The resultant climate forcings could be of great concern as boreal forests store a large proportion of carbon in their permafrost, soils, and wetlands (Sabine et al., 2004).

For temperate forests, models indicate greater cooling and decreased surface air temperature due to evapotranspiration when compared to agriculture and pastureland, although the differences are less pronounced (Liu et al., 2005). Temperate forests also depict better resilience to droughts as they retain moisture better due to greater rooting depths. Models unequivocally depict that with greater disturbances, these forests would also become net carbon emitters rather than sinks; however, their impact on temperature is uncertain (Albani et al., 2006). This is due to mixed effects of competing processes, such as reduced albedo during winter and summer evapotranspiration which could further be impacted by deforestation-induced higher albedo and reduced evapotranspiration (Bala et al., 2007).

In summary, vegetation-climate models depict overall varying positive feedback between vegetation-regulated carbon cycles and global warming-induced increases in temperatures (Bonan, 2008). Uncertainties can be attributed to differentiated responses of plant functional types to increasing carbon concentrations. With rising carbon concentrations, NPP rates are estimated to increase through enhanced photosynthesis (CO_2 fertilisation). However, these effects are not projected to be long-term when considering limiting factors such as nitrogen availability, signifying reduced carbon storage in the long term (Huntingford et al., 2022). Climate models are fundamental to informing policies and actionable change for mitigating climate change (Nabuurs et al., 2022). Uncertainties and certain limitations of climate-vegetation coupled models, such as realistic representation of biosphere processes, are still challenging. Despite this, models remain pivotal in developing and scaling up projected responses of biophysical systems to changing climate, as well as drafting climate scenarios under changing vegetation and land use dynamics. There is a need for a more comprehensive monitoring of ecosystem changes through spatially distributed ecological flux towers; advancements in spatial resolution and temporal coverage of remote sensing datasets; improved parameterisation of models through experiment and observation-based data; more robust recording of atmospheric carbon; and syntheses of long-term ecological datasets (Reichstein et al., 2007). Enhancement of ground monitoring and resultant datasets that better represent ecological heterogeneity at different successional stages across varying topographic and bioclimatic regions are projected to improve the confidence of model results significantly. Inferences from previous models have already outlined certain policy directives, such as emphasising the importance of tropical forests over greater uncertainties around temperate forest-climate interactions, but these need to be revised continuously (Bala et al., 2007). With improved long-term datasets, the inclusion of biogeophysical processes beyond albedo, evapotranspiration, and carbon cycling, increasing the scope of modelling to other GHGs and aerosols and their interactions with forest ecosystems will be significant steps in evolving our understanding of the climate-forest nexus.

14.3 CLIMATE MITIGATION THROUGH FORESTS: CURRENT OUTLOOK AND FUTURE PROSPECTS

The Intergovernmental Panel on Climate Change's (IPCC's) sixth assessment report indicates that the Agriculture, Forestry and Other Land Use (AFOLU) sector contributes over 23% of total GHG emissions globally (IPCC, 2023). Their latest assessments also report that deforestation and forest degradation-related emissions alone constitute ~45% of total AFOLU emissions; however, this is projected to gradually reduce with improving forest conservation measures. Drivers of GHG emissions from forests can be direct, such as conversion of natural ecosystems for agriculture, infrastructure development, and urbanisation; industrial resource extraction; climate-induced or anthropogenic disturbances in fire regimes; timber and fuelwood extraction; livestock rearing and management; cash crop cultivation and increasing use of high yielding varieties and associated fertiliser use; along with indirect drivers emanating from associated governance, cultural, economic, technological, and demographic factors (FAO, 2020).

Protection of forest carbon sinks and generation of carbon offsets within the AFOLU sector have already reduced total GHG emissions by 1–2%. However, research suggests that this can be comprehensively upscaled to address 25–30% of the total emission reductions needed to limit global mean temperature rise to 1.5–2°C delineated in IPCC's low-cost pathway for climate change mitigation (IPCC, 2023; Psistaki et al., 2024). IPCC's fifth and sixth assessment reports estimate that the protection, restoration, and improved management of forests and related ecosystems, when scaled appropriately, can not only address the mitigation targets mentioned above but also address it in a cost-effective manner. Roe et al. (2019) have reviewed annual mitigation potentials of protecting, restoring, and managing forests using sectoral and integrated financial models. They assess that input costs of less than 20 USD, 50 USD, and 100 USD can help mitigate 2.4–2.9 tCO_{2eq}, 3.1–3.3 tCO_{2eq}, and 4.2–7.3 tCO_{2eq} annually, respectively. Furthermore, the forest sector not only has substantial scope to ameliorate global warming but also provides numerous co-benefits that other mitigation strategies such as technology transfers do not provide, for example, improved agricultural productivity, food security, sustained timber supply,

NTFPs, fodder, and other indirect ecosystem services such as sustained water availability, improved air quality, and enhanced soil productivity (Karlsson et al., 2020).

14.3.1 Forest-Based Climate Change Mitigation Measures

There are several ways in which forests and related ecosystems can be effectively incorporated into climate change mitigation. The most direct measure is to reduce deforestation and forest degradation, conserving existing carbon pools in vegetation and soils while enhancing pools in degraded ecosystems. This would require a slew of measures, from legislating laws and policies for protected areas, better enforcement of existing laws, improving forest-related governance measures, developing sustainable land tenure policies and practices, encouraging community-based forest management, and implementing forest certification (FC) (Barber et al., 2014; Smith et al., 2020). Protecting existing carbon sinks provides a range of co-benefits, such as biodiversity conservation, air and water filtration, and improved water and nutrient cycling, at a cost substantially lower than afforestation and reforestation. Reducing deforestation and degradation can help achieve an emissions reduction of up to 4.5 Gt CO_{2eq} per year at the cost of less than 100 USD per tCO_{2eq} per year, and thus, it has been a pronounced inclusion in several Nationally Determined Contributions (NDCs) under the Paris Agreement (Seddon et al., 2019).

Afforestation, reforestation, and restoration (ARR) activities are active ways to convert land use/land cover to forest, with afforestation taking place on land that historically did not have forest cover, while reforestation refers to activities that occur on land that previously had forest cover. However, an often-overlooked component is restoration, which aims to improve the ecological structure and function of existing forests and vegetative cover through activities that support natural regeneration and ecological recovery. All three activities have a significant mitigation potential; however, reforestation and restoration are preferable as they don't alter the landscape in a major way and are aimed at reversing changes which are usually anthropogenic (Jones et al., 2022). Scientifically and ecologically sound restoration can improve climate resilience, enhance ecosystem services such as water provisioning, microclimate regulation, soil erosion reduction, increase timber, and non-timber forest production as well as provide alternate income and livelihood options. While afforestation potentially offers the same co-benefits, there is a significant chance for adverse effects due to the planting of inappropriate and non-native species, misidentification of suitable areas, and creation of monoculture stands. However, when executed in suitably identified areas, afforestation has immense potential to reduce runoff and soil erosion, enhance soil water retention and water yield, and improve soil productivity, thereby reducing land degradation (Ellison et al., 2017).

Improved Forest Management (IFM) is another key approach to addressing climate change, and it involves scientifically established practices that augment existing forest management. This includes but is not limited to reduced impact harvesting practices, sustainable silvicultural rotation, mixed stand management, continuous cover forestry, etc. A crucial advantage of IFM practices is the boosted monetary incentive it can provide by generating high-quality carbon credits (Ekholm, 2020). Lastly, although fire regimes are a naturally occurring phenomenon, anthropogenic fires and fire events exacerbated by global warming imply that forest fires have become a key source of GHG emissions. Decadal studies project that improved fire management through the prevention, early detection, control, management, and restriction of forest fires can sequester an additional 0.48 Gt CO_{2eq} per year while also reducing air pollution, preventing habitat loss, soil erosion, and land degradation, as well as improving forage and fodder quality (Arora & Melton, 2018).

Recognising the plethora of forest-derived climate change adaptation and mitigation services, forest governance and related instruments have evolved across the globe to build climate resilience in two broad ways – regulatory approaches and incentive-based mechanisms.

14.3.2 Approaches for Forest-Based Climate Change Mitigation

Several national laws and policies, along with international instruments, have been designed to address mitigation and adaptation towards climate change following the United Nations Framework Convention on Climate Change (UNFCCC). Regulatory approaches largely consist of national and subnational jurisdictions with appropriate stakeholder engagement to protect forests. Economic incentives refer to financial and market-based mechanisms through which climate change mitigation and adaptation can be addressed. The third pivotal Conference of Parties (CoPs) to UNFCCC at Kyoto in 1997 led to the inception of the Clean Development Mechanism (CDM), allowing countries and private entities to meet their emission reduction targets by trading carbon offsets through various financial mechanisms (Böhringer, 2003). CDM gave rise to compliance and voluntary carbon markets, leading to the generation of a significant amount of carbon offsets and credits (IPCC, 2023). This was supplemented in CoP 11 and 13 with the introduction of other incentive-based mitigation programmes such as REDD+ and the role of conservation, sustainable management of forests, and enhancement of forest carbon stocks in developing countries.

14.3.2.1 Forest Protection/Management Regulations

The most frequent and direct manner in which governments and international bodies leverage forests for climate change is by legislating protected status for natural forests, vegetative cover, and habitats, thereby protecting existing forest ecosystem services, including their roles as carbon sinks (FAO, 2020). This has historically been the most prevalent approach and has a multitude of co-outcomes, such as addressing habitat degradation, biodiversity loss, and ecosystem service preservation. According to the World Database on Protected Areas maintained and updated by IUCN and UNEP, 21% of the total forest areas are currently under protected status. This includes 39% of primary old-growth forests, which are the largest terrestrial non-soil carbon sinks having remarkable capacity for carbon storage and 46% of globally intact forests, i.e. forest tracts that have not experienced any anthropogenic activities yet (Protected Forest Indicator, 2024). The importance of legal protection for forests is underlined by international agreements such as the Kunming-Montreal Global Biodiversity Framework's 23 action-oriented global targets for urgent action over the decade for addressing biodiversity loss. This agreement includes an ambitious 30 × 30 target, which aims to bring 30% of all terrestrial and marine areas under protected area status by 2030 (Convention on Biological Diversity, 2022).

Regulations to avoid land use conversion of forests have largely succeeded in slowing down deforestation-linked emissions in biodiversity hotspots such as Amazonian Brazil (Nepstad et al., 2014). Bebber and Butt (2017) have assessed the impact of protected status on carbon emissions and report that legal protection reduced emissions by over 29% during 2000–2012 in tropical areas. In particular, OECD (Organisation for Economic Co-operation and Development) countries have successfully enforced strong statutory regulations that protect endangered species and habitats, promote sustainable harvest techniques, compensate deforestation with planting, conserve natural species composition, and encourage nature-based solutions. However, this has been a challenge in developing and least-developed countries, with studies from the Amazon suggesting a reversal in trends of deforestation and emissions reduction from forest areas due to inconsistent forest policies borne out of socio-economic priorities and weak governance (Azevedo et al., 2017; Pelicice, 2019).

14.3.2.2 Community-Based Forest Management

Assigning protection and management responsibilities to stakeholder communities and indigenous peoples is increasingly gaining traction to protect forests whilst empowering marginalised livelihoods (Burivalova et al., 2019; Yadava & Sinha, 2020). By enabling communities to become active custodians of forests, not only can their meticulously developed traditional and indigenous knowledge be used for sustainable forest management (SFM) but also provide them with property and/or land tenure rights over forests. Additionally, if the forest protection practices can be streamlined to align with the standards of carbon markets and registries, communities can continue to reap forest resource benefits through participatory management while also getting economic benefits through carbon offsets (Blackman & Veit, 2018; Duchelle et al., 2018). This would go a considerable way in alleviating socio-economic stress amongst marginalised and highly vulnerable forest-dependent populations that would otherwise be vulnerable to land use conversion for agriculture, mining, urban expansion, etc. While IPCC reports several case studies of emissions reduction through community-based forest management, particularly in the tropics, there is a need to formally quantify the economic benefits generated at regional and global scales to make a stronger evidence-based case for improving community engagement in forest management (Nabuurs et al., 2022).

14.3.2.3 Sustainable Forest Management and Forest Certification

An assessment from the United Nations Forum on Forests estimates that over a quarter of the world's population, a substantial proportion of which are bracketed under extreme poverty, depend on forests for their livelihoods, income, and subsistence (UNDESA-UNFF, 2021). FAO's State of the Forest Reports (FAO, 2020, 2024) have outlined several crucial socio-economic dependencies on forests over the years, such as their contribution to freshwater (75% of freshwater is sourced from forested watersheds) and revenue generation (1.15 billion ha of forests worldwide are managed solely for the production of timber/non-timber goods). Meanwhile, the assessments also report that pressures on forests for these services have increased while the extent and health of forests have gradually deteriorated, for example, while the absolute areas under production forests have largely remained the same in the last three decades, global wood removals have increased by 385 million cubic metres between 2010 and 2018 (FAO, 2022). Furthermore, with rising global threats such as climate change, biodiversity loss, and environmental degradation, there is an expected shift towards decarbonisation and circular economies. This would have cascading implications for forests as increased pricing and demands for raw materials sourced from mineral, oil-based derivatives, and metals are projected to be partially offset by wood and NTFP-based substitutions, particularly for the raw materials in the construction sector. Such trends are bound to put tremendous strain on forest resources, and hence, SFM has rapidly gained importance over the last decades (International Tropical Timber Organization (ITTO), 2016).

SFM refers to the responsible management of forested areas, resources, ecosystems, and services in order to ensure their qualitative and quantitative availability to all relevant stakeholders for future generations across a host of environmental, social, and economic values (World Commission on Environment and Development, 1987). It comprises policies and regulations that support forest managers to implement sustainable management principles through forest inventories, certification, management plans, monitoring, and stakeholder engagement.

Since its inception, progress towards SFM has been evaluated in two broad ways – through protocols such as NFI and/or as certifications for goods and services extracted from forests. Protocols have evolved as part of several regional and international processes, such as the Montreal Process (MP), Forest Europe (FE), International Tropical Timber Organisation (ITTO), and FAO, with the most popular approach being the Criteria and Indicator (C&I) framework for SFM (ITTO, 2016). The approach employs NFIs using a wide range of indicators under various ecological, economic, and social criteria backed by national/subnational regulatory, jurisdictional, and institutional policies. While the C&Is may vary across the frameworks, they broadly pertain to aspects of forest extent; forest ecosystem health and vitality; forest productivity; protective functions of forests (soil and water); socio-economic linkages; forest biodiversity and governance, legal and institutional mechanisms for enabling SFM (Forest Europe, 2015; ITTO, 2016; Montreal Process Working Group, 2015). Along with SFM, FC co-evolved in the 1990s as a mechanism of independently certifying the trade and quality of products sourced from sustainably managed forests. It emerged in the form of voluntary certification schemes led by organisations such as the Forest Stewardship Council (FSC) and the Programme for the Endorsement of Forest Certification (PEFC), the two most widely subscribed to certification standards globally. Since their conceptualisation, SFM and FC have undergone several technical, regulatory, governance, and implementation-based revisions as our understanding of scientific management of forests, forest dependencies, and their challenges has evolved. Subsequently, as SFM frameworks began to be adopted globally, market influences over certification-based organisations became apparent, with schemes such as the Sustainable Forest Initiative (SFI) focusing on institutional aspects and FSC on socio-cultural aspects. Meanwhile, SFM protocols such as the ITTO, FE, and MP have converged along ecological C&Is due to a globally increasing impetus on ecosystem protection (Kadam et al., 2021; Linser & Wolfslehner, 2022; Linser et al., 2018; Rodriguez Franco & Conje, 2023).

A key area of progress in SFM has been in the measurement of forest resource stock under NFIs with the number of reporting countries having doubled in the last decade (FAO, 2022; MacDicken et al., 2015). While it is pertinent to note that over three-quarters of the NFI, reporting countries use a C&I-based framework, they largely belong to high- and middle-income countries, with low-income countries unable to fund such comprehensive monitoring (Linser & Wolfslehner, 2022; MacDicken et al., 2015; Managi et al., 2019). Similarly, while the area under certified forests has risen tremendously over the last 30 years, the bulk of these forests are restricted to North America, Europe, and high-income countries in Asia. Furthermore, these schemes seem to majorly cover boreal and temperate forests, with tropical and particularly sub-tropical forests not being certified in proportion to their extent (FAO, 2020).

Overall, there has been a radical shift in the sectoral scope of forest management plans from an early emphasis on production to a more holistic approach at present encompassing conservation and regulatory facets. However, there is still a great need to strengthen their frequency of monitoring and reporting, with lower-income countries struggling to finance monitoring at scheduled periodicities and frequencies (Kadam et al., 2021; MacDicken et al., 2015). The frequency and scope of SFM monitoring under different criteria, robust measurement of indicators, and their implementation efficacy are undergoing continuous transformations (Kadam et al., 2021; Linser & Wolfslehner, 2022; Linser et al., 2018; MacDicken et al., 2015). While the presence and extensive coverage of various SFM protocols & FC schemes is a positive indicator of global commitment towards SFM, measurement of contextual facets of SFM and monitoring of qualitative indicators remains a greatly debated subject (Sabogal et al., 2013; Siry et al., 2018; Teketay et al., 2010). Even in countries with C&I frameworks, indicators requiring robust technical enquiries and financial capacities for certain aspects, such as above/below ground carbon, impact evaluation of reduced harvest procedures, issues of gender inclusion and resource ownership, and stakeholder participation need to be improved (ITTO, 2016; Teketay et al., 2010). At the same time, the integration of remote sensing techniques and improved socio-ecological datasets are projected to enhance forest monitoring (Bosela et al., 2024). Furthermore, climate policies and biodiversity frameworks helped streamline SFM and FC frameworks in the past, an amalgamation of SFM with international forest schemes such as REDD+, carbon markets as well as corporate sustainability commitments are some expected trends in the future.

14.3.3 Incentive-Based Approaches for Climate Mitigation

14.3.3.1 Carbon Markets

The idea behind emissions trading and carbon markets was inspired by reduction-based targets to reverse ozone layer depletion following the Montreal Protocol. Formalised at the third CoP to UNFCCC in Kyoto in 1997, ratifying signatories to the Kyoto Protocol committed to certain emission targets against a pre-agreed baseline (Newell et al., 2014). Verification and validation systems were set up to estimate these reductions while allowing countries to meet targets through trade upon failure. These markets facilitated the trading of carbon credits, with each credit representing 1 tCO_{2eq} that has either been prevented from being emitted or sequestered through activities such as afforestation, reforestation, and/or restoration (Böhringer, 2003).

There are two types of carbon markets – compliance-based and voluntary carbon markets. The former comprises agencies that govern the carbon trading that occurs amongst both private entities and governments to meet their emission reduction targets set out by legally binding or self-declared goals such as the Kyoto targets, Paris Accords, or those self-declared in their NDCs to the UNFCCC. The voluntary carbon market exists for carbon trading between private entities which are not legally obligated to meet emission reduction targets but do so of their own volition (Michaelowa et al., 2019). As is the case for any trading in the market, carbon credits need to be authenticated in terms of their estimation, quality, and pricing; therefore, governing bodies and standards were required to regulate this market. Subsequently, several carbon registries have developed detailed scientific methodologies to quantify reductions, ensuring that the offsets generated are real, additional, verifiable, and permanent. These qualities are crucial in characterising the quality of generated offsets and grading their discernible impact on reducing emissions. Additionality and permanence are particularly two very important criteria, the former qualifying that offsets generated under a carbon project would not have occurred in a business-as-usual scenario. Permanence refers to the principle that offsets generated would be permanently stored in sinks (for periods significant enough that they don't contribute to global warming) and not be readily emitted back into the atmosphere. Several carbon standards and registries are now in place, such as the European Union Emissions Trading Scheme (compliance), Voluntary Carbon Standards by Verra (voluntary), and the Gold Standard by WWF (voluntary). They ensure that projects and activities registered, evaluated, and verified by them generate offsets that adhere to the principles of additionality and permanence. These were developed after much deliberation between the scientific community and project developers while being continuously updated to keep up with the latest advancements in climate science (Lovell, 2010).

A sizeable proportion of carbon credits or offsets generated from both compliance and voluntary carbon markets stem from AFOLU activities. A review under IPCC AR6 of carbon offsetting projects between 2010 and 2019 indicates that of a total of 7.9 Gt CO_{2eq} offsets generated by the AFOLU sector during the period, about 80% originate from the forest-based activities, which accounts for 1.2% of total global emissions and 5.5% of the emissions from the AFOLU sector (Nabuurs et al., 2022). The largest share of these offsets came from REDD+ activities (particularly in the Amazon) followed by compliance markets and then by voluntary markets. About 248.9 Mt CO_{2eq} came from regulatory carbon market programmes such as California's Cap-and-Trade compliance carbon programme, Australia's Emissions Reductions Funds Registry, and New Zealand's Environmental Protection Authority (Nabuurs et al., 2022). Hamrick and Gallant (2018) reviewed the offsets generated by the voluntary markets between 2008 and 2018, reporting a total of 95.3 Mt CO_{2eq} offsets largely from Latin America, Africa, and Asia and a smaller proportion from North America with very small/negligible contribution from the European voluntary market.

Despite the substantial offsets generated through forest-based activities, there have been some criticisms of AFOLU-based carbon projects surrounding uncertainties in their estimations as well as a debate around their net benefits (Krug, 2018). Projects are often unable to clearly demonstrate additionality due to a lack of standardisation between different carbon registries and their methodologies (Mason &

Plantinga, 2013). This issue is further exacerbated by the increasing number of carbon standards, particularly in the voluntary markets. Despite the tremendous mitigation potential of forests, financing for carbon offsetting through forest activities has been well short of the figure required (USD 0.7 billion raised as opposed to USD 400 billion required based on IPCC AR6) to limit temperature rise to 2°C before 2100 (Nabuurs et al., 2022). Another barrier to upscaling forest-based carbon offsetting projects is the implication it could have for marginalised people by limiting agricultural expansion and driving up food prices (FAO, 2020). There are also institutional barriers around a lack of transparency, poor governance (particularly surrounding land tenure and ownership), and a lack of technical capacity in developing countries where these projects are most often feasible (Zelli et al., 2017). Lastly, monitoring, reporting, and verification are still challenging when it comes to offset estimation. Even with the advancements in satellite datasets and forest cover data products such as those curated by Hansen et al. (2013), techniques are still evolving when it comes to monitoring and assessing forest degradation and restoration. However, with the evolution of LiDAR and SAR datasets, corresponding improvements in their image processing techniques, and incorporation of artificial intelligence and neural networks in data processing, along with their growing spatio-temporal coverage, these barriers are projected to gradually dissolve (Cunningham et al., 2013; Solberg et al., 2015; Vatandaşlar & Abdikan, 2022).

The inclusion of forest-based targets in most NDCs signifies the overwhelming consensus on their importance in meeting emission reduction targets. The carbon market is still an evolving space, with ever-improving assessment methodologies and reporting formats (Michaelowa et al., 2019; Newell et al., 2014). It has significant potential in incentivising emissions reductions by channelling public and private finance towards mitigation efforts. This is exemplified by the fact that emerging economies such as Brazil, India, and Türkiye are in the process of developing their national carbon registries and markets (Singh et al., 2024; World Bank, 2024). The World Bank (World Bank, 2024) reports that carbon taxes and emissions trading systems cover approximately 24% of global GHG emissions, and while this proportion has been relatively stable over the past few years, revenues from carbon markets reached USD 100 billion in 2023 illustrating their growing financial impact. However, challenges persist around the standardisation of different carbon registries and integrity concerns, particularly in the voluntary carbon markets. With strengthening market credibility, transparency, and advancements in the robustness of monitoring, verification, and reporting of carbon credits, carbon markets continue to remain at the forefront of nature-based solutions in achieving our climate goals.

14.3.3.2 Potential and Challenges in REDD+

Attributing monetary value to forest ecosystem services has long been outlined by forest experts to have tremendous potential in incentivising their conservation and sustainable management. Although payment for ecosystem services (PES) was conceptualised as early as the 1970s, the theoretical basis for it has undergone several iterations over the decades (Gómez-Baggethun et al., 2010). With the advent of financial mechanisms post-Kyoto for emissions reductions, REDD+ emerged as a leading functional PES mechanism that engaged relevant stakeholders in conserving forests and therein-derived services. The concept began with assigning economic value to emissions reduced through forest conservation and protection by preventing deforestation and forest degradation. Subsequently, it expanded over time to include enhancing forest carbon stocks, promoting science-based SFM and increasing community engagement (Parrotta et al., 2022). At present, REDD+ is implemented on a country, region, or project basis wherein the conservation of forests and prevention of their degradation is encouraged through remuneration based on the potentially avoided emissions in a business-as-usual scenario (Pistorius, 2012). It is a results-based scheme that provides financial benefits for meeting certain targets such as a reduced rate of deforestation, avoided deforestation, and avoided degradation (Angelsen et al., 2018).

To date, REDD+ is the largest PES mechanism generating the most offsets from the AFOLU sector. According to IPCC's evaluation period for AR6, of a total of 7.9 Gt CO_{2eq} offsets generated between 2005 and 2018, over 7.5 Gt CO_{2eq} offsets were generated from REDD+ activities alone. These have mostly been from the Amazon (>90% from Brazil, Guyana, and Argentina) with the rest from other parts of Latin America, Asia, Africa, and Oceania (Houghton & Nassikas, 2018; Nabuurs et al., 2022). With the development of robust monitoring, inventorying, and verifying methodologies, REDD+ now covers over 50 countries, with payments being received by them for conserving forests (Parrotta et al., 2022). The effectiveness of REDD+ has also been documented with studies underscoring the influence these projects had in significantly slowing down deforestation in Brazil and Guyana (Roopsind et al., 2019; Simonet et al., 2019).

The mitigation potential of REDD+ activities is quite substantial, particularly considering the current state of forests and global carbon fluxes. Roe et al. (2021) have estimated the forest and other terrestrial ecosystems to account for 3.5–6.6 Gt CO_{2eq} of the total mitigation potential of land-based activities. REDD+ can directly address 3–5.6 Gt CO_{2eq} of this total if all REDD+ projects and activities in developing countries are carried out and verified (Parrotta et al., 2022). These figures are determined from the forest cover and deforestation rate reference levels submitted by 54 countries to UNFCCC and are quite noteworthy as these account for over 70% of global deforestation (FAO, 2020, 2024).

The optimism behind the future of REDD+ in meeting its potential for mitigation stems from much-improved forest monitoring and incentivisation at regional, national, and subnational levels. Remote sensing-based monitoring and mapping of forest carbon stocks has advanced tremendously due to radical improvements in Earth observation technology, satellite image acquisition, and accessibility to global remote sensing data (Hubau et al., 2020). This is complemented by an exponential increase in computational capacities with cloud computing, machine learning, and artificial intelligence. Improvements in methodologies for measuring deforestation have also contributed to this, such as a shift from pixel-based deforestation estimation to sample-based extrapolations due to larger errors being attributed to post-classification-based change detection techniques (Olofsson et al., 2014; Penman & Hoekman, 2014). Our ability to quantify forest resources and ecosystem functions is rapidly progressing with this being exemplified by the rising number of independent global forest and tree cover datasets such as the Global Forest Change datasets and the Tropical Moist Forest data (Hansen et al., 2013; Vancutsem et al., 2020). This is further mirrored by the fact that countries are using satellite remote sensing ever more than before for natural resource management, including forest mapping and monitoring (55 countries in 2005 to 99 in 2020; FAO, 2020).

While improvements in ecosystem monitoring and reduced rates of deforestation have been the impetus behind promoting REDD+, the programme has also seen its fair share of challenges. Uncertainties in carbon estimation, exclusion of carbon pools beyond aboveground biomass (soils, deadwood, and litter), omission of gases apart from CO_2, and measurement of degradation are often argued when debating the credibility of REDD+ offsets (Morita & Matsumoto, 2023). For example, there is large variability in emissions assessments emanating from degradation which range between 25 and 65% of total forest-related emissions (Baccini et al., 2017; FAO, 2020; Pearson et al., 2017). Reference levels are crucial in monitoring results in REDD+ but most countries do not submit reference levels pertaining to degradation or PLUS activities. The former is challenging due to the need for a sufficiently dense time series of high-resolution satellite imagery over a long period and the latter is due to complications in stocking assessments beyond aboveground carbon, e.g., soil carbon estimations require intensive field sample inventories (Lee et al., 2018). Impact evaluation studies of REDD+ payments from parts of the world outside Amazonia are scarce and there is still ambiguity in determining their influence in maintaining and enhancing the quality of forest structure, biodiversity, and carbon stocks rather than a primary focus on forest extent. This is reflected in reports on REDD+ payments with over 95% of the projects reporting interventions restricted to limiting deforestation and degradation and only 5% reporting results for enhancing stocks (Parrotta et al., 2022). There are also a few caveats in existing REDD+ reporting. Degradation in peat forests and soils such as those in the Congo and the Amazon has been observed to result in significant emissions; hence, their exclusion from emissions due to degradation is precarious (Draper et al., 2014; Mitchard, 2018). Finally, uncertainties in estimation are either not reported or have not been comprehensively evaluated, potentially leading to overestimating emissions reductions from REDD+ activities (FAO, 2020; Neeff, 2021; Yanai et al., 2020). Nonetheless, there is little debate surrounding the most discernible impact of REDD+ which is the significant slowing down of deforestation with over 46% of reporting countries declaring a significant decrease in deforestation rates from 2010–2015 to 2015–2020 periods (significantly greater than 16% of non-REDD+ countries reporting the same; FAO, 2020). Regardless of arguments around specific quantities of emissions reductions and coupled uncertainties, REDD+ most definitely has not only contributed to emissions reductions but also empowered relevant stakeholders to take ownership of forested landscapes while demonstrating the sway financial incentives can have on sustaining forest resources.

14.4 CONCLUSION

Forests have coupled responses to shifting climate patterns whilst also having regulatory effects through intricate biogeophysical feedback mechanisms. Anthropogenic climate change has been demonstrated to significantly impact these climate-forest dynamics. The immense potential forests have in mitigating GHG emissions and ameliorating the adverse effects of climate change has become irrefutable with developments in our scientific understanding of these interlinkages. This has been achieved through a rigorous evolution of ensemble time continuous physical process-based models and developments in physical theories and is supported by comprehensive ecological and climate datasets. Empirically supported conservation and management of forest ecosystems have been demonstrated to be financially viable nature-based climate solutions with several co-benefits. Discussions on enhancing climate resilience are increasingly being centred on forests, either through protected status, engagement of stakeholders, and SFM or through incentives-based mechanisms such as the CDM, voluntary carbon markets, and REDD+. While the aforementioned approaches have faced strategic challenges surrounding their efficacy and implementation, their continuous refinements will be imperative in adapting to an increasingly warming planet.

REFERENCES

Adão, T., Hruška, J., Pádua, L., Bessa, J., Peres, E., Morais, R., & Sousa, J. J. (2017). Hyperspectral imaging: A review on UAV-based sensors, data processing and applications for agriculture and forestry. *Remote Sensing*, *9*(11), 1110.

Albani, M., Medvigy, D., Hurtt, G. C., & Moorcroft, P. R. (2006). The contributions of land-use change, CO_2 fertilization, and climate variability to the Eastern US carbon sink. *Global Change Biology*, *12*(12), 2370–2390.

Albrich, K., Rammer, W., Thom, D., & Seidl, R. (2018). Trade-offs between temporal stability and level of forest ecosystem services provisioning under climate change. *Ecological Applications*, *28*(7), 1884–1896.

Angelsen, A., Martius, C., De Sy, V., Duchelle, A. E., Larson, A. M., & Pham, T. T. (2018). *Transforming REDD+: Lessons and new directions*. CIFOR.

Arora, V. K., & Boer, G. J. (2005). A parameterization of leaf phenology for the terrestrial ecosystem component of climate models. *Global Change Biology*, *11*(1), 39–59.

Arora, V. K., & Melton, J. R. (2018). Reduction in global area burned and wildfire emissions since 1930s enhances carbon uptake by land. *Nature Communications*, *9*(1), 1326.

Azevedo, A. A., Rajão, R., Costa, M. A., Stabile, M. C., Macedo, M. N., Dos Reis, T. N., Alencar, A., Soares-Filho, B. S., & Pacheco, R. (2017). Limits of Brazil's Forest Code as a means to end illegal deforestation. *Proceedings of the National Academy of Sciences*, *114*(29), 7653–7658.

Aziz, G., Minallah, N., Saeed, A., Frnda, J., & Khan, W. (2024). Remote sensing based forest cover classification using machine learning. *Scientific Reports*, *14*(1), 69.

Aznar-Sánchez, J. A., Belmonte-Ureña, L. J., López-Serrano, M. J., & Velasco-Muñoz, J. F. (2018). Forest ecosystem services: An analysis of worldwide research. *Forests*, *9*(8), 453.

Baccini, A., Walker, W., Carvalho, L., Farina, M., Sulla-Menashe, D., & Houghton, R. A. (2017). Tropical forests are a net carbon source based on aboveground measurements of gain and loss. *Science*, *358*(6360), 230–234.

Bala, G., Caldeira, K., Wickett, M., Phillips, T. J., Lobell, D. B., Delire, C., & Mirin, A. (2007). Combined climate and carbon-cycle effects of large-scale deforestation. *Proceedings of the National Academy of Sciences*, *104*(16), 6550–6555.

Barber, C. P., Cochrane, M. A., Souza, C. M., Jr, & Laurance, W. F. (2014). Roads, deforestation, and the mitigating effect of protected areas in the Amazon. *Biological Conservation*, *177*, 203–209.

Baumgartner, R. J. (2019). Sustainable development goals and the forest sector—A complex relationship. *Forests*, *10*(2), 152.

Bebber, D. P., & Butt, N. (2017). Tropical protected areas reduced deforestation carbon emissions by one third from 2000–2012. *Scientific Reports*, *7*(1), 14005.

Betts, R. A., Cox, P. M., Collins, M., Harris, P. P., Huntingford, C., & Jones, C. D. (2004). The role of ecosystem-atmosphere interactions in simulated Amazonian precipitation decrease and forest dieback under global climate warming. *Theoretical and Applied Climatology*, *78*, 157–175.

Betts, R. A., Cox, P. M., & Woodward, F. I. (2000). Simulated responses of potential vegetation to doubled-CO_2 climate change and feedbacks on near-surface temperature. *Global Ecology and Biogeography*, *9*(2), 171–180.

Blackman, A., & Veit, P. (2018). Titled Amazon indigenous communities cut forest carbon emissions. *Ecological Economics*, *153*, 56–67.

Blanco, J. A., & Lo, Y. H. (2023). Latest trends in modelling forest ecosystems: New approaches or just new methods?. *Current Forestry Reports*, *9*(4), 219–229.

Böhringer, C. (2003). The Kyoto protocol: A review and perspectives. *Oxford Review of Economic Policy*, *19*(3), 451–466.

Bonan, G. (2015). Ecological climatology: Concepts and applications (3rd ed.). Cambridge University Press. https://doi.org/10.1017/CBO9781107339200.

Bonan, G. B. (2008). Forests and climate change: Forcings, feedbacks, and the climate benefits of forests. *Science*, *320*(5882), 1444–1449.

Bonan, G. B., Pollard, D., & Thompson, S. L. (1992). Effects of boreal forest vegetation on global climate. *Nature*, *359*(6397), 716–718.

Bosela, M., Larocque, G. R., Baycheva, T., Valbuena, R., & Lier, M. (2024). Criteria and indicators of sustainable forest management. In *Ecological forest management handbook* (pp. 356–385). CRC Press.

Boyd, D. S., & Danson, F. M. (2005). Satellite remote sensing of forest resources: Three decades of research development. *Progress in Physical Geography*, *29*(1), 1–26.

Burivalova, Z., Allnutt, T. F., Rademacher, D., Schlemm, A., Wilcove, D. S., & Butler, R. A. (2019). What works in tropical forest conservation, and what does not: Effectiveness of four strategies in terms of environmental, social, and economic outcomes. *Conservation Science and Practice*, *1*(6), e28.

Calders, K., Jonckheere, I., Nightingale, J., & Vastaranta, M. (2020). Remote sensing technology applications in forestry and REDD+. *Forests*, *11*(2), 188.

Canadell, J. G., & Raupach, M. R. (2008). Managing forests for climate change mitigation. *Science*, *320*(5882), 1456–1457.

Chen, S., Woodcock, C. E., Bullock, E. L., Arévalo, P., Torchinava, P., Peng, S., & Olofsson, P. (2021). Monitoring temperate forest degradation on Google Earth Engine using Landsat time series analysis. *Remote Sensing of Environment*, *265*, 112648.

Convention on Biological Diversity. (2022). *Kunming-Montreal Global Biodiversity Framework*. Secretariat of the Convention on Biological Diversity. https://www.cbd.int/article/cop15-final-text-kunming-montreal-gbf

Coops, N. C. (2015). Characterizing forest growth and productivity using remotely sensed data. *Current Forestry Reports*, *1*, 195–205.

Cunningham, K. W., Little, J., & Montgomery, M. N. (2013). Satellite remote sensing & carbon market transparency. *Advances in Forestry Letters*, *2*(3), 37–41.

De Frenne, P., Lenoir, J., Luoto, M., Scheffers, B. R., Zellweger, F., Aalto, J., Ashcroft, M. B., Christiansen, D. M., Decocq, G., De Pauw, K., Govaert, S., Greiser, C., Gril, E., Hampe, A., Jucker, T., Klinges, D. H., Koelemeijer, I. A., Lembrechts, J. J., Marrec, R., … Hylander, K. (2021). Forest microclimates and climate change: Importance, drivers and future research agenda. *Global Change Biology*, *27*(11), 2279–2297.

DeLucia, E. H., Drake, J. E., Thomas, R. B., & Gonzalez-Meler, M. (2007). Forest carbon use efficiency: Is respiration a constant fraction of gross primary production?. *Global Change Biology*, *13*(6), 1157–1167.

De Matteis, A. (2019). Decomposing the anthropogenic causes of climate change. *Environment, Development and Sustainability*, *21*, 165–179.

Draper, F. C., Roucoux, K. H., Lawson, I. T., Mitchard, E. T. A., Honorio Coronado, E. N., Lähteenoja, O., Torres Montenegro, L., Valderrama Sandoval, E., Zaráte, R., & Baker, T. R. (2014). The distribution and amount of carbon in the largest peatland complex in Amazonia. *Environmental Research Letters*, *9*(12), 124017.

Dubaya, R. O., Sheldon, S. L., Clark, D. B., Hofton, M. A., Blair, J. B., Hurtt, G. C., & Chazdon, R. L. (2010). Estimation of tropical forest height and biomass dynamics using lidar remote sensing at La Selva, Costa Rica. *Journal of Geophysical Research: Biogeosciences*, *115*(G2).

Duchelle, A. E., Simonet, G., Sunderlin, W. D., & Wunder, S. (2018). What is REDD+ achieving on the ground?. *Current Opinion in Environmental Sustainability*, *32*, 134–140.

Edwards, P. N. (2011). History of climate modeling. *Wiley Interdisciplinary Reviews: Climate Change*, *2*(1), 128–139.

Ekholm, T. (2020). Optimal forest rotation under carbon pricing and forest damage risk. *Forest Policy and Economics*, *115*, 102131.

Ellison, D., Morris, C. E., Locatelli, B., Sheil, D., Cohen, J., Murdiyarso, D., Gutierrez, V., van Noordwijk, M., Creed, I. F., Pokorny, J., Gaveau, D., Spracklen, D. V., Bargués Tobella, A., Ilstedt, U., Teuling, A. J., Gebrehiwot, S. G., Sands, D. C., Muys, B., Verbist, B., … Sullivan, C. A. (2017). Trees, forests and water: Cool insights for a hot world. *Global Environmental Change*, *43*, 51–61.

FAO. (2020). *Global Forest Resources Assessment 2020: Main report*. Rome. https://doi.org/10.4060/ca9825en

FAO. (2022). *Global forest sector outlook 2050: Assessing future demand and sources of timber for a sustainable economy – Background paper for The State of the World's Forests 2022* [FAO Forestry Working Paper No. 31]. FAO.

FAO. (2024). *The State of the World's Forests 2024 – Forest-sector innovations towards a more sustainable future*. Rome. https://doi.org/10.4060/cd1211en

Fisher, J. B., Lee, B., Purdy, A. J., Halverson, G. H., Dohlen, M. B., Cawse-Nicholson, K., Wang, A., Anderson, R. G., Aragon, B., Arain, M. A., Baldocchi, D. D., Baker, J. M., Barral, H., Bernacchi, C. J., Bernhofer, C., Biraud, S. C., Bohrer, G., Brunsell, N., Cappelaere, B., … Hook, S. (2020). ECOSTRESS: NASA's next generation mission to measure evapotranspiration from the international space station. *Water Resources Research*, *56*(4), e2019WR026058.

Foley, J. A., Levis, S., Costa, M. H., Cramer, W., & Pollard, D. (2000). Incorporating dynamic vegetation cover within global climate models. *Ecological Applications*, *10*(6), 1620–1632.

Forest Europe. (2015). *State of Europe's forests 2015*. Ministerial Conference on the Protection of Forests in Europe. https://foresteurope.org/wp-content/uploads/2022/02/soef_21_12_2015.pdf

Frazier, A. E., Renschler, C. S., & Miles, S. B. (2013). Evaluating post-disaster ecosystem resilience using MODIS GPP data. *International Journal of Applied Earth Observation and Geoinformation*, *21*, 43–52.

Friedlingstein, P., O'Sullivan, M., Jones, M. W., Andrew, R. M., Gregor, L., Hauck, J., Le Quéré, C., Luijkx, I. T., Olsen, A., Peters, G. P., Peters, W., Pongratz, J., Schwingshackl, C., Sitch, S., Canadell, J. G., Ciais, P., Jackson, R. B., Alin, S. R., Alkama, R., … Zheng, B. (2022). Global carbon budget 2022. *Earth System Science Data*, *14*(11), 4811–4900.

Gatti, L. V., Basso, L. S., Miller, J. B., Gloor, M., Gatti Domingues, L., Cassol, H. L. G., Tejada, G., Aragão, L. E. O. C., Nobre, C., Peters, W., Marani, L., Arai, E., Sanches, A. H., Corrêa, S. M., Anderson, L., Von Randow, C., Correia, C. S. C., Crispim, S. P., & Neves, R. A. L. (2021). Amazonia as a carbon source linked to deforestation and climate change. *Nature*, *595*(7867), 388–393. https://doi.org/10.1594/PANGAEA.926834

Gómez-Baggethun, E., De Groot, R., Lomas, P. L., & Montes, C. (2010). The history of ecosystem services in economic theory and practice: From early notions to markets and payment schemes. *Ecological Economics*, *69*(6), 1209–1218.

Gruber, N., & Galloway, J. N. (2008). An Earth-system perspective of the global nitrogen cycle. *Nature*, *451*(7176), 293–296.

Hamrick, K., & Gallant, M. (2018). Voluntary carbon markets insights: 2018 outlook and first quarter trends (Full version). Forest Trends' Ecosystem Marketplace. https://www.forest-trends.org/wp-content/uploads/2018/09/VCM-Q1-Report_Full-Version-2.pdf.

Hansen, M. C., Potapov, P. V., Moore, R., Hancher, M., Turubanova, S. A., Tyukavina, A., Thau, D., Stehman, S. V., Goetz, S. J., Loveland, T. R., Kommareddy, A., Egorov, A., Chini, L., Justice, C. O., & Townshend, J. R. G. (2013). High-resolution global maps of 21st-century forest cover change. *Science*, *342*(6160), 850–853.

Hashimoto, H., Nemani, R. R., White, M. A., Jolly, W. M., Piper, S. C., Keeling, C. D., Myneni, R. B., & Running, S. W. (2004). El Niño–Southern Oscillation–induced variability in terrestrial carbon cycling. *Journal of Geophysical Research: Atmospheres*, *109*(D23).

Houghton, R. A., & Nassikas, A. A. (2018). Negative emissions from stopping deforestation and forest degradation, globally. *Global Change Biology*, *24*(1), 350–359.

Hubau, W., Lewis, S. L., Phillips, O. L., Affum-Baffoe, K., Beeckman, H., Cuní-Sanchez, A., Daniels, A. K., Ewango, C. E. N., Fauset, S., Mukinzi, J. M., Sheil, D., Sonké, B., Sullivan, M. J. P., Sunderland, T. C. H., Taedoumg, H., Thomas, S. C., White, L. J. T., Abernethy, K. A., Adu-Bredu, S., … Zemagho, L. (2020). Asynchronous carbon sink saturation in African and Amazonian tropical forests. *Nature*, *579*(7797), 80–87.

Huete, A. R. (1988). A soil-adjusted vegetation index (SAVI). *Remote Sensing of Environment*, *25*(3), 295–309. https://doi.org/10.1016/0034-4257(88)90106-X

Huete, A. R. (2012). Vegetation indices, remote sensing and forest monitoring. *Geography Compass*, *6*(9), 513–532.

Huete, A. R., Didan, K., Miura, T., Rodriguez, E. P., Gao, X., & Ferreira, L. G. (2002). Overview of the radiometric and biophysical performance of the MODIS vegetation indices. *Remote Sensing of Environment*, *83*(1–2), 195–213. https://doi.org/10.1016/S0034-4257(02)00096-2

Huntingford, C., Burke, E. J., Jones, C. D., Jeffers, E. S., & Wiltshire, A. J. (2022). Nitrogen cycle impacts on CO_2 fertilisation and climate forcing of land carbon stores. *Environmental Research Letters*, *17*(4), 044072.

Intergovernmental Panel on Climate Change (IPCC). (2023). *Climate change 2023: Synthesis report*. Contribution of Working Groups I, II and III to the Sixth Assessment Report of the Intergovernmental Panel on Climate Change (Core Writing Team, H. Lee, & J. Romero, Eds.). IPCC. https://doi.org/10.59327/IPCC/AR6-9789291691647

International Tropical Timber Organization (ITTO). (2016). *Criteria and indicators for the sustainable management of tropical forests* [Policy Development Series No. 21]. ITTO. https://www.itto.int

Ishtiaque, A., Myint, S. W., & Wang, C. (2016). Examining the ecosystem health and sustainability of the world's largest mangrove forest using multi-temporal MODIS products. *Science of the Total Environment*, *569*, 1241–1254.

Jackson, R. B., Randerson, J. T., Canadell, J. G., Anderson, R. G., Avissar, R., Baldocchi, D. D., Bonan, G. B., Caldeira, K., Diffenbaugh, N. S., Field, C. B., Hungate, B. A., Jobbágy, E. G., Kueppers, L. M., Nosetto, M. D., & Pataki, D. E. (2008). Protecting climate with forests. *Environmental Research Letters*, *3*(4), 044006.

Jones, J., Ellison, D., Ferraz, S., Lara, A., Wei, X., & Zhang, Z. (2022). Forest restoration and hydrology. *Forest Ecology and Management*, *520*, 120342.

Kadam, P., Dwivedi, P., & Karnatz, C. (2021). Mapping convergence of sustainable forest management systems: Comparing three protocols and two certification schemes for ascertaining the trends in global forest governance. *Forest Policy and Economics, 133*, 102614.

Karlsson, M., Alfredsson, E., & Westling, N. (2020). Climate policy co-benefits: A review. *Climate Policy, 20*(3), 292–316.

Kellogg, K., Hoffman, P., Standley, S., Shaffer, S., Rosen, P., Edelstein, W., Dunn, C., Baker, C., Barela, P., Shen, Y., Guerrero, A. M., Xaypraseuth, P., Sagi, V. R., Harinath, N., Kumar, R., Bhan, R., & Sarma, C. V. H. S. (2020). NASA-ISRO synthetic aperture radar (NISAR) mission. In *2020 IEEE aerospace conference* (pp. 1–21). IEEE.

Kloster, S., Mahowald, N. M., Randerson, J. T., Thornton, P. E., Hoffman, F. M., Levis, S., Lawrence, P. J., Feddema, J. J., Oleson, K. W., & Lawrence, D. M. (2010). Fire dynamics during the 20th century simulated by the Community Land Model. *Biogeosciences, 7*(6), 1877–1902.

Krieger, D. J. (2001). Economic value of forest ecosystem services: A review (Report). The Wildernes Society. https://www.sierraforestlegacy.org/Resources/Conservation/FireForestEcology/ForestEconomics/EcosystemServices.pdf.

Krug, J. H. (2018). Accounting of GHG emissions and removals from forest management: A long road from Kyoto to Paris. *Carbon Balance and Management, 13*, 1–11.

Kucharik, C. J., Barford, C. C., El Maayar, M., Wofsy, S. C., Monson, R. K., & Baldocchi, D. D. (2006). A multiyear evaluation of a Dynamic Global Vegetation Model at three AmeriFlux forest sites: vegetation structure, phenology, soil temperature, and CO_2 and H_2O vapor exchange. *Ecological Modelling, 196*(1–2), 1–31.

Kurz, W. A., Dymond, C. C., Stinson, G., Rampley, G. J., Neilson, E. T., Carroll, A. L., Ebata, T., & Safranyik, L. (2008). Mountain pine beetle and forest carbon feedback to climate change. *Nature, 452*(7190), 987–990.

Lafleur, P. M. (2009). Connecting atmosphere and wetland: Trace gas exchange. *Geography Compass, 3*(2), 560–585.

Lausch, A., Erasmi, S., King, D. J., Magdon, P., & Heurich, M. (2016). Understanding forest health with remote sensing-part I—A review of spectral traits, processes and remote-sensing characteristics. *Remote Sensing, 8*(12), 1029.

Lausch, A., Erasmi, S., King, D. J., Magdon, P., & Heurich, M. (2017). Understanding forest health with remote sensing-part II—A review of approaches and data models. *Remote Sensing, 9*(2), 129.

Lechner, A. M., Foody, G. M., & Boyd, D. S. (2020). Applications in remote sensing to forest ecology and management. *One Earth, 2*(5), 405–412.

Lee, D., Skutsch, M., & Sandker, M. (2018). *Challenges with measurement and accounting of the plus in REDD+*. Climate and Land Use Alliance.

Leite-Filho, A. T., Soares-Filho, B. S., Davis, J. L., Abrahão, G. M., & Börner, J. (2021). Deforestation reduces rainfall and agricultural revenues in the Brazilian Amazon. *Nature Communications, 12*(1), 2591.

Levis, S. (2010). Modeling vegetation and land use in models of the Earth System. *Wiley Interdisciplinary Reviews: Climate Change, 1*(6), 840–856.

Linser, S., & Wolfslehner, B. (2022). National implementation of the Forest Europe indicators for sustainable Forest management. *Forests, 13*(2), 191.

Linser, S., Wolfslehner, B., Bridge, S. R. J., Gritten, D., Johnson, S., Payn, T., Prins, K., Raši, R., & Robertson, G. (2018). 25 years of criteria and indicators for sustainable forest management: How intergovernmental C&I processes have made a difference. *Forests, 9*(9), 578.

Liu, H., Randerson, J. T., Lindfors, J., & Chapin, F. S, III. (2005). Changes in the surface energy budget after fire in boreal ecosystems of interior Alaska: An annual perspective. *Journal of Geophysical Research: Atmospheres, 110*(D13).

Lorenz, K., & Lal, R. (2010). *Carbon sequestration in forest ecosystems*. https://doi.org/10.1007/978-90-481-3266-9

Lovell, H. C. (2010). Governing the carbon offset market. *Wiley Interdisciplinary Reviews: Climate Change*, *1*(3), 353–362.

Luo, Y., Hui, D., & Zhang, D. (2006). Elevated CO_2 stimulates net accumulations of carbon and nitrogen in land ecosystems: A meta-analysis. *Ecology*, *87*(1), 53–63.

MacDicken, K. G., Sola, P., Hall, J. E., Sabogal, C., Tadoum, M., & de Wasseige, C. (2015). Global progress toward sustainable forest management. *Forest Ecology and Management*, *352*, 47–56.

Managi, S., Wang, J., & Zhang, L. (2019). Research progress on monitoring and assessment of forestry area for improving forest management in China. *Forestry Economics Review*, *1*(1), 57–70.

Mansourian, S., Belokurov, A., & Stephenson, P. J. (2009). The role of forest protected areas in adaptation to climate change. *Unasylva*, *60*(231–232), 63–69.

Maselli, F., Papale, D., Puletti, N., Chirici, G., & Corona, P. (2009). Combining remote sensing and ancillary data to monitor the gross productivity of water-limited forest ecosystems. *Remote Sensing of Environment*, *113*(3), 657–667.

Mason, C. F., & Plantinga, A. J. (2013). The additionality problem with offsets: Optimal contracts for carbon sequestration in forests. *Journal of Environmental Economics and Management*, *66*(1), 1–14.

May, P. B., Schlund, M., Armston, J., Kotowska, M. M., Brambach, F., Wenzel, A., & Erasmi, S. (2024). Mapping aboveground biomass in Indonesian lowland forests using GEDI and hierarchical models. *Remote Sensing of Environment*, *313*, 114384.

Pelicice, F. M. (2019). Weak democracies, failed policies, and the demise of ecosystems in poor and developing nations. *Tropical Conservation Science*, *12*, https://doi.org/10.1177/1940082919839902.

Michaelowa, A., Shishlov, I., & Brescia, D. (2019). Evolution of international carbon markets: Lessons for the Paris Agreement. *Wiley Interdisciplinary Reviews: Climate Change*, *10*(6), e613.

Migolet, P., Goïta, K., Pambo, A. F. K., & Mambimba, A. N. (2022). Estimation of the total dry aboveground biomass in the tropical forests of Congo Basin using optical, LiDAR, and radar data. *GIScience & Remote Sensing*, *59*(1), 431–460.

Mitchard, E. T. (2018). The tropical forest carbon cycle and climate change. *Nature*, *559*(7715), 527–534.

Montreal Process Working Group. (2015). *Criteria and indicators for the conservation and sustainable management of temperate and boreal forests.* Montreal Process Working Group. https://www.montrealprocess.org

Moorcroft, P. R. (2006). How close are we to a predictive science of the biosphere?. *Trends in Ecology & Evolution*, *21*(7), 400–407.

Morita, K., & Matsumoto, K. I. (2023). Challenges and lessons learned for REDD+ finance and its governance. *Carbon Balance and Management*, *18*(1), 8.

Nabuurs, G.-J., Mrabet, R., Abu Hatab, A., Bustamante, M., Clark, H., Havlík, P., House, J., Mbow, C., Ninan, K. N., Popp, A., Roe, S., Sohngen, B., & Towprayoon, S. (2022). Chapter 9: Agriculture, forestry and other land uses (AFOLU). In P. R. Shukla, J. Skea, R. Slade, A. Al Khourdajie, R. van Diemen, D. McCollum, M. Pathak, S. Some, P. Vyas, R. Fradera, M. Belkacemi, A. Hasija, G. Lisboa, S. Luz, & J. Malley (Eds.), *Climate change 2022: Mitigation of climate change. Contribution of working group III to the sixth assessment report of the intergovernmental panel on climate change.* Cambridge University Press. https://doi.org/10.1017/9781009157926.009

Neeff, T. (2021). What is the risk of overestimating emission reductions from forests–and what can be done about it?. *Climatic Change*, *166*(1), 26.

Nepstad, D., McGrath, D., Stickler, C., Alencar, A., Azevedo, A., Swette, B., Bezerra, T., DiGiano, M., Shimada, J., Seroa da Motta, R., Armijo, E., Castello, L., Brando, P., Hansen, M. C., McGrath-Horn, M., Carvalho, O., & Hess, L. (2014). Slowing Amazon deforestation through public policy and interventions in beef and soy supply chains. *Science*, *344*(6188), 1118–1123.

Newell, R. G., Pizer, W. A., & Raimi, D. (2014). Carbon markets: Past, present, and future. *Annual Review of Resource Economics*, *6*(1), 191–215.

Ninan, K. N., & Inoue, M. (2014). Valuing forest ecosystem services: What we know and what we don't. *Valuing Ecosystem Services*, *93*, 189–226.

Olofsson, P., Foody, G. M., Herold, M., Stehman, S. V., Woodcock, C. E., & Wulder, M. A. (2014). Good practices for estimating area and assessing accuracy of land change. *Remote Sensing of Environment*, *148*, 42–57.

Pan, Y., Birdsey, R. A., Phillips, O. L., Houghton, R. A., Fang, J., Kauppi, P. E., Keith, H., Kurz, W. A., Ito, A., Lewis, S. L., Nabuurs, G.-J., Shvidenko, A., Hashimoto, S., Lerink, B., Schepaschenko, D., Castanho, A., & Murdiyarso, D. (2024). The enduring world forest carbon sink. *Nature*, *631*(8021), 563–569.

Pandey, P. C., & Arellano, P. (Eds.). (2022). *Advances in remote sensing for forest monitoring*. John Wiley & Sons.

Parrotta, J., Mansourian, S., Wildburger, C., & Grima, N. (Eds.). (2022). Forests, climate, biodiversity and people: Assessing a decade of REDD+ (IUFRO World Series Vol. 40). International Union of Forest Research Organizations (IUFRO). https://www.fs.usda.gov/research/publications/wo/wo_2022_parrotta_002.pdf.

Pearson, T. R., Brown, S., Murray, L., & Sidman, G. (2017). Greenhouse gas emissions from tropical forest degradation: An underestimated source. *Carbon Balance and Management*, *12*, 1–11.

Peng, C. (2000). From static biogeographical model to dynamic global vegetation model: A global perspective on modelling vegetation dynamics. *Ecological Modelling*, *135*(1), 33–54.

Penman, J., & Hoekman, D. H. (2014). *Integrating remote-sensing and ground-based observations for estimation of emissions and removals of greenhouse gases in forests: Methods and guidance from global forest observation initiative*. GEO-GFOI MGD.

Pistorius, T. (2012). From RED to REDD+: The evolution of a forest-based mitigation approach for developing countries. *Current Opinion in Environmental Sustainability*, *4*(6), 638–645.

Poortinga, A., Tenneson, K., Shapiro, A., Nquyen, Q., San Aung, K., Chishtie, F., & Saah, D. (2019). Mapping plantations in Myanmar by fusing Landsat-8, Sentinel-2 and Sentinel-1 data along with systematic error quantification. *Remote Sensing*, *11*(7), 831.

Protected Forest Indicator. (2024). *Global forest review*. World Resources Institute. Retrieved January 20, 2025, from https://research.wri.org/gfr/forest-designation-indicators/protected-forests

Psistaki, K., Tsantopoulos, G., & Paschalidou, A. K. (2024). An overview of the role of forests in climate change mitigation. *Sustainability*, *16*(14), 6089.

Purves, D., & Pacala, S. (2008). Predictive models of forest dynamics. *Science*, *320*(5882), 1452–1453.

Qi, J., Chehbouni, A., Huete, A. R., Kerr, Y. H., & Sorooshian, S. (1994). A modified soil adjusted vegetation index. *Remote Sensing of Environment*, *48*(2), 119–126. https://doi.org/10.1016/0034-4257(94)90134-1

Quillet, A., Peng, C., & Garneau, M. (2010). Toward dynamic global vegetation models for simulating vegetation–climate interactions and feedbacks: Recent developments, limitations, and future challenges. *Environmental Reviews*, *18*(NA), 333–353.

Ramachandran, N., Saatchi, S., Tebaldini, S., d'Alessandro, M. M., & Dikshit, O. (2023). Mapping tropical forest aboveground biomass using airborne SAR tomography. *Scientific Reports*, *13*(1), 6233.

Randerson, J. T., Liu, H., Flanner, M. G., Chambers, S. D., Jin, Y., Hess, P. G., Pfister, G., Mack, M. C., Treseder, K. K., Welp, L. R., Chapin, F. S., Harden, J. W., Goulden, M. L.,

Lyons, E., Neff, J. C., Schuur, E. A. G., & Zender, C. S. (2006). The impact of boreal forest fire on climate warming. *Science, 314*(5802), 1130–1132.

Reiche, J., Hamunyela, E., Verbesselt, J., Hoekman, D., & Herold, M. (2018). Improving near-real time deforestation monitoring in tropical dry forests by combining dense Sentinel-1 time series with Landsat and ALOS-2 PALSAR-2. *Remote Sensing of Environment, 204*, 147–161.

Reichstein, M., Papale, D., Valentini, R., Aubinet, M., Bernhofer, C., Knohl, A., Laurila, T., Lindroth, A., Moors, E., Pilegaard, K., & Seufert, G. (2007). Determinants of terrestrial ecosystem carbon balance inferred from European eddy covariance flux sites. *Geophysical Research Letters, 34*(1).

Rodriguez Franco, C., & Conje, J. (2023). The evolution of the dialogue and perspectives on sustainable forest management with special emphasis on the United States of America. *Journal of Sustainable Forestry, 42*(8), 747–791.

Roe, S., Streck, C., Beach, R., Busch, J., Chapman, M., Daioglou, V., Deppermann, A., Doelman, J., Emmet-Booth, J., Engelmann, J., Fricko, O., Frischmann, C., Funk, J., Grassi, G., Griscom, B., Havlik, P., Hanssen, S., Humpenöder, F., Landholm, D., … Lawrence, D. (2021). Land-based measures to mitigate climate change: Potential and feasibility by country. *Global Change Biology, 27*(23), 6025–6058.

Roe, S., Streck, C., Obersteiner, M., Frank, S., Griscom, B., Drouet, L., Fricko, O., Gusti, M., Harris, N., Hasegawa, T., Hausfather, Z., Havlík, P., House, J., Nabuurs, G.-J., Popp, A., Sanz Sánchez, M. J., Sanderman, J., Smith, P., Stehfest, E., & Lawrence, D. (2019). Contribution of the land sector to a 1.5°C world. *Nature Climate Change, 9*(11), 817–828.

Rondeaux, G., Steven, M., & Baret, F. (1996). Optimization of soil-adjusted vegetation indices. *Remote Sensing of Environment, 55*(2), 95–107. https://doi.org/10.1016/0034-4257(95)00186-7

Roopsind, A., Sohngen, B., & Brandt, J. (2019). Evidence that a national REDD+ program reduces tree cover loss and carbon emissions in a high forest cover, low deforestation country. *Proceedings of the National Academy of Sciences, 116*(49), 24492–24499.

Rosenzweig, C., Karoly, D., Vicarelli, M., Neofotis, P., Wu, Q., Casassa, G., Menzel, A., Root, T. L., Estrella, N., Seguin, B., & Tryjanowski, P. (2008). Attributing physical and biological impacts to anthropogenic climate change. *Nature, 453*(7193), 353–357.

Rouse, J. W., Haas, R. H., Schell, J. A., & Deering, D. W. (1974). Monitoring vegetation systems in the Great Plains with ERTS. *NASA Special Publication, 351*, 309–317.

Sabine, C. L., Heimann, M., Artaxo, P., Bakker, D. C., Chen, C. T. A., Field, C. B., Gruber, N., Le Quere, C., Prinn, R. G., Richey, J. E., Lankao, P. R., Sathaye, J. A., & Valentini, R. (2004). Current status and past trends of the global carbon cycle. In *The global carbon cycle: Integrating humans, climate and the natural world* (Vol. 62, pp. 17–44). Island Press.

Sabogal, C., Guariguata, M. R., Broadhead, J., Lescuyer, G., Savilaakso, S., Essoungou, N., & Sist, P. (2013). Multiple-use forest management in the humid tropics: Opportunities and challenges for sustainable forest management (FAO Forestry Paper No. 173). Food and Agriculture Organization of the United Nations; Center for International Forestry Research. https://agritrop.cirad.fr/570548/1/document_570548.pdf.

Scheiter, S., Langan, L., & Higgins, S. I. (2013). Next-generation dynamic global vegetation models: Learning from community ecology. *New Phytologist, 198*(3), 957–969.

Schröter, D., Brussaard, L., De Deyn, G., Poveda, K., Brown, V. K., Berg, M. P., Wardle, D. A., Moore, J., & Wall, D. H. (2004). Trophic interactions in a changing world: Modelling aboveground–belowground interactions. *Basic and Applied Ecology, 5*(6), 515–528.

Seddon, N., Sengupta, S., García-Espinosa, M., Hauler, I., Herr, D., & Rizvi, A. R. (2019). Nature-based solutions in nationally determined contributions: Synthesis and recommendations for enhancing climate ambition and action by 2020. IUCN and University of Oxford. https://portals.iucn.org/library/efiles/documents/2019-030-En.pdf.

Sellers, P. J., Dickinson, R. E., Randall, D. A., Betts, A. K., Hall, F. G., Berry, J. A., Collatz, G. J., Denning, A. S., Mooney, H. A., Nobre, C. A., Sato, N., Field, C. B., & Henderson-Sellers, A. (1997). Modeling the exchanges of energy, water, and carbon between continents and the atmosphere. *Science, 275*(5299), 502–509.

Simonet, G., Subervie, J., Ezzine-de-Blas, D., Cromberg, M., & Duchelle, A. E. (2019). Effectiveness of a REDD+ project in reducing deforestation in the Brazilian Amazon. *American Journal of Agricultural Economics, 101*(1), 211–229.

Singh, M., Sinha, B., Bisaria, J., Thomas, T., & Srivastava, P. (2022). Understanding synergies and tradeoffs between forests, water, and climate change. *Wiley Interdisciplinary Reviews: Water, 9*(6), e1614.

Singh, N., Singh, H., Arora, C., & Chaturvedi, V. (2024). *Role of financial players in the Indian carbon market* (Issue brief). Council on Energy, Environment and Water (CEEW). https://www.ceew.in/sites/default/files/role-of-financial-players-in-indian-carbon-credit-trading-and-offest-markets.pdf

Sinha, S., Jeganathan, C., Sharma, L. K., & Nathawat, M. S. (2015). A review of radar remote sensing for biomass estimation. *International Journal of Environmental Science and Technology, 12*(5), 1779–1792.

Siry, J. P., Cubbage, F. W., Potter, K. M., & McGinley, K. (2018). Current perspectives on sustainable forest management: North America. *Current Forestry Reports, 4*, 138–149.

Smith, P., Calvin, K., Nkem, J., Campbell, D., Cherubini, F., Grassi, G., Korotkov, V., Le Hoang, A., Lwasa, S., McElwee, P., Nkonya, E., Saigusa, N., Soussana, J. F., Taboada, M. A., Manning, F. C., Nampanzira, D., Arias-Navarro, C., Vizzarri, M., House, J., … Arneth, A. (2020). Which practices co-deliver food security, climate change mitigation and adaptation, and combat land degradation and desertification?. *Global Change Biology, 26*(3), 1532–1575.

Snyder, P. K., Delire, C., & Foley, J. A. (2004). Evaluating the influence of different vegetation biomes on the global climate. *Climate Dynamics, 23*, 279–302.

Solberg, S., Gizachew, B., Næsset, E., Gobakken, T., Bollandsås, O. M., Mauya, E. W., Olsson, H., Malimbwi, R., & Zahabu, E. (2015). Monitoring forest carbon in a Tanzanian woodland using interferometric SAR: A novel methodology for REDD+. *Carbon Balance and Management, 10*, 1–14.

Stern, D. I., & Kaufmann, R. K. (2014). Anthropogenic and natural causes of climate change. *Climatic Change, 122*, 257–269.

Teketay, D., Lemenih, M., Bekele, T., Yemshaw, Y., Feleke, S., Tadesse, W., Moges, Y., Hunde, T., & Nigussie, D. (2010). Forest resources and challenges of sustainable forest management and conservation in Ethiopia. In *Degraded forests in Eastern Africa* (pp. 19–63). Routledge.

Thom, D., Rammer, W., & Seidl, R. (2017). The impact of future forest dynamics on climate: Interactive effects of changing vegetation and disturbance regimes. *Ecological Monographs, 87*(4), 665–684.

Tol, R. S. (2018). The economic impacts of climate change. *Review of Environmental Economics and Policy*. 12(1), 4–25. https://doi.org/10.1093/reep/rex027

Turner, D. P., Ritts, W. D., Cohen, W. B., Gower, S. T., Running, S. W., Zhao, M., Costa, M. H., Kirschbaum, A. A., Ham, J. M., Saleska, S. R., & Ahl, D. E. (2006). Evaluation of MODIS NPP and GPP products across multiple biomes. *Remote Sensing of Environment, 102*(3–4), 282–292.

United Nations. (2015). *Transforming our world: The 2030 agenda for sustainable development*. United Nations. https://sustainabledevelopment.un.org/content/documents/21252030%20Agenda%20for%20Sustainable%20Development%20web.pdf

United Nations Department of Economic and Social Affairs, United Nations Forum on Forests Secretariat. (2021). *The global forest goals report 2021*. United Nations. https://www.un.org/esa/forests/wp-content/uploads/2021/04/Global-Forest-Goals-Report-2021.pdf

Upadhyay, V., & Kumar, A. (2018). Hyperspectral remote sensing of forests: Technological advancements, opportunities and challenges. *Earth Science Informatics, 11*(4), 487–524.

Vancutsem, C., Achard, F., Pekel, J.-F., Vieilledent, G., Carboni, S., Simonetti, D., Gallego, J., Aragão, L. E. O. C., Nasi, R. (2021). Long-term (1990–2019) monitoring of forest cover changes in the humid tropics. *Science Advances*, 7(10), eabe1603. https://doi.org/10.1126/sciadv.abe1603.

Vatandaşlar, C., & Abdikan, S. (2022). Carbon stock estimation by dual-polarized synthetic aperture radar (SAR) and forest inventory data in a Mediterranean forest landscape. *Journal of Forestry Research, 33*(3), 827–838.

Wheeler, T., & Von Braun, J. (2013). Climate change impacts on global food security. *Science, 341*(6145), 508–513.

World Bank. (2024). *State and trends of carbon pricing 2024*. World Bank. https://hdl.handle.net/10986/41544

World Commission on Environment and Development (U. N.). (1987). *Our common future*. Oxford University Press.

Xiao, J., Chevallier, F., Gomez, C., Guanter, L., Hicke, J. A., Huete, A. R., Ichii, K., Ni, W., Pang, Y., Rahman, A. F., Sun, G., Yuan, W., Zhang, L., & Zhang, X. (2019). Remote sensing of the terrestrial carbon cycle: A review of advances over 50 years. *Remote Sensing of Environment, 233*, 111383.

Yadava, R. N., & Sinha, B. (2020). Vulnerability assessment of forest fringe villages of Madhya Pradesh, India for planning adaptation strategies. *Sustainability, 12*(3), 1253.

Yanai, R. D., Wayson, C., Lee, D., Espejo, A. B., Campbell, J. L., Green, M. B., Zukswert, J. M., Yoffe, S. B., Aukema, J. E., Lister, A. J., Kirchner, J. W., & Gamarra, J. G. P. (2020). Improving uncertainty in forest carbon accounting for REDD+ mitigation efforts. *Environmental Research Letters, 15*(12), 124002.

Zelli, F., Möller, I., & Asselt, H. V. (2017). Institutional complexity and private authority in global climate governance: The cases of climate engineering, REDD+ and short-lived climate pollutants. *Environmental Politics, 26*(4), 669–693.

Zolkos, S. G., Goetz, S. J., & Dubayah, R. (2013). A meta-analysis of terrestrial aboveground biomass estimation using lidar remote sensing. *Remote Sensing of Environment, 128*, 289–298.

15 Synergistic Impacts of Climate and Land Use Land Cover (LULC) Changes on the Hydrology of Forested Landscapes

Sumantra Basu, Shubhi Kulshrestha, and Sagrika Kalta

15.1 INTRODUCTION

Water plays an important role in agricultural food production and livelihood generation. According to the World Bank, by 2050, the Earth will require feeding 9 billion people with the same amount of land and water used in 2012. Due to an increase in urbanization and economic development, land use land cover (LULC) is changing rapidly. Such changes in LULC have a deep impact on hydrology. Particularly in dry regions, water availability is very scarce due to low rainfall, high temperature, wind velocity, and potential evapotranspiration. The impact of climate change and human-induced land use on hydrological processes has garnered significant attention and presented a significant challenge for numerous researchers (Liu et al., 2013). Forests supply human water in the densely populated eastern USA, and an estimate suggests that the southeastern forests provide surface drinking water to approximately 2130 communities with 48.7 million people (Caldwell et al., 2014, 2016). Change in LULC and climate is complicated and affects water availability, quality, and overall ecosystem functions. LULC changes include afforestation, deforestation, urbanization, agricultural land expansion, and other alterations in forested landscapes, whereas, through changes in temperature, precipitation patterns, and extreme weather events, climate change influences forest hydrology. Two major factors contributing to the degradation of forested mountain watersheds, which provide clean, reliable, and abundant fresh water essential for supporting ecosystems and an increasing human population, are climate change and alterations in forest ecosystems (Caldwell et al., 2016; Postel & Richter, 2012; Viviroli et al., 2007). Developing countries such as India are facing rapid increases in population due to changes in LULC through deforestation, urbanization, and extension of agricultural lands that

DOI: 10.1201/9781003588580-15

impact the hydrological cycle in several river basins of India. A report on global land cover changes between 2000 and 2017 reveals an 86% change in land cover in India with an 82% change in agricultural land and a 4% change in forests (Chen et al., 2019a; IPCC, 2019). Thus, a detailed understanding and evaluation of the change in climate and LULC and their impact on hydrological processes are crucial for decision-makers in planning policies, environmental conservation plans, allocation of water resources, river restoration, and protection of riparian ecosystems (Chen et al., 2019b; Chu et al., 2013; Naha et al., 2021).

15.2 FUNDAMENTALS OF HYDROLOGICAL PROCESSES AND WATER RESOURCES

One-third of the Earth's surface is covered with water, and about 97% of this water, which is salty in nature, with a volume of 141 billion m^3, is found in the seas and oceans. The remaining 3% is considered freshwater. Poles and ice masses store approximately 87%–90% of this freshwater, rendering it unusable for direct use. Humans can only use 0.08% of the total freshwater. Groundwater stores 95%–96% of freshwater resources, while lakes, rivers, and streams hold the remaining 0.4%–0.5% (Figure 15.1). Therefore, humans can only utilize a very small portion of Earth's total water resources (Keleş, 2019).

Hydrology processes on Earth systems—the atmosphere, lithosphere, and biosphere—focus on water movement, distribution, and storage. The hydrology, which characterizes the continuous cycle of water movement in its solid, liquid, and gaseous states, drives these processes. The solar energy that moves the water cycle allows the release of water into the atmosphere through evapotranspiration. Evaporated steam ascends into the atmosphere, cooling and condensing into water vapor. Precipitation serves as the primary input of water into terrestrial systems and is critical for recharging surface and groundwater resources (Dingman, 2015; Trenberth et al., 2007). The land surface manages the interaction of precipitation, which can infiltrate, run across surfaces, or evaporate. Once infiltrated and distributed across the landscape, it may find its way to streams, rivers, and even the ocean. Topography,

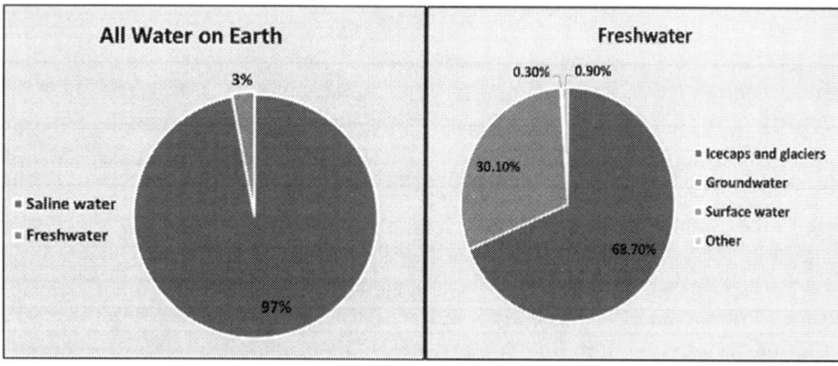

FIGURE 15.1 Distribution of Earth's water resources.

type of land improvement, and human activities state different yields for all three drainage systems, mainly because nature does not do the same kinds of changes to the environment as man. Topography, land cover, and human activities influence runoff, which plays a crucial role in determining water availability and quality in various ecosystems (Ward & Trimble, 2004). The most superficial asset in soil and rock is subsurface hydrology. The soil pore spaces contain two distinct zones: the unsaturated zone, which contains both air and water, and the slightly saturated zone, which fills all the soil pore spaces with water. Aquifers, acting as natural reservoirs, store the latter, known as groundwater. The most important reason for this is the linkage of surface water with groundwater. It sustains the flow of rivers, wetlands, and other ponds, especially during periods of drought (Sophocleous, 2002).

Hydrological processes tightly interlink water resources, ensuring their availability and sustainability. Climate, geographical, and geological factors influence the uneven distribution of global freshwater resources. This is primarily caused by human actions such as climate change, deforestation, urbanization, and agriculture, which disrupt the dramatic shift in the water cycle. Such disruptions have led to issues such as altered river flows, depletion of groundwater reserves, and increased incidence of extreme hydrological events, including floods and droughts (Falkenmark & Rockström, 2006; Oki & Kanae, 2006). The regulation of hydrological processes is essential for ecosystem services, including nutrient cycling, habitat provision, and climate regulation.

15.3 IMPORTANCE OF FOREST ECOSYSTEMS AND HYDROLOGY

Forest ecosystems provide several production- and protection-related hydrological services, including quality and continuous water yield, flood control, soil protection, landslide protection, avalanche protection, and rockfall protection. These forests are a crucial ecosystem, providing the majority of the water available on Earth. However, these services are influenced by various factors such as the structure and composition of forest types, tree species, age class distribution, leaf area, biomass, basal area, crown closure, and stages of stand development (Keleş, 2019). Forest ecosystems also play a crucial role in understanding the hydrological processes that influence the distribution, storage, movement, and quality of water in a watershed. These matters involving the hydrological cycle are termed forest hydrology, primarily dealing with how forests can influence the hydrological cycle using their processes such as transpiration, interception, and shading (Ellison et al., 2017). The transpiration process, which increases water loss through roots and across leaves, contributes significantly to this impact by increasing moisture convergence in the atmosphere. This process can significantly alter subsequent weather patterns by modifying the atmospheric response to surface moisture. Conversely, in drier atmospheres, increased transpiration may reduce atmospheric moisture convergence and runoff. Resolving and understanding this dichotomy of hydrological deforestation impacts could potentially lead to the prescription of ecological restorations. Forests affect water quality and erosion control. The plants help to keep the topsoil stable, which cuts down on erosion and stops sediment from washing away. However, events like wildfires can reduce this effect, as seen in the 2020 Cameron Peak Fire in Colorado. This wildfire led to more erosion and sediment entering water bodies, negatively impacting water quality and increasing sediment levels. These issues make water treatment harder. Such situations highlight why effective forest management is vital for water quality (Robichaud et al., 2020). Several natural and anthropogenic processes lead to the degradation of forest ecosystems, which negatively impacts the hydrological cycle by reducing their capacity to provide various hydrological and ecological services. These negative impacts could be in the form of biodiversity loss, production loss, and reduction in forest quality. Restoration of forest ecosystems by increasing forest cover, structure, function, and species composition helps in reviving the hydrology of a degraded forest through partitioning of precipitation into evapotranspiration and streamflow (Jones et al., 2022). Jones et al. (2022) studied the hydrological effect on three types of forests, i.e., mature and old-growth, managed forest plantations, and the early stages of native forest succession. It was found that the mature and old-growth forests that are often considered as the reference model for restoration have high evapotranspiration and consistent water yield, whereas the managed forest plantations provide low water yield, especially during dry periods. The water yield from the early stages of native forest succession may be greater with an increase in low flows than those from managed plantations. It was also observed that there are some hydrological benefits if the native species and natural processes are followed in forest restoration as compared to other restoration processes.

Forests play a particularly critical role in enhancing infiltration, reducing surface runoff, and contributing to local and regional precipitation through evapotranspiration. The destruction of forest ecosystems disrupts these processes, exacerbating soil erosion, altering hydrological cycles, and reducing the availability of clean water (Ellison et al., 2017; Sheil & Murdiyarso, 2009). Forestry activities, aimed at increasing water production, alter the ecosystem's total evaporation, leading to an increase in discharge. This, in turn, increases the concentration of nutrients and sediment in streams (Sun et al., 2005). As the sustainable management of forested ecosystems impacts the provision of freshwater and other hydrological services, knowing the interaction between forest ecosystems and water is very important for the sustainable management of forest ecosystems and water.

15.4 IMPACTS OF CHANGE IN CLIMATE AND LAND USE LAND COVER ON HYDROLOGY OF FORESTED CATCHMENT

LULC changes are ongoing processes, typically influenced by both natural and human-driven factors. Climate and LULC changes significantly impact watershed hydrology. Studies show that climate change generally has a more substantial effect on hydrology than LULC changes (Talib & Randhir, 2017; van Roosmalen et al., 2009). A study was conducted in a large-scale agricultural catchment to simulate changes in the groundwater system and its discharge to rivers under two different climate scenarios, and it was found that in one of the scenarios, groundwater recharge increased significantly, which in turn amplified stream discharge and the seasonal dynamics (van Roosmalen et al., 2009). LULC changes, particularly urbanization, can increase surface runoff and peak flow volumes (Lamichhane & Shakya, 2019; Talib & Randhir, 2017). The combined effects of climate and LULC changes can intensify hydrological processes, leading to increased vulnerability to floods, especially during spring (Farjad et al., 2017). Combined modelling methods that include climate models, LULC projections, and hydrological models are necessary to fully grasp how climate, LULC, and hydrology all affect each other (Farjad et al., 2017; Lamichhane & Shakya, 2019). Climate change has the most substantial effect on the hydrology of a large-scale agricultural catchment, while other factors like land use change have a lesser impact. Combined changes in climate and land use can significantly impact the hydrology of forested watersheds, including increased surface runoff and stream discharge (Birhanu et al., 2019).

Lamichhane and Shakya (2019) looked at how changes in climate and land use affect hydrology. They used the Kathmandu Valley watershed in central Nepal as an example. To simulate hydrology and look at how it changes under different representative concentration pathway (RCP) scenarios (RCP4.5 and RCP8.5), they made a hydrological model using the Soil and Water Assessment Tool (SWAT). According to the modelling results, the RCP4.5 scenario projects an increase of 37%, 21%, and 12%, respectively, in river runoff due to climate change, LULC changes, and the combined effects of both factors.

The change in LULC of an area can significantly change the hydrology of that area by modifying the hydrological processes such as baseflow, surface runoff, evapotranspiration (ET), water holding capacity of the soil, interception, and groundwater recharge (Rogger et al., 2017), whereas climate change is reshaping precipitation patterns, temperature trends, and the occurrence of extreme weather events, which in turn affects hydrological cycles. Research indicates that rising atmospheric temperatures increase the atmosphere's ability to hold moisture, resulting in more intense and unpredictable rainfall events. This situation, often termed "hydroclimate whiplash," leads to alternating phases of heavy rainfall and drought, making water management more challenging (Diffenbaugh et al., 2015).

One of the major causes of LULC change is rapid anthropogenic activities (Rogger et al., 2017). Changes in LULC, including urbanization, deforestation, and agricultural expansion, have a profound impact on hydrological responses. Urban development increases impervious surfaces, resulting in greater runoff and heightened flood risks (Jacobson, 2011). A study examining a changing watershed predicted that LULC changes could significantly affect streamflow and flood volumes by the years 2050 and 2080 (El Khoury et al., 2021). Deforestation and land degradation diminish the land's ability to retain water, worsening erosion and reducing groundwater recharge. On the other hand, afforestation and reforestation initiatives can improve water infiltration and storage, although the specific results depend on factors such as species choice and local climate conditions (Filoso et al., 2017). Interactions between climate change and changes in LULC can have cumulative effects on hydrological systems. For example, research on the Ravi River Basin has shown that both climate variability and land use changes significantly impact river flows (Ali et al., 2022). Also, research in the Tana River Basin highlighted that combined climate and land use challenges could alter hydrological processes affecting water availability and quality (LGebremicael et al., 2019). Moreover, higher temperatures lead to increased evapotranspiration rates, which diminish soil moisture and streamflow.

15.4.1 Changes in Precipitation Patterns and Intensity

Climate change can lead to more extreme rainfall events, including both heavier storms and longer dry periods. Global warming directly impacts precipitation patterns. As temperatures rise, evaporation increases, leading to greater surface drying and more intense, prolonged droughts. However, the air's capacity to hold water also increases by approximately 7% for every 1°C of warming, resulting in higher levels of moisture in the atmosphere. This increased moisture supply fuels storms—whether they are individual thunderstorms, extratropical rain or snowstorms, or tropical cyclones—leading to more intense precipitation events. This can increase the frequency and intensity of floods and droughts, altering the timing and volume of streamflow (Trenberth, 2011). Warmer temperatures can cause a shift from snow to rain in colder climates, which can reduce snowpack storage and lead to earlier peak flows, potentially increasing the risk of flooding and altering seasonal water availability (Adam et al., 2009).

Rising temperatures can increase evaporation and transpiration rates from both vegetation and soil, potentially reducing surface runoff and altering groundwater recharge. This can affect streamflow and water supply during dry periods. Global terrestrial evapotranspiration (ET) consists of transpiration (T), soil evaporation (E), and interception (I), with their relative contributions varying across climate regimes. Transpiration dominates ET in wet regions, while soil evaporation is more significant in dry areas (Zhang et al., 2017). Estimates of global T/(E+T+I) range widely from 24% to 90% (Wei et al., 2017), with one study suggesting T accounts for 57.2% of global terrestrial ET (Wei et al., 2017). Another study estimates T at 80%, E at 7%, I at 11%, and snow sublimation at 2% of total ET (Miralles et al., 2011). Climate drivers differently influence ET components, with radiation being crucial in wet regions and precipitation in dry areas (Zhang et al., 2017). Accurate partitioning of ET is essential for understanding hydrological cycles and improving climate models (Lawrence et al., 2007). Notably, Australia contributes disproportionately to global soil evaporation variance despite its small land area (Zhang et al., 2017).

15.4.2 Changes in Vegetation and Forest Cover

Conversion of forests to agriculture or urban areas leads to reduced evapotranspiration, increased surface runoff, and potentially higher sedimentation in water bodies (Lockaby et al., 2013). Deforestation can also reduce the ability of the catchment to regulate floodwaters, leading to higher peak flows and a decrease in groundwater recharge. The restoration of forest cover can improve water retention, stabilize soil, and enhance groundwater recharge, thus mitigating the effects of flooding (van Meerveld & Seibert, 2025).

15.4.3 Altered Soil Properties due to Climate and Land Use Change

Altered soil properties due to climate and LULC changes can have profound effects on the hydrology of forested catchments (Babaremu et al., 2024). Soil plays a critical role in regulating water flow, storage, and quality. Changes in soil properties, whether through direct alterations from land use changes or indirect effects from climate change, can significantly influence runoff, infiltration, erosion, and overall water dynamics. Reduced infiltration causes the soil to store less water, resulting in faster runoff during rain events, thereby increasing flood risks and diminishing groundwater recharge (Archer et al., 2020). Additionally, soils with poor infiltration characteristics are unable to support vegetation, which further exacerbates erosion and degradation.

15.4.4 Soil Organic Matter and Carbon Content

Land use changes, particularly deforestation and the conversion of forests to agricultural land, can lead to a loss of soil organic matter (SOM) and carbon (Amoakwah et al., 2022). Forest soils typically have high organic content due to the accumulation of decaying plant material, which is crucial for soil structure, moisture retention, and fertility. Loss of SOM leads to reduced soil structure, which can decrease soil aggregation and increase the likelihood of compaction and erosion. Soils with lower organic matter have reduced water retention capacity, meaning less water is available for infiltration and groundwater recharge (Seiler & Gat, 2007). This can result in more rapid runoff and an increased likelihood of flooding.

15.4.5 Soil pH and Nutrient Availability

Agricultural practices, especially the overuse of chemical fertilizers and irrigation, can alter the pH and nutrient composition of soils. Acidic soils can inhibit the availability of essential nutrients and affect the health of vegetation, which in turn impacts soil structure and hydrological processes. Altered soil pH and nutrient levels can affect the growth and health of vegetation, influencing the soil's ability to absorb and retain water. Shifts in temperature and rainfall patterns can also influence soil chemistry and nutrient cycling. Warmer temperatures may enhance the rate of chemical weathering and nutrient cycling, potentially increasing nutrient leaching during intense rainfall events, which can contribute to soil degradation and water quality issues (Biswas et al., 2018).

15.4.6 Changes in Water Quality

One of the most significant impacts of land use change on water quality is increased sedimentation. Clearing forests or disturbing land for agriculture or urban development increases the susceptibility of soil erosion. Rainfall can wash away topsoil, which carries with it not only soil particles but also pollutants such as nutrients, pesticides, and organic matter (Weldeslassie et al., 2018). Increased sediment load in rivers, lakes, and reservoirs can degrade water quality by making it turbid (cloudy). High turbidity can decrease sunlight penetration, affecting aquatic plants that rely on light for photosynthesis (Kirk, 1985). Sediment can also smother aquatic habitats, including spawning grounds for fish, and clog water treatment systems. Moreover, sediment frequently carries contaminants with it, which can further pollute water sources. Pesticides and herbicides can be toxic to aquatic life, including fish, amphibians, and invertebrates. They can also pose health risks to humans, especially when they enter drinking water supplies. Furthermore, these chemicals can disrupt the delicate balance of aquatic ecosystems, affecting species diversity and food webs (Relyea, 2005).

15.5 CHALLENGES AND ADAPTIVE MANAGEMENT CONSIDERATIONS

The synergistic effects of climate and LULC changes on the hydrology of forested landscapes pose several challenges. Forests play a crucial role in carbon storage. Deforestation or degradation due to land use changes can release carbon into the atmosphere, contributing to climate change, which in turn affects hydrological processes. Changes in forest cover and water availability can impact biodiversity in forested catchments, leading to changes in species composition and ecosystem services, such as water purification, carbon storage, and flood regulation (Seidl et al., 2019; Zafirah et al., 2017).

Changes in precipitation and temperature patterns, coupled with land use changes, can alter the timing of streamflow, shifting peak flow periods earlier or later in the year, with implications for water management (Tomer & Schilling, 2009). Increased rainfall intensity and reduced infiltration capacity due to land use changes may contribute to more frequent or intense floods in forested catchments, particularly in the absence of natural floodplain buffering. Deforestation and climate-induced droughts can reduce groundwater recharge, leading to lower baseflow levels, which can affect streamflow during dry periods (Kløve et al., 2014).

Climate and LULC changes significantly impact watershed hydrology, altering processes like evapotranspiration, runoff, and water yield (Hyandye et al., 2018; Talib & Randhir, 2017). These changes can intensify the water cycle and introduce seasonal variations in watershed systems (Talib & Randhir, 2017). Future climate projections suggest that warmer temperatures and altered precipitation patterns, in conjunction with LULC changes, could potentially reduce water availability in certain watersheds (Hyandye et al., 2018; Lamichhane & Shakya, 2019). The impacts of integrated LULC and climate changes on hydrology are not simply additive, highlighting the complexity of these interactions (Lamichhane & Shakya, 2019). To address these challenges, researchers propose various adaptation and mitigation strategies, including land and water management practices like filter strips, which can help increase water availability (Hyandye et al., 2018). Understanding these dynamic changes is crucial for developing effective mitigation and adaptation options to increase watershed resilience (Liu et al., 2017; Talib & Randhir, 2017).

15.5.1 Sustainable Forest Management

Sustainable forest management can help mitigate some of the negative hydrological impacts. Techniques like selective logging, reforestation, and afforestation can reduce runoff, improve soil stability, and enhance water retention capacity. Implementation of water conservation practices, such as watershed protection and sustainable land use practices, can help improve resilience to climate and land use changes and ensure the long-term health of the catchment (Bunch et al., 2011). The main goals of LULC change detection and monitoring are to look at spatial representations from two different points in time and compare them, making sure that any differences aren't caused by things that aren't related to the variables of interest, like differences in orbital or platform altitudes. They also want to find changes that happen because of differences in the variables that are of interest. This goal emphasizes the importance of making accurate comparisons between spatial data from two different points in time. The data may come from various sources, such as satellite imagery or aerial photographs.

These variations in data acquisition can introduce distortions that are unrelated to the actual land cover or environmental changes. To observe only the changes caused by the variables of interest (such as deforestation, urban expansion, or climate-induced shifts), the first goal is to minimize these irrelevant differences (Oliver & Morecroft, 2014). The second goal is to focus on identifying and measuring actual changes resulting from the specific variables under investigation. After eliminating irrelevant variations, the focus shifts to detecting changes that occur because of the variables of interest—whether those variables are related to human activities (e.g., urbanization, agriculture) or natural processes (e.g., erosion, forest regeneration) (Wang et al., 2023). Quantifying the magnitude of change that occurred is useful for assessing the impact of the change, whether it's beneficial or detrimental, and in predicting future trends. Change detection, monitoring, and updating primarily utilize two types of techniques: map-to-map comparisons and image-to-image comparisons. Map-to-map comparisons involve identifying differences between two maps from different time periods. This method requires images (either digital or analog) from two separate dates to be classified by interpreters, after which the maps are compared. In the past, this comparison was done manually using a light table and was the only method available for spatially referenced change detection. However, with the advent of GIS technology, map-to-map comparisons can now be conducted digitally, either by analyzing the area in each polygon type or on a pixel-by-pixel basis.

Digital remote sensing and multispectral scanner data have been used by researchers to find changes in urban-suburban areas (Jensen & Toll, 1982; Yang et al., 2003). SPOT data has also been used to track land conversion at the edge between cities and rural areas (Zhang et al., 2021). Combining SPOT panchromatic and multispectral data to make images that are better to look at has been shown to be very useful for figuring out land use and defining urban growth (Pacifici et al., 2009). Gao et al. (2017) used Landsat data to differentiate crop types and monitor crop growth over time, while Lee et al. (2010) manually distinguished between various crop types using SPOT data. To find changes in land use (Coppin et al., 2004), researchers have also used image differencing (Alqurashi & Kumar, 2013), principal components analysis (Deng et al., 2008), and tasseled cap transformation (Healey et al., 2005).

15.5.2 Integrated Watershed Management

This approach involves managing both the land and water resources within a watershed holistically. It incorporates land use planning, soil conservation, and water management practices to enhance ecosystem services and mitigate hydrological disruptions. Integrated Watershed Management (IWM) is crucial for addressing the impacts of climate and LULC changes on hydrological processes in forested landscapes. Studies have shown that both climate and LULC changes can significantly alter watershed hydrology, including increased runoff, peak flows, and flood vulnerability (Farjad et al., 2017; Lamichhane & Shakya, 2019). These changes are not simply additive, highlighting the need for integrated assessment approaches (Lamichhane & Shakya, 2019). Adaptive management and decision analysis techniques can help address the complexities of IWM by improving ecosystem understanding, facilitating stakeholder participation, and managing uncertainty (Ohlson, 1999). Modelling tools, such as hydrological models combined with climate and land use projections, are valuable for simulating future scenarios and informing management strategies (Farjad et al., 2017; Talib & Randhir, 2017). To enhance watershed resilience, restoration strategies and ongoing monitoring are recommended to mitigate the impacts of climate and LULC changes on hydrological processes (Ohlson, 1999; Talib & Randhir, 2017). The adaptive management framework of IWM is a good and adaptable way to deal with the effects that changes in climate and land use have on the water flow in forested areas. By continuously monitoring and adjusting management strategies in response to evolving conditions, IWM can help ensure the sustainability of both water resources and ecosystems. Adaptive IWM can lessen the effects of climate and LULC changes by getting stakeholders to work together, restoring ecosystems, and using climate-resilient methods. This makes watersheds healthier and more resilient in the long term.

Forest restoration can significantly influence the water management results. Choosing the right species, applying various silvicultural methods, and taking past land use into account is important. Even though it is understood that restoration influences water behavior, detailed research is required in understanding how different restoration techniques influence this aspect (Filoso et al., 2017). Climate change is shifting precipitation patterns, leading to prolonged droughts in some areas while others experience severe flooding. These changes complicate water management and threaten the supply of clean water (Diffenbaugh et al., 2015). Additionally, changes in land use, such as deforestation and urbanization, worsen soil erosion and elevate sediment levels in rivers, negatively affecting water quality and aquatic ecosystems (Jacobson, 2011). The combination of climate change and LULC further depletes groundwater recharge. Additionally, rising temperatures due to climate change boost evapotranspiration rates, further depleting groundwater resources (Filoso et al., 2017). Adaptation strategies must include sustainable land use planning. Permeable pavements, urban green spaces, and buffer strips are examples of green infrastructure that can be implemented to reduce runoff, boost groundwater recharge, and enhance water quality (Jacobson, 2011). Afforestation and reforestation are essential to stabilize soils, reduce erosion, and boost water retention. Choosing the right tree species is essential for ensuring water yield and ecosystem recovery (Filoso et al., 2017). IWM is a critical component of adaptation, merging climate and land use data for adaptive watershed planning, which includes flood control measures, erosion prevention, and efficient water allocation (Ali et al., 2022).

15.2 CONCLUSION

Several research studies highlighted particular shortcomings in simulating discharges in small forested watersheds, according to Marin et al. (2020), who analyzed a number of studies. Consequently, in order to pinpoint particular risks to water supplies throughout time, modelling activities are crucial. Additionally, this technique will assist watershed managers in adopting sustainable measures to mitigate the effects of climate and land use change. This strategy can improve community resilience to natural disasters, as well as the objectives of the Water Framework Directive (2007/60/EC) and the Sustainable Development Goals (SDGs).

For economic growth and environmental preservation, hydrological studies are crucial. Research like this helps policymakers figure out how much forest cover is necessary to safeguard water supplies and encourage economic growth in a given area. Additionally, operational activities, including watershed restoration, best forest management practices, and forest regeneration initiatives, might benefit greatly from the data provided by this research. Several research studies have demonstrated the beneficial effects of trees on reducing surface runoff and peak streamflow. Reduced surface and peak flows are projected to generate less soil erosion and flood frequency in wooded ecosystems compared to harvested areas. Because forest influences on hydrology vary with size, it is important to be careful when generalizing findings from one scale to another. Recent hydrological research has greatly expanded our understanding of the connections between forest ecosystems and water. However, in order to fill in certain important gaps in our understanding and provide assistance for sustainable forest resource management, additional research is required.

REFERENCES

Adam, J. C., Hamlet, A. F., & Lettenmaier, D. P. (2009). Implications of global climate change for snowmelt hydrology in the twenty-first century. *Hydrological Processes: An International Journal*, *23*(7), 962–972.

Ali, R., Hassan, M. A., & Wang, W. (2022). Effects of climate and land-use changes on river flows in a Himalayan basin. *Science of The Total Environment*, *846*, 157282.

Alqurashi, A., & Kumar, L. (2013). Investigating the use of remote sensing and GIS techniques to detect land use and land cover change: A review. *Advances in Remote Sensing*. *2*(2), 193–204.

Amoakwah, E., Lucas, S. T., Didenko, N. A., Rahman, M. A., & Islam, K. R. (2022). Impact of deforestation and temporal land-use change on soil organic carbon storage, quality, and lability. *PLoS ONE*, *17*(8), e0263205.

Archer, N. A., Bell, R. A., Butcher, A. S., & Bricker, S. H. (2020). Infiltration efficiency and subsurface water processes of a sustainable drainage system and consequences to flood management. *Journal of Flood Risk Management*, *13*(3), e12629.

Babaremu, K., Taiwo, O., & Ajayi, D. (2024). Impacts of land use and land cover changes on hydrological response. *TWIST*, *19*(1), 256–267.

Birhanu, A., Masih, I., Van Der Zaag, P., Nyssen, J., & Cai, X. (2019). Impacts of land use and land cover changes on hydrology of the Gumara catchment, Ethiopia. *Physics and Chemistry of the Earth, Parts a/b/c*, *112*, 165–174.

Biswas, B., Qi, F., Biswas, J. K., Wijayawardena, A., Khan, M. A. I., & Naidu, R. (2018). The fate of chemical pollutants with soil properties and processes in the climate change paradigm—A review. *Soil Systems*, *2*(3), 51.

Bunch, M. J., Morrison, K. E., Parkes, M. W., & Venema, H. D. (2011). Promoting health and well-being by managing for social–ecological resilience: The potential of integrating ecohealth and water resources management approaches. *Ecology and Society*, *16*(1).

Caldwell, P., Muldoon, C., Ford-Miniat, C., Cohen, E., Krieger, S., Sun, G., McNulty, S., & Bolstad, P. V. (2014). Quantifying the role of national forest system lands in providing surface drinking water supply for the Southern United States [Gen. Tech. Rep. SRS-197]. Asheville, NC: US Department of Agriculture Forest Service, Southern Research Station. 135 p., 197, 1-135.

Caldwell, P. V., Miniat, C. F., Elliott, K. J., Swank, W. T., Brantley, S. T., & Laseter, S. H. (2016). Declining water yield from forested mountain watersheds in response to climate change and forest mesophication. *Global Change Biology*, *22*(9), 2997–3012.

Chen, C., Park, T., Wang, X., Piao, S., Xu, B., Chaturvedi, R. K., Fuchs, R., Brovkin, V., Ciais, P., Fensholt, R., Tømmervik, H., Bala, G., Zhu, Z., Nemani, R, R., & Myneni, R. B. (2019a). China and India lead in greening of the world through land-use management. *Nature Sustainability*, *2*(2), 122–129.

Chen, Y., Xu, C. Y., Chen, X., Xu, Y., Yin, Y., Gao, L., & Liu, M. (2019b). Uncertainty in simulation of land-use change impacts on catchment runoff with multi-timescales based on the comparison of the HSPF and SWAT models. *Journal of Hydrology*, *573*, 486–500.

Chu, M. L., Knouft, J. H., Ghulam, A., Guzman, J. A., & Pan, Z. (2013). Impacts of urbanization on river flow frequency: A controlled experimental modelling-based evaluation approach. *Journal of Hydrology*, *495*, 1–12.

Coppin, P., Jonckheere, I., Nackaerts, K., Muys, B., & Lambin, E. (2004). Review article digital change detection methods in ecosystem monitoring: A review. *International Journal of Remote Sensing*, *25*(9), 1565–1596.

Deng, J. S., Wang, K., Deng, Y. H., & Qi, G. J. (2008). PCA-based land-use change detection and analysis using multitemporal and multisensor satellite data. *International Journal of Remote Sensing*, *29*(16), 4823–4838.

Diffenbaugh, N. S., Swain, D. L., & Touma, D. (2015). Anthropogenic warming has increased drought risk in California. *Proceedings of the National Academy of Sciences*, *112*(13), 3931–3936.

Dingman, S. L. (2015). *Physical hydrology*. Waveland Press.

El Khoury, D., Coenders-Gerrits, M., & van der Ent, R. (2021). Impacts of land-use changes on watershed hydrology in the semi-arid tropics. *Hydrological Processes*, *35*(10), e14308.

Ellison, D., Futter, M. N., & Bishop, K. (2017). On the forest cover–water yield debate: From demand- to supply-side to thinking. *Global Change Biology*, *23*(2), 516–527.

Falkenmark, M., & Rockström, J. (2006). The new blue and green water paradigm: Breaking new ground for water resources planning and management. *Journal of Water Resources Planning and Management*, *132*(3), 129–132.

Farjad, B., Gupta, A., Razavi, S., Faramarzi, M., & Marceau, D. J. (2017). An integrated modelling system to predict hydrological processes under climate and land-use/cover change scenarios. *Water*, *9*(10), 767.

Filoso, S., Bezerra, M. O., Weiss, K. C. B., & Palmer, M. A. (2017). Impacts of forest restoration on water yield: A systematic review. *PLoS ONE*, *12*(8), e0183210.

Gao, F., Anderson, M. C., Zhang, X., Yang, Z., Alfieri, J. G., Kustas, W. P., Mueller, R., Johnson, D. M., & Prueger, J. H. (2017). Toward mapping crop progress at field scales through fusion of Landsat and MODIS imagery. *Remote Sensing of Environment*, *188*, 9–25.

Healey, S. P., Cohen, W. B., Zhiqiang, Y., & Krankina, O. N. (2005). Comparison of Tasseled Cap-based Landsat data structures for use in forest disturbance detection. *Remote Sensing of Environment*, *97*(3), 301–310.

Hyandye, C. B., Worqul, A., Martz, L. W., & Muzuka, A. N. (2018). The impact of future climate and land use/cover change on water resources in the Ndembera watershed and their mitigation and adaptation strategies. *Environmental Systems Research*, *7*(1), 24.

IPCC. (2019). *Special report on global warming of 1.5°C (SR15)*. Available at: https://www.ipcc.ch/sr15

Jacobson, C. R. (2011). Identification and quantification of the hydrological impacts of imperviousness in urban catchments: A review. *Journal of Environmental Management*, *92*(6), 1438–1448.

Jensen, J. R., & Toll, D. L. (1982). Detecting residential land-use development at the urban fringe. *Photogrammetric Engineering and Remote Sensing*, *48*.

Jones, J., Ellison, D., Ferraz, S., Lara, A., Wei, X., & Zhang, Z. (2022). Forest restoration and hydrology. *Forest Ecology and Management*, *520*, 120342.

Keleş, S. (2019). An assessment of hydrological functions of forest ecosystems to support sustainable forest management. *Journal of Sustainable Forestry*, *38*(4), 305–326.

Kirk, J. T. (1985). Effects of suspensoids (turbidity) on penetration of solar radiation in aquatic ecosystems. *Hydrobiologia*, *125*, 195–208.

Kløve, B., Ala-Aho, P., Bertrand, G., Gurdak, J. J., Kupfersberger, H., Kværner, J., Muotka, T., Mykrä, H., Preda, E., Rossi, P., Uvo, C. B., Velasco, E., & Pulido-Velazquez, M. (2014). Climate change impacts on groundwater and dependent ecosystems. *Journal of Hydrology*, *518*, 250–266.

Lamichhane, S., & Shakya, N. M. (2019). Integrated assessment of climate change and land use change impacts on hydrology in the Kathmandu Valley watershed, Central Nepal. *Water*, *11*(10), 2059.

Lawrence, D. M., Thornton, P. E., Oleson, K. W., & Bonan, G. B. (2007). The partitioning of evapotranspiration into transpiration, soil evaporation, and canopy evaporation in a GCM: Impacts on land–atmosphere interaction. *Journal of Hydrometeorology*, *8*(4), 862–880.

Lee, W. S., Alchanatis, V., Yang, C., Hirafuji, M., Moshou, D., & Li, C. (2010). Sensing technologies for precision specialty crop production. *Computers and Electronics in Agriculture*, *74*(1), 2–33.

LGebremicael, T. G., Mohamed, Y. A., Betrie, G. D., van der Zaag, P., & Teferi, E. (2019). The impact of land use change and climate variability on the hydrology of the Lake Tana Basin, Ethiopia. *Hydrology and Earth System Sciences*, *23*(1), 2905–2920.

Liu, S., Bond-Lamberty, B., Boysen, L. R., Ford, J. D., Fox, A., Gallo, K., Hatfield, J., Henebry, G. M., Huntington, T. G., Liu, Z., Loveland, T. R., Norby, R. J., Sohl, T., Steiner, A. L., Yuan, W., Zhang, Z., & Zhao, S. (2017). Grand challenges in understanding the interplay of climate and land changes. *Earth Interactions*, *21*(2), 1–43.

Liu, Y., Zhang, X., Xia, D., You, J., Rong, Y., & Bakir, M. (2013). Impacts of land-use and climate changes on hydrologic processes in the Qingyi River watershed, China. *Journal of Hydrologic Engineering*, *18*(11), 1495–1512.

Lockaby, G., Nagy, C., Vose, J. M., Ford, C. R., Sun, G., McNulty, S., Caldwell, P., Cohen, E., & Myers, J. M.. (2013). Forests and water. In: Wear, D. N.; Greis, J. G. (Eds.), *The Southern forest futures project: Technical report* (pp. 309–339., 178, 309–339) [Gen. Tech. Rep. SRS-GTR-178]. USDA-Forest Service, Southern Research Station.

Marin, M., Clinciu, I., Tudose, N. C., Ungurean, C., Adorjani, A., Mihalache, A. L., Davidescu. A. A., Davidescu. S. O., Dinca. L., & Cacovean, H. (2020). Assessing the vulnerability of water resources in the context of climate changes in a small forested watershed using SWAT: A review. *Environmental Research*, *184*, 109330.

Miralles, D. G., De Jeu, R. A. M., Gash, J. H., Holmes, T. R. H., & Dolman, A. J. (2011). Magnitude and variability of land evaporation and its components at the global scale. *Hydrology and Earth System Sciences*, *15*(3), 967–981.

Naha, S., Rico-Ramirez, M. A., & Rosolem, R. (2021). Quantifying the impacts of land cover change on hydrological responses in the Mahanadi River basin in India. *Hydrology and Earth System Sciences*, *25*(12), 6339–6357.

Ohlson, D. W. (1999). *Exploring the application of adaptive management and decision analysis to integrated watershed management* [Doctoral dissertation, University of British Columbia].

Oki, T., & Kanae, S. (2006). Global hydrological cycles and world water resources. *Science*, *313*(5790), 1068–1072.

Oliver, T. H., & Morecroft, M. D. (2014). Interactions between climate change and land use change on biodiversity: Attribution problems, risks, and opportunities. *Wiley Interdisciplinary Reviews: Climate Change*, *5*(3), 317–335.

Pacifici, F., Chini, M., & Emery, W. J. (2009). A neural network approach using multi-scale textural metrics from very high-resolution panchromatic imagery for urban land-use classification. *Remote Sensing of Environment*, *113*(6), 1276–1292.

Postel, S., & Richter, B. (2012). *Rivers for life: Managing water for people and nature*. Island Press.

Relyea, R. A. (2005). The impact of insecticides and herbicides on the biodiversity and productivity of aquatic communities. *Ecological Applications*, *15*(2), 618–627.

Robichaud, P. R., Lewis, S. A., & Wagenbrenner, J. W. (2020). Post-fire mulching for runoff and erosion mitigation: Field trials, effectiveness, and applicability. *International Journal of Wildland Fire*, *29*(8), 733–745.

Rogger, M., Agnoletti, M., Alaoui, A., Bathurst, J. C., Bodner, G., Borga, M., Chaplot, V., Gallart, F., Glatzel, G., Hall, J., Holko, L., Horn, R., Kiss, A., Kohnová, S., Leitinger, G., Lennartz, B., Parajka, J., Perdigão, R., Peth, S., & Blöschl, G. (2017). Land use change impacts on floods at the catchment scale: Challenges and opportunities for future research. *Water Resources Research*, *53*(7), 5209–5219.

Seidl, R., Albrich, K., Erb, K., Formayer, H., Leidinger, D., Leitinger, G., Tappeiner, U., Tasser, E., & Rammer, W. (2019). What drives the future supply of regulating ecosystem services in a mountain forest landscape?. *Forest Ecology and Management*, *445*, 37–47.

Seiler, K. P., & Gat, J. R. (2007). *Groundwater recharge from run-off, infiltration and percolation* (Vol. 55). Springer Science & Business Media.

Sheil, D., & Murdiyarso, D. (2009). How forests attract rain: An examination of a new hypothesis. *Bioscience*, *59*(4), 341–347.

Sophocleous, M. (2002). Interactions between groundwater and surface water: The state of the science. *Hydrogeology Journal*, *10*, 52–67.

Sun, G., McNulty, S. G., Lu, J., Amatya, D. M., Liang, Y., & Kolka, R. K. (2005). Regional annual water yield from forest lands and its response to potential deforestation across the southeastern United States. *Journal of Hydrology*, *308*(1-4), 258–268.

Talib, A., & Randhir, T. O. (2017). Climate change and land use impacts on hydrologic processes of watershed systems. *Journal of Water and Climate Change*, *8*(3), 363–374.

Tomer, M. D., & Schilling, K. E. (2009). A simple approach to distinguish land-use and climate-change effects on watershed hydrology. *Journal of Hydrology*, *376*(1–2), 24–33.

Trenberth, K. E. (2011). Changes in precipitation with climate change. *Climate Research*, *47*(1–2), 123–138.

Trenberth, K. E., Smith, L., Qian, T., Dai, A., & Fasullo, J. (2007). Estimates of the global water budget and its annual cycle using observational and model data. *Journal of Hydrometeorology*, *8*(4), 758–769.

van Meerveld, I., & Seibert, J. (2025). Reforestation effects on low flows: Review of public perceptions and scientific evidence. *Wiley Interdisciplinary Reviews: Water*, *12*(1), e1760.

van Roosmalen, L., Sonnenborg, T. O., & Jensen, K. H. (2009). Impact of climate and land use change on the hydrology of a large-scale agricultural catchment. *Water Resources Research*, *45*(7).

Viviroli, D., Dürr, H. H., Messerli, B., Meybeck, M., & Weingartner, R. (2007). Mountains of the world, water towers for humanity: Typology, mapping, and global significance. *Water Resources Research*, *43*(7).

Wang, G., Peng, W., Zhang, L., & Zhang, J. (2023). Quantifying the impacts of natural and human factors on changes in NPP using an optimal parameters-based geographical detector. *Ecological Indicators*, *155*, 111018.

Ward, A. D., & Trimble, S. W. (2004). *Environmental hydrology* (2nd ed.). CRC Press.

Wei, Z., Yoshimura, K., Wang, L., Miralles, D. G., Jasechko, S., & Lee, X. (2017). Revisiting the contribution of transpiration to global terrestrial evapotranspiration. *Geophysical Research Letters*, *44*(6), 2792–2801.

Weldeslassie, T., Naz, H., Singh, B., & Oves, M. (2018). *Chemical contaminants for soil, air and aquatic ecosystem. Modern age environmental problems and their remediation*, (pp. 1–22). Cham: Springer International Publishing..

Yang, L., Xian, G., Klaver, J. M., & Deal, B. (2003). Urban land-cover change detection through sub-pixel imperviousness mapping using remotely sensed data. *Photogrammetric Engineering & Remote Sensing*, *69*(9), 1003–1010.

Zafirah, N., Nurin, N. A., Samsurijan, M. S., Zuknik, M. H., Rafatullah, M., & Syakir, M. I. (2017). Sustainable ecosystem services framework for tropical catchment management: A review. *Sustainability*, *9*(4), 546.

Zhang, J., Yuan, X., Tan, X., & Zhang, X. (2021). Delineation of the urban-rural boundary through data fusion: Applications to improve urban and rural environments and promote intensive and healthy urban development. *International Journal of Environmental Research and Public Health*, *18*(13), 7180.

Zhang, Y., Chiew, F. H., Peña-Arancibia, J., Sun, F., Li, H., & Leuning, R. (2017). Global variation of transpiration and soil evaporation and the role of their major climate drivers. *Journal of Geophysical Research: Atmospheres*, *122*(13), 6868–6881.

Index

O

P

R

S

T

V

W

Made in the USA
Las Vegas, NV
19 September 2026

6da50e5e-3174-4ee8-8111-4363428cc3b9R01